Catalysis
by Nonmetals

PHYSICAL CHEMISTRY
A Series of Monographs

ERNEST M. LOEBL, *Editor*
*Department of Chemistry, Polytechnic Institute of Brooklyn
Brooklyn, New York*

1. W. Jost: Diffusion in Solids, Liquids, Gases, 1952
2. S. Mizushima: Structure of Molecules and Internal Rotation, 1954
3. H. H. G. Jellinek: Degradation of Vinyl Polymers, 1955
4. M. E. L. McBain and E. Hutchinson: Solubilization and Related Phenomena, 1955
5. C. H. Bamford, A. Elliott, and W. E. Hanby: Synthetic Polypeptides, 1956
6. George J. Janz: Thermodynamic Properties of Organic Compounds — Estimation Methods, Principles and Practice, revised edition, 1967
7. G. K. T. Conn and D. G. Avery: Infrared Methods, 1960
8. C. B. Monk: Electrolytic Dissociation, 1961
9. P. Leighton: Photochemistry of Air Pollution, 1961
10. P. J. Holmes: Electrochemistry of Semiconductors, 1962
11. H. Fujita: The Mathematical Theory of Sedimentation Analysis, 1962
12. K. Shinoda, T. Nakagawa, B. Tamamushi, and T. Isemura: Colloidal Surfactants, 1963
13. J. E. Wollrab: Rotational Spectra and Molecular Structure, 1967
14. A. Nelson Wright and C. A. Winkler: Active Nitrogen, 1968
15. R. B. Anderson: Experimental Methods in Catalytic Research, 1968
16. Milton Kerker: The Scattering of Light and Other Electromagnetic Radiation, 1969
17. Oleg V. Krylov: Catalysis by Nonmetals — Rules for Catalyst Selection, 1970

In preparation

Alfred Clark: The Theory of Adsorption and Catalysis

CATALYSIS BY NONMETALS

RULES FOR CATALYST SELECTION

OLEG V. KRYLOV
Institute of Chemical Physics
Academy of Sciences of the USSR
Moscow, USSR

TRANSLATED BY

MICHAEL F. DELLEO, JR.
UNITED STATES ARMY
FORT CARSON, COLORADO

GEORGE DEMBINSKI
UNION CARBIDE CORPORATION
TARRYTOWN, NEW YORK

JOHN HAPPEL
DEPARTMENT OF CHEMICAL ENGINEERING
NEW YORK UNIVERSITY
BRONX, NEW YORK

ALVIN H. WEISS
DEPARTMENT OF CHEMICAL ENGINEERING
WORCESTER POLYTECHNIC INSTITUTE
WORCESTER, MASSACHUSETTS

1970

ACADEMIC PRESS   New York and London

COPYRIGHT © 1970, BY ACADEMIC PRESS, INC.
ALL RIGHTS RESERVED.
NO PART OF THIS BOOK MAY BE REPRODUCED IN ANY FORM,
BY PHOTOSTAT, MICROFILM, RETRIEVAL SYSTEM, OR ANY
OTHER MEANS, WITHOUT WRITTEN PERMISSION FROM
THE PUBLISHERS.

ACADEMIC PRESS, INC.
111 Fifth Avenue, New York, New York 10003

United Kingdom Edition published by
ACADEMIC PRESS, INC. (LONDON) LTD.
Berkeley Square House, London W1X 6BA

LIBRARY OF CONGRESS CATALOG CARD NUMBER: 70-107578

PRINTED IN THE UNITED STATES OF AMERICA

Contents

Preface ix

PART I

General Methods of Catalyst Selection

INTRODUCTION 3

1. PROPERTIES OF A SOLID AND CATALYTIC ACTIVITY IN OXIDATION–REDUCTION REACTIONS

   1.1. Electroconductivity 6
   1.2. Work Function 15
   1.3. Width of the Forbidden Zone 21
   1.4. Doped Levels 36
   1.5. The Difference of Electronegativities and the Effective Charges of Atoms 40
   1.6. The Role of d-Electrons in the Catalytic Properties of a Solid 47
   1.7. The Color of Solids 66

2. SOLID PROPERTIES AND CATALYTIC ACTIVITY IN ACID–BASE REACTIONS

   2.1. Proton-Acid and -Base Properties of a Surface 70
   2.2. Coordination Capacity of Metals 81
   2.3. Dielectric Constant 89

3. Catalytic Activity and Structure of Solids

| 3.1. | Crystalline Lattice Parameter | 93 |
| 3.2. | Type of Crystal Lattice | 103 |

PART II

Principles of Selection of Catalysts for Various Reactions

Introduction   113

4. Decomposition of Alcohols and Acids

| 4.1. | Decomposition of Alcohols | 115 |
| 4.2. | Decomposition of Acids | 134 |

5. Dehydrogenation and Hydrogenation Reactions

5.1.	Dehydrogenation of Hydrocarbons	140
5.2.	Hydrogenation	148
5.3.	Decomposition of Inorganic Hydrides	152

6. Hydrogen–Deuterium Exchange and Other Simple Reactions   157

7. Reactions of Oxidation and Decomposition of Oxygen-Containing Compounds

7.1.	Simple Oxidation Reactions	168
7.2.	Oxidation of Organic Compounds	178
7.3.	The Decomposition of Oxygen-Containing Substances	185

8. Acid–Base Reactions

8.1.	Addition and Removal of Water and Hydrogen Halides	198
8.2.	The Reactions of Condensation and Polycondensation	207
8.3.	Isomerization Reactions	212
8.4.	Cracking and Alkylation of Hydrocarbons	215

9. Polymerization Reactions   219

Conclusions   240

Appendix PHYSICAL PROPERTIES OF SOME NONMETALLIC COMPOUNDS
           USED AS CATALYSTS   247

REFERENCES   259

Subject Index   279

Preface

The development of scientific principles for the selection of catalysts appears to be the central problem of the science of catalysis. Nevertheless, until now, the selection of catalysts was typically carried out empirically. Successes obtained in the investigation of the mechanism of chemisorption and catalysis permit a series of rules to be developed, which significantly reduce the time required for catalyst selection for a specific reaction.

Both of these subjects have been touched upon to an extent in a series of collections and monographs published in recent years and discussed in International Congresses on Catalysis in Philadelphia (1956), in Paris (1960), and in Amsterdam (1964). In 1964, in Moscow, an All Union conference on the scientific bases of selection of catalysts also took place.

In Part I of this monograph, an endeavor is made to relate a series of properties of a solid (such as type of conductivity, the width of a forbidden zone of a semiconductor, work function, charge and radius of ions, electronegativity of atoms, acid–base properties of a surface, lattice parameters, and types of lattices) with catalytic activity in such a manner as to develop a method of catalyst selection. Principally, the problems of selection of one-component nonmetallic catalysts are discussed.* Questions of the selectivity of catalysts are not considered. The book is developed using the principles of heterogeneous catalysis.

---

* A recently published monograph by G. C. Bond is devoted to catalysis by metallic catalysts ("Catalysis by Metals," Academic Press, New York, 1962).

However, in those cases where the mechanism of a reaction is believed to be identical in homogeneous and heterogeneous media, the topic of selection of homogeneous catalysts is also discussed.

Consideration is given to questions of the elementary mechanism of catalysis when steps occur in which this is necessary for the development of methods of catalyst selection.

In Part II, an abundance of literary material is examined for oxide, sulfide, and other catalyst selection methods for the specific reactions oxidation, hydrogenation, dehydrogenation, isomerization, polymerization, etc. These rules are related to the properties of solids examined in Part I. It is shown which theories of catalysis are the best correlations to explain data. A review of the literature is complete to mid 1964.

The author hopes that the monograph will be useful to both the scientific worker and the practical man working in the field of catalysis.

It should be emphasized that a single theory of catalyst selection does not exist at the present time. The present book does not lay claim to creation of a theory founded on the basis of rigorous quantum-mechanical treatment of bonds of chemisorbed particles with a surface. Explanation based on these principles is so controversial that the subjective presentation of the author is indicated. All remarks will be accepted by the author with appreciation.

The author expresses sincere gratitude to F. F. Volkenshtein, G. M. Zhabrova, J. J. Ioffe, L. R. Margolis, and S. Z. Roginskii for their valuable advice and comments, and for having read the manuscript.

O. V. KRYLOV

**Catalysis
by Nonmetals**

PART I • **GENERAL METHODS OF CATALYST SELECTION**

Introduction

Experiment shows that in the majority of cases the chemical composition of the principal component of a catalyst exerts considerably greater influence on its catalytic activity than the method of preparation. For example, neither the method of preparation nor doping will succeed in increasing the activity of quartz or NaCl to a level that they are able to function at room temperature like metal catalysts (Pt, Pd, Ni). Naturally, the question arises, according to which of its characteristics does one select a catalyst; and with which of its known properties does one correlate its catalytic activity?

One must distinguish two types of properties of a solid. To one type are related the properties, which, according to existing theoretical ideas, directly determine the catalytic activity (e.g., the number of current carriers, work function, and crystalline lattice parameter). The other type of property (e.g., color, and melting point) is that whose alteration is perhaps functionally associated with the alteration of catalytic activity. Alterations of the first type of properties simultaneously determine both alteration of the properties of the second type and catalytic activity.

For the practical problem of catalyst selection, properties of the first and second types are equally valuable. A choice between them is determined on the basis of quantitative information. In the literature, one can find examples of successful application of methods of catalyst selection by examination of such properties as heats of fusion [1] and sublimation [2], compressibility, and a series of others. For creation of a theory of catalyst selection, it is important to delineate those properties of a material that directly determine catalytic activity. The applicability of particular forms of existing theories of catalysis for this determination will be shown.

One can explain the majority of cases of catalysis by an intermediate chemical reaction of the reactants with the catalyst. Therefore, in the selection of catalysts, very general chemical ideas and analogies can be found to be valuable. The nature of a few active species can be deduced by chemical examination [3].

The position of an element in the periodic table of D. I. Mendeleev, (i.e., the electronic shell structure of atoms and ions) determines, in fact, all the basic chemical properties as well as the series of physical properties of a substance. Therefore, the comparison of catalytic activity of solids with their position in the periodic table of elements results in the development of a series of rules for catalyst selection.

There have been experiments to relate the catalytic activity of an element directly with its atomic number. Atomic or molecular weight and volume of the elements and of the structural groups combined in a catalyst have also been used. A series of such experiments is explained in the well-known monograph of Berkman et al. [4]. The general disadvantage of comparing such a series on the basis of its formalism is the absence of theoretical concepts for explaining various correlations. Nevertheless, one can still draw from such comparisons a series of conclusions that are essential for catalyst selection. The conclusion appears to be common and indisputable concerning the high catalytic activity of transition metals in reactions of oxidation, hydrogenation, and dehydrogenation. For reactions of catalytic cracking and dehydration, the oxides of those elements found in the upper right part of the periodic table are the most active. Data show that typical catalytic poisons are located principally at the lower right of the table of Mendeleev [5]. Patents are often useful in establishing that the elements of a certain group of the periodic table can be employed as catalysts for a given reaction.

The greatest successes in relating catalytic activity of an element with its position in the periodic table have been obtained for metallic catalysts. Very clear rules establishing the relationship of catalytic activity with solid properties have been found (e.g., with crystalline lattice parameter, number of d-electrons, etc.). Note, however, that the problem of reducing the time for metallic catalyst selection is not very serious, because the number of requisite experiments is not great. Upon switching to binary compounds, the number of possible experiments increases by a factor of 10^4. If different valence states appearing in these elements are taken into account, along with various crystalline modifications, the number is even greater. Further increase of the number of elements in a catalyst even more seriously hampers selection by the method of relating catalytic properties with position of elements in the periodic table of elements of Mendeleev.

INTRODUCTION

For general orientation in catalyst selection, a useful classification of a catalytic process is according to the mechanism of operation of the catalyst. Thus, according to Roginskii [6, 7], the majority of catalytic reactions can be separated into two types—oxidation–reduction (electronic) and acid–base (ionic). Reactions pertaining to the first type are those of oxidation, reduction, hydrogenation, dehydrogenation, decomposition of unstable oxygen-containing compounds, and others. These reactions are catalyzed by solids possessing free or easily excited electrons, i.e., metals and semiconductors. The mechanism of these reactions is a characteristic transfer of an electron in an elementary act of catalysis from the catalyst to the reacting substance or vice versa.[1]

Catalytic cracking, hydration, dehydration, hydrolysis, many reactions of isomerization, polymerization, and a series of other reactions pertaining to the second type are accelerated under the influence of acid or base. For this type of reaction, the elementary act is the characteristic transfer of protons or the production of a heteropolar donor–acceptor pair.

Other investigators [9–11] use analogous types of classification for organic reactions. This classification corresponds to the classification of homogeneous reactions in organic chemistry according to the mechanism of the elementary step on homolytic and heterolytic catalysts [12].

Methods of catalyst selection will be examined in detail by us, as a rule, separately for each of the two types of reactions.

---

[1] According to Volkenshtein [8], the idea of "electron transfer" in catalysis and chemisorption is incorrect. It is more correct to say, concerning a specific step, that there is a localization of an electron or a hole of a semiconductor on an adsorbed particle.

1 • Properties of a Solid and Catalytic Activity in Oxidation–Reduction Reactions

1.1. ELECTROCONDUCTIVITY

As early as 1928, in the works of Ioffe [13] and Roginskii and Schultz [14], ideas were expressed concerning the connection of the catalytic activity of a solid with the number of conduction electrons in it. In the opinion of Ioffe [13], a catalytic reaction can proceed on elementary surface defects of the crystalline lattice of a semiconductor. In 1933 Roginskii [6] showed the preponderance of semiconductors among catalysts of the oxidation–reduction class. Wagner and Hauffe [15] experimentally demonstrated the change of electrical conductivity of NiO while the catalytic reactions of oxidation of CO and decomposition of N_2O were proceeding on its surface. Electronic ideas in catalysis on semiconductors grew especially rapidly in the 1950's [1, 16–19].

We shall examine the connection between electronic properties of the bulk of a solid semiconductor and its adsorptive capacity according to Volkenshtein [16].

There are three possibilities for the type of bond of a chemisorbed particle (atom, molecule) with a solid surface: (1) "weak" bond, (2) "strong" acceptor bond and (3) "strong" donor bond. In the first case, an electron of the chemisorbed particle is drawn close to a cation of the lattice or an electron of the anion of the lattice is drawn close to the chemisorbed particle. The latter remains electrically neutral. In the second case, an electron of the particle adsorbed on the cation interacts with a free electron of the semiconductor, thus bringing about a chemical

1.1. ELECTROCONDUCTIVITY

bond with the lattice. In the third case, an atom or molecule is adsorbed on an anion of the lattice and enters into an interaction with a free hole of the semiconductor.

A chemisorbed particle produces local energetic levels in the forbidden zone (Fig. 1a). The transition of an electron to acceptor level A corre-

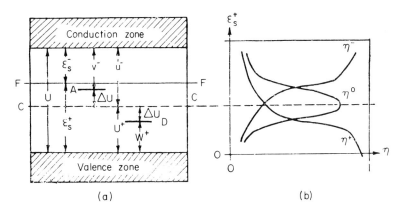

FIG. 1. (a) A zone diagram of a semiconductor and (b) the change of concentration of adsorbed particles during a change in position of the Fermi level FF in the forbidden zone of a semiconductor [16].

sponds to the production of a "strong" acceptor bond; corresponding to the production of a "strong" donor bond is the withdrawal of an electron from donor level D, that is, the transition of a hole to it. The relative coverage of particles on a surface, found in state "weak" (η^0), "strong" acceptor (η^-), and "strong" donor (η^+) are described by the formulas below:

$$\eta_0 = \frac{1}{1 + 2\exp\left(-\frac{\Delta U}{kT}\right)\operatorname{ch}\frac{\epsilon_s^+ - U^+}{kT}}$$

$$\eta^- = \frac{\exp\left(-\frac{\Delta U}{kT} + \frac{\epsilon_s^+ - U^+}{kT}\right)}{1 + 2\exp\left(-\frac{\Delta U}{kT}\right)\operatorname{ch}\frac{\epsilon_s^+ - U^+}{kT}} \qquad (1)$$

$$\eta^+ = \frac{\exp\left(-\frac{\Delta U}{kT} - \frac{\epsilon_s^+ - U^+}{kT}\right)}{1 + 2\exp\left(-\frac{\Delta U}{kT}\right)\operatorname{ch}\frac{\epsilon_s^+ - U^+}{kT}}$$

In these formulas (see also Fig. 1a) ϵ_s^- and $\epsilon_s^+ \equiv$ the distance from the Fermi level FF (level of electrochemical potential)[1] to the conduction zone or, correspondingly, to the valence zone; $U \equiv$ the width of the forbidden zone ($U = \epsilon_s^- + \epsilon_s^+ = U^- + U^+$); $v^- \equiv$ the distance from the acceptor local level A to the zone of conduction; $w^+ \equiv$ the distance from donor level D to the valence zone. It is obvious that $\eta^0 + \eta^- + \eta^+ = 1$.

From formulas (1) it is evident that the relative coverage of various forms of chemisorbed particles on the surface of a semiconductor in the presence of a fixed electronic equilibrium is determined by the location of the Fermi level. In Fig. 1b is shown the change of magnitude of η^0, η^-, and η^+ during a shift of the Fermi level in the forbidden zone from the valence zone to the conduction zone according to formulas (1). If it is assumed that the catalytic activity is determined by the coverage of some type of chemisorbed particles, one can derive, based on the location of the Fermi level, the equation for the rate of a chemical reaction.

Analogous conclusions concerning the connection between catalytic activity and the location of the Fermi level in a semiconductor (not introducing, however, the idea of three types of bonds) were reached by other previously mentioned investigators involved in the development of an electronic theory of catalysis.[2] Upon a shift of the Fermi level from the valence zone to the conduction zone (Fig. 1b), catalytic activity can increase, decrease, or go through a maximum.

An increase in the concentration of electrons in a semiconductor (i.e., electronic conductivity) displaces the Fermi level FF upward—closer to the conduction zone; increasing the concentration of holes (i.e., hole conductivity) displaces the Fermi level downward—closer to the valence zone. It follows, on the basis of the above discussion, that a dependence of catalytic activity on the magnitude and charge of electronic conductivity results.

Garner [21, 22] and Hauffe [17] introduced the idea of acceptor and donor reactions. According to their presentation, the acceptor reactions are accompanied by the transfer of an electron in the limiting step of a reaction from the catalyst to the adsorbed particles; and, in donor reactions, a transfer effects a hole (i.e., the transfer of an electron proceeds in the opposite direction). According to more rigorous analyses [8, 16], based on theories relating the Fermi level to catalytic activity, acceptor

---

[1] The probability of the filling of the Fermi levels by electrons is $\frac{1}{2}$. In semiconductors and dielectrics the Fermi level is located in the forbidden zone, in metals in the permitted zone.

[2] In recent times the notion [16] of the role of weak one-electron chemisorption in catalysis has been criticized [20].

1.1. ELECTROCONDUCTIVITY

reactions accelerate during the displacement of the Fermi level upward (see Fig. 1) (i.e., accelerations by electrons), donor reactions accelerate according to the degree of lowering of the Fermi level (i.e. acceleration by holes). Hence, one can draw immediate conclusions regarding methods of catalyst selection: In order to increase the rate of an acceptor reaction, it is necessary to introduce a dopant into a semiconductor that increases its electronic conductivity (for example, Ga_2O_3 in ZnO); for an increase of rate of a donor reaction introduce a dopant that increases its hole conductivity (for example, Li_2O in NiO).

In the opinion of Garner [21] and others, the reaction of H_2–D_2 exchange on a ZnO catalyst (an n-type electronic semiconductor) appears to be an acceptor reaction (in the sense of transfer of an electron from the adsorbent to the adsorbate). The introduction of a donor dopant in ZnO (an oxide of a trivalent metal such as Al_2O_3 or Ga_2O_3) increases the concentration of free electrons and the rate of H_2–D_2 exchange; the introduction of an acceptor dopant (an oxide of a univalent metal such as Li_2O) lowers the concentration of free electrons and the rate of exchange. In the opinion of Hauffe [23], this is demonstrated by the participation of the electrons of a semiconductor in the limiting step of a reaction.

$$H_{2\,gas} \underset{}{\overset{fast}{\rightleftarrows}} 2H^+_{ads} + 2e$$

$$D_{2\,gas} \underset{}{\overset{fast}{\rightleftarrows}} 2D^+_{ads} + 2e$$

$$H^+_{ads} + D^+_{ads} + 2e \underset{}{\overset{slow}{\rightleftarrows}} HD_{gas}$$

The decomposition of hydrogen peroxide [24] on ZnO and other catalysts also appears to be an acceptor reaction.

The decomposition of N_2O on the hole semiconductor (p-type) NiO [25] appears to be donor reaction. The introduction of Li_2O in NiO increases, but the introduction of In_2O_3 reduces its catalytic activity. The mechanism of this reaction is as follows:

$$N_2O \overset{fast}{\longrightarrow} N_2 + O^- + p$$

$$2O^- + 2p \overset{slow}{\longrightarrow} O_2$$

where p is the symbol for a hole. In the limiting step of the reaction—the desorption of oxygen—the reaction of O^- with free holes occurs.

According to Dell *et al.* [26], the limiting step of this reaction is

$$N_2O + O^{2-} + 2p \to N_2 + O_2$$

which also proceeds with participation of free holes of the semiconductor.

Later works showed that the connection of catalytic and chemisorptive properties of a semiconductor with its electronic properties (quantity and character of introduced dopant or magnitude and charge of electroconductivity) is not as simple as it first seemed [27].

For example, in the reaction of H_2–D_2 exchange, Kuchaev and Boreskov did not observe a change in activity of an elemental germanium catalyst upon the introduction into it of both donor and acceptor dopes. During a study of the chemisorption of H_2 on ZnO [29] it was found that addition of Li_2O or Ga_2O_3 to ZnO, for all practical purposes, did not influence either the rate or activation energy of the process. This contradicts the above discussed mechanism of Hauffe for H_2–D_2 exchange. Analogous results were obtained [30] in a study of the effect of adding Ga_2O_3 and Al_2O_3 on the catalytic activity of ZnO for dehydrogenation of butylene to divinyl.

For the other discussed reaction (decomposition of N_2O), it was shown [31] that the addition of Li_2O and Ga_2O_3 to ZnO and the addition of Al_2O_3 and WO_3 to TiO_2 lowered the catalytic activity of ZnO and TiO_2. However, according to the above discussed ideas about the role of the Fermi level, donor and acceptor additions must exhibit contrasting influence on chemisorption and catalysis since the rate of reaction is proportional to the amount of one of the charged chemisorbed forms η^- or η^+ [see Formulas (1)]. The author explains these anomalies by diffusion phenomena in the pores of the compounds under study.

Often, one investigator will refer to a specific reaction as a donor type, another to the same reaction as an acceptor type. In particular, a considerable amount of contradictory data was obtained during the study of the oxidation of CO on oxide catalysts. This reaction is often used for the verification of specific electronic ideas. According to Schwab and Block [32], the reaction of oxidation of CO appears to be of the donor type; according to Parravano [33] it is an acceptor type. The former found that the addition of Li_2O in NiO increased activity but the addition of Cr_2O_3 reduced it. Parravano obtained contradictory data. The data of Keier *et al.* [34], concurred with the results of Parravano. French workers, Coue *et al.* [35] investigating recently the same reaction on the same catalyst—NiO, in general, did not discover the effect of acceptor and donor additions of Li_2O and Ga_2O_3 on the catalytic activity of NiO.

During the study of the chemisorption of CO and O_2 on ZnO and NiO, it was discovered [36, 37] that the introduction of Li_2O reduced the

1.1. ELECTROCONDUCTIVITY

electroconductivity of the electronic semiconductor, ZnO and increased the electroconductivity of the hole semiconductor, NiO in accordance with the electronic theory of semiconductors. At the same time Li_2O accelerates the chemisorption of O_2—an acceptor of electrons—on ZnO and NiO and suppresses the chemisorption of CO—a donor of electrons —which does not immediately follow from the above mentioned ideas.

Garner et al. [38] found that dehydrogenation of alcohols on oxide semiconductors is an acceptor reaction. According to the data of Frolov et al. [39] this reaction, studied on Ge, is a donor reaction. In Fig. 2 is shown the dependence of the activation energy for dehydro-

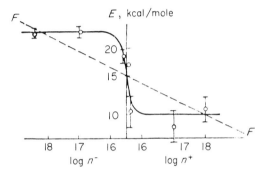

FIG. 2. The change of activation energy of dehydrogenation of C_2H_5OH on Ge in relation to concentration of basic carriers of current: donor ($\log n^-$) and acceptor ($\log n^+$) dopes. The change of position of the Fermi level FF is represented by the dashed line.

genation of C_2H_5OH on the concentration of the principle current carriers in the catalyst (Ge), alloyed by doping with donors (Sb) and acceptors (Ga). The transition from electron to hole type of conduction produces a sharp decrease in activation energy. Similar discontinuities in the curves of dependence of activation energy on number and charge of current carriers were found in other mentioned works.

The stated contradictions in experimental data alone do not provide a conclusion about the inaccuracy of the original hypothesis that there is a relationship between catalytic activity and position of the Fermi level in a semiconductor. Each of these contradictions can be explained. For example, the different effect of dopes in the semiconductor on the course of the oxidation of CO or dehydrogenation of alcohols, having been observed by some authors, could be explained by different limiting steps at different reaction conditions, [16, 40]. One must also consider that, in the process of a reaction, a change in the chemical composition of a catalyst can take place (e.g., its oxidation or reduction) that is accompanied by a change in its electronic structure. Such a change, in a specific case, can lead to some type of relationship of catalytic activity with quantity and character of the dopant. In addition, one usually compares the catalytic activity of a semiconductor with the quantity of introduced donor or acceptor dopants but not directly with the position of the Fermi level, which is impossible to determine from the data usually quoted in tables.

Experimental data often give more complex dependencies than those that follow from the elementary electronic ideas mentioned above. Further on, some reasons for this will be examined in detail. Not all the deviations in experimental data succeed in being explained at this time. This, in one stroke, reduces the forecasting capability of the electronic theory in regard to methods of catalyst selection.

It is necessary to note that significant changes of energy of activation of a catalytic reaction, observed during the introduction of donor and acceptor dopes, almost always are compensated by a cymbate change of the pre-exponential factor k_0 in the Arrhenius equation. This leads to the fact that the observed changes of rate-constants of reactions are not great. For example, in the work [39] on dehydrogenation of C_2H_5OH and iso-C_3H_7OH on Ge, the specimens n-Ge (with addition of Sb) and p-Ge (with addition of Ga) were typically characterized by an activity five to six times greater. There were cases, e.g., during the study of hydrogenation of C_2H_4 and C_3H_6 on n- and p-InAs and InSb [41], when the more active n-specimens had a noticeably greater activation energy. Thus, the introduction of a dope into semiconductors, in contrast to the introduction of a dope into metals, has not as yet enabled one to obtain large order-of-magnitude changes in catalytic activity. Only in very rare cases is one successful in changing the activity more than one to two orders of magnitude. This was obtained, for example, in the work [34] on the oxidation of CO on NiO, where catalysis was carried out at near room temperature.

One of the authors of the electronic theory, Volkenshtein, determinedly speaks out against research to relate the catalytic activity of semiconductors with their type of conductivity if the discussions concern semiconductors of different chemical nature [16, p. 116]. Other not so categorical authors [7, 42–44] indicate that there is the existence of such a connection in a number of cases. Dell et al. [26], using literature data, showed that oxide semiconductors catalyzing the decomposition of N_2O are placed according to activity (according to the temperature at the beginning of the reaction, t) in the following order: NiO, Cu_2O, and CoO ($t < 400°C$); CaO, CuO, MgO, and CeO_2 ($t = 400$–$550°C$); Fe_2O_3, TiO_2, Cr_2O_3, Al_2O_3, and ZnO ($t > 550°C$). The p-semiconductors

1.1. ELECTROCONDUCTIVITY

are the more active catalysts, followed by insulators and n-semiconductors, respectively. These large differences in reaction initiation temperatures correspond to large differences in catalytic activity that do not occur during the study of a specific catalyst with a different quantity of introduced dopes. In this simplified model there are exceptions. One is CuO—a natural semiconductor falling in the second group. The other, listed among the few active catalysts that fall in the third group, is Cr_2O_3—a natural semiconductor that readily exhibits p-semiconduction [45]. In addition, Al_2O_3 more correctly should be included among insulators, although by special preparation a "black" aluminum oxide can be obtained [46] which conducts electrons.

In later works [47] it was shown that MgO also belongs to the slightly active catalysts for decomposition of N_2O ($>550°C$). Still less active are the other insulators: BeO, GeO_2, and SiO_2. Based on these data, Stone [48] suggests that the series obtained should be made more precise. The general tendency of catalytic activity diminishing during the transition from p- to n-semiconductors is observed, but insulators possibly are less active than n-semiconductors. In addition, the oxides of the transition metals appear to be the more active catalysts. Therefore, it is difficult to say whether their high catalytic activity is caused by the hole character of the semiconductor or by the presence of d-electrons.

Regular changes of catalytic activity of oxides in oxidation–reduction reactions can be explained by a change of the type of conductivity during the change of chemical structure of the basic oxide that forms the catalyst. The type of conductivity of a semiconductor can be measured according to the sign of thermoelectromotive force and the sign of the Hall effect [17]. For the oxides, the type of conductivity can be measured by the direction of the dependence of electroconductivity on oxygen pressure. On oxides—n-semiconductors—conductivity is reduced with an increase of oxygen pressure, but for p-semiconductors, it increases. As a rule, oxides of higher valences are prone to a partial reduction that leads to the appearance in the lattice of a stoichiometric excess of the metal and semiconduction of the n-type. Oxides of lower oxidation states are prone to partial oxidation, which results in a stoichiometric excess of oxygen and semiconduction of the p-type.

If one examines the oxides of metals in the basic valence states, one can find that regular changes of type of conduction are dependent upon the location of a metal in the periodic table. In Fig. 3 are pictured the stable oxides of metals for which there is information in the literature on the type of conductivity. With respect to alkaline earth oxides, information in the literature is contradictory. There is one explanation, e.g., that CaO, SrO, and BaO are hole semiconductors [17]; but according to other data [19] they are electron semiconductors. Cr_2O_3, ZrO_2, and ThO_2 can have n- and p-conductivity. Oxides of transition elements at the beginning of large periods have n-conductivity; at the end, usually p-conductivity. For example, in the fourth period, TiO_2, V_2O_5, Cr_2O_3, MnO, and Fe_2O_3 are n-semiconductors; CoO, NiO, and Cu_2O are p-semiconductors; but ZnO is again an n-semiconductor. In the fifth and sixth periods this dependence is also observed, although the given data are less significant. Upon a change of oxidation state, the type of conductivity may change. For example, CrO_2 and CrO_3 are n-semiconductors; but Cr_2O_3 is frequently a p-semiconductor. MnO, CoO, NiO, and FeO appear to be p-semiconductors, but CuO is an n-semiconductor.

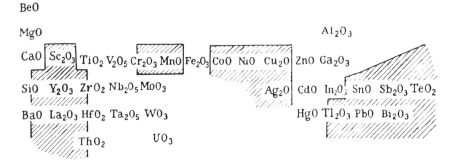

FIG. 3. Dependence of the type of conductivity of oxides on the position of the metal in the periodic table.

In the crosshatched area are those oxides which in the presence of low temperatures are predominantly of p-type conductivity. In the noncrosshatched area are those with n-type conductivity.

It is impossible, of course, to derive an absolute meaning from the observations presented. The type of conductivity, as a rule, is determined at low temperature and at pressure close to atmospheric. At catalytic conditions, the type of conductivity can be anything. In some cases a surface layer and bulk of an oxide have a different sign of conductivity. Even for such a well-studied electron semiconductor as ZnO, it was found that at high temperatures and oxygen pressures [50] and also during the addition of some oxides [51], ZnO can become hole-conducting.

The methods for changing the type of conductivity of sulfides, selenides, and tellurides to some extent reiterate the methods for oxides of these same metals. However, there is a distinction. For example, ZnO and ZnS are frequently obtained with electron conductivity, but ZnSe and ZnTe are frequently obtained with hole conductivity. For nonoxide semiconductors, a simple connection between type of conductivity and position of the element in the periodic table has not yet been successfully established (possibly because of a lack of data). For elementary semiconductors (Ge, Si, Se, As, and Te) and binary semiconductors, which are close in electronic and crystalline structure to the element (for example, GaAs and InSb), it is equally easy to obtain both electronic and hole types.

We notice that Eqs. (1) give a connection of the magnitudes of η^0, η^-, and η^+ not with the electroconductivity, but with the location of the Fermi level. In the majority of works, a connection between catalytic activity and magnitude and sign of electroconductivity is found. More recently the product of three factors has appeared

$$\sigma = e(n^-\mu^- + n^+\mu^+) \qquad (2)$$

where e is the charge of an electron; n^- and n^+ are the concentrations of free current carriers—electrons and holes, respectively; μ^+ and μ^- are the mobilities of electrons and holes.

The Fermi level determines the magnitudes of n^- and n^+. If the mobilities of an electron and a hole have nearly the same magnitude, then from Eqs. (1) and (2) one can draw a conclusion about the relation between electroconductivity and catalytic activity. This, however, is not always so. For example, in oxides of transition metals (Section 1.6) $\mu^+ \gg \mu^-$. In this case the oxide can have hole conductivity, although the concentration of free electrons is higher than that of free holes; and from such a relation between catalytic activity and electroconductivity one can arrive at false conclusions. At present nothing is known about the influence of mobilities of current carriers on catalytic activity. Data are presented below (Section 1.3) on the properties of the imperfect semiconductor, Ga_2Se_3. Its catalytic activity, as well as the width of its forbidden zone, has a value between ZnSe and GaAs. At the same time, the mobility of the current carriers in Ga_2Se_3 is on the order of one to two times less than in GaAs and ZnSe. Thus, mobility apparently does not exhibit a direct influence on catalytic activity.

1.2. Work Function

It was shown previously that the chemisorptive and catalytic activities of a semiconductor can be determined by the position of the Fermi level. The magnitudes of ϵ_s^- and ϵ_s^+ in Eqs. (1) specify the position of the Fermi level on the surface of a crystal, which, in the general case, differs from its position in the bulk phase [16]. Chemisorbed particles on surfaces result in either an excess of electrons or of holes (i.e., a surface charge). This charge is balanced in the semiconductor layer near the surface by the formation of a bulk charge equal in magnitude but opposite in sign. As a result, the energy levels near the surface are distorted, and the position of the Fermi level upon formation of a negative charge on the surface can be described by

$$\epsilon_s^- = \epsilon_v^- + \Delta\epsilon \qquad \epsilon_s^+ = \epsilon_v^+ - \Delta\epsilon \qquad (3)$$

and during formation of a positive charge

$$\epsilon_s^- = \epsilon_v^- - \Delta\epsilon \qquad \epsilon_s^+ = \epsilon_v^+ + \Delta\epsilon \qquad (4)$$

where ϵ_v^- and ϵ_v^+ are the position of the Fermi level in the bulk of the crystal; $\Delta\epsilon$ is the magnitude of the distortion of the zones.

The distortion of the zones during the charging of a surface, according to Volkenshtein [16], is represented in Fig. 4.

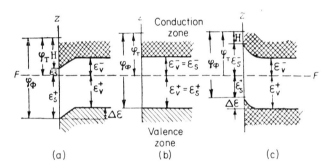

Fig. 4. Distortion of zones on the surface of a semiconductor according to Volkenshtein [16]: (a) a positively charged surface, (b) a neutral surface, and (c) a negatively charged surface. The surface coincides with the z axis.

The value of ϵ_v is determined by measurement of electroconductivity, as shown in Section 1.1. One can determine the value of the zone distortion that occurs upon adsorption by its relationship to the work of escape of a semiconductor electron. The value can be determined by measuring thermoelectronic emission or by measuring the change of contact difference of potentials [52]. Thus, the value of the work function is obtained; this is the energy of transfer of an electron from the Fermi level in a vacuum, i.e., the statistically average work of the removal of an electron from a crystal. This is also called the thermoelectric or thermodynamic work function φ_T. From Fig. 4 it is seen that φ_T is represented as follows:

$$\varphi_T = \epsilon_s^- + H \qquad (5)$$

1.2. WORK FUNCTION

In its turn

$$H = H_0 \pm \mu$$

where the value H_0 is determined by the chemical nature of the semiconductor (sometimes H_0 is called the affinity of a crystal to an electron); μ is the drop in potential caused by the polarization of adsorbed molecules.

Usually $\mu \ll \epsilon_s^- + H_0$. From Eqs. (3)–(5) it follows that φ_T will be a property of a simple substance at high temperatures. When the deviation of the zones is not great, the semiconductor becomes a natural conductor with its Fermi level near the middle of the forbidden zone. In this case $\epsilon_s^- = U/2$ (where U is the width of the forbidden zone) and $\varphi_T \approx H + U/2$.

The photoelectric work function φ_Φ will be the property which more accurately indicates the properties of a simple substance at low temperatures.

$$\varphi_\Phi = \varphi_T + \epsilon_s^+ = H + U$$

It is determined by the wavelength of a surface photo effect and characterizes the energy of electrons, drawn out of the upper edge of the valence zone.

During adsorption, the thermoelectronic work function is changed in magnitude

$$\Delta\varphi_T = \Delta\epsilon + \Delta\mu \tag{6}$$

where $\Delta\mu$ is the change of dipole moment of the adsorbate.

The photoelectric work function is changed in magnitude $\Delta\varphi_\Phi = \Delta\mu$. Measuring $\Delta\varphi_T$ and $\Delta\varphi_\Phi$ separately, one can determine $\Delta\epsilon$ [8].

Often, $\Delta\varphi_T$ is close to the magnitude of $\Delta\epsilon$. In accordance with Eqs. (1), (3), and (4), the magnitude of $\Delta\epsilon$ (or $\Delta\varphi_T$) can influence directly the equilibrium capacity of active particles on a surface, i.e., catalytic activity. The influence of $\Delta\varphi_T$ on the activation energy of a reaction can also be obtained, assuming, according to Roginskii [53], that $\Delta\varphi_T$ influences not the equilibrium configuration of the adsorbed molecules, but the charge of the activated complex. In the general case, the deformation of the zones, $\Delta\epsilon$, can be specified not only by charges on the adsorption surface planes but also by nonadsorption surface conditions (the levels of Tamm, surface defects). The smaller the size of the crystal, the greater the role of $\Delta\epsilon$ in the total value of ϵ_s. Finally, at a sufficiently small size, it can happen that $\Delta\epsilon$ will completely determine ϵ_s, i.e., the properties of the surface; and there will be no dependence on the properties of the bulk of the basic crystalline lattice. This case is the so-called "quasi-insulated surface" [16]. Theoretical analysis [54] shows that for oxide semiconductors such as Cu_2O, V_2O_5, and ZnO, independence of electronic properties of the surface from the bulk occurs upon breaking the crystals to a value of 10^{-5} cm, i.e., upon attainment of sufficiently large specific surfaces (10–100 m²/gm). For the hole specimens Ge, AlSb, and InSb, the influence of (crystal) breaking on electronic properties was studied experimentally [55]. The surface properties begin to predominate over bulk properties at sizes of crystals less than 1 μ. In these semiconductors there is a chemical bond that is completely, or to a considerable extent, covalent. The breaking of a chemical bond (for example, at the crack of a crystal) creates "free valences," which produce surface levels. On semiconductors (e.g., Ge), they are created by s-orbitals which are energetically more favorable than p-orbitals and can accept an electron, i.e., they appear to be acceptor levels. However—for electronic specimens in this case—the rectification, not the deformation of the zones, must occur during (crystal) breaking, i.e., compensation of the donor levels by the surface acceptor levels and approach to natural conduction.

In the final analysis, the value of $\Delta\epsilon$ is also determined by the properties of the lattice; however, the dependence is complex and is in terms of an explicit function which has not yet been evaluated. It has even been shown [56] that, during displacement of the Fermi level up or down in the bulk of the crystal (see Fig. 4), the Fermi level on the surface is displaced in the same direction, providing that the surface levels are not concurrently changed:

$$d\epsilon_s^+/d\epsilon_v^+ = d\epsilon_s^-/d\epsilon_v^- \geqslant 0 \qquad (7)$$

Data are available [57, 58] which suggest that even at room temperature the surface states do not completely shield the bulk for semiconductors of the type A^{IV}, $A^{III}B^{V}$ and those near to them, having a small surface (less than 1 m²). With an increase of temperature, i.e., an approach to temperatures of catalysis, the role of surface states drops. Even for substances with a small width of forbidden zone, such as InAs and InSb, it has been shown that in the case of crystals of size greater than 10^{-4} cm, the temperature dependence of the width of the forbidden zone is more important for change of position of the Fermi level than are the surface states [59]. The changes of φ during adsorption are usually not great and do not exceed several tens of electron volts. As an example, Fig. 5 shows the change of the work function, $\Delta\varphi$ during adsorption of isopropyl alcohol on ZnO (according to the data of Enikeev et al. [60]). The value of the work function $\Delta\varphi$ passes through a maximum (simultaneously with this, the value of the contact potential V_k passes through a minimum) upon the alcohol's occupying of $\sim 10\%$ of the surface of the

1.2. WORK FUNCTION

adsorbent. Enikeev et al. [60] explain the left part of the curve by strong irreversible adsorption causing the deviation of the zones, $\Delta\epsilon$; the right part—by a weak reversible adsorption, leading to the polarization of the adsorbed molecules μ, balancing $\Delta\epsilon$. The total change of $\Delta\varphi$ in this case amounts altogether to 110 mV, and that corresponds to 2.5 kcal/mole. During the study of the adsorption of oxygen on ZnO [64] the maximum value of $\Delta\epsilon$ appeared equal to 0.46 eV.

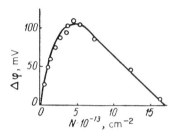

FIG. 5. The dependence of the change of the work function $\Delta\varphi$ on the degree of coverage of the surface of ZnO by isopropyl alcohol [60].

Significantly larger changes in $\Delta\varphi$ were observed upon the introduction of dope into the bulk of the solid. In Fig. 6 is shown the dependence of the change of activation energy of adsorption of O_2 on ZnO on the value of $\Delta\varphi$ [53, 62]. Pure specimens of ZnO and specimens of ZnO with additions of Li_2O and $ZnSO_4$ were studied. The change of φ

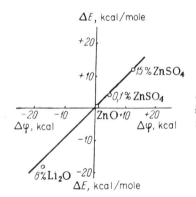

FIG. 6. The change of energy of activation of adsorption of O_2 and the work function upon introduction into ZnO of donor and acceptor impurities [53, 62].

(in kilocalories) exactly equaled ΔE (the angle of the slope of the line in Fig. 6 is equal to 45°). Although the interval of $\Delta\varphi$ and ΔE amounts to 1.5 eV (greater than 30 kcal), strangely, the rates only change in the ranges of one to two orders of magnitude because of strong compensative effect of E and log k_0 (where k_0 is the preexponential factor in the Arrhenius equation. During the study of the oxidation of propylene on Cu_2O, as well as on pure and impure compounds of Li, Pb, Fe, Cl, S, and P, the change of the logarithm of the rate of reaction also responded to the change of the work function $\Delta\varphi$ [63].

The principal components, determining the work function, according to Eqs. (3)–(5), ϵ_v and H, depend on the properties of the crystalline lattice. In particular, the magnitude of ϵ_v^- for electronic semiconductors is approximately zero. In this case, disregarding $\Delta\epsilon$ and μ, we obtain approximately $\varphi_T \approx H$. Analogously, for natural semiconductors, $\varphi_T \approx H + U/2$; for hole semiconductors $\varphi_T \approx H + U$ [where U is the width of the forbidden zone (see Section 1.3)]; and this width also appears to be a constant of the lattice. Consequently, n-semiconductors must have a work function which is less than that for p-semiconductors. Information about the absolute values of the work function φ for the most prevalent semiconductor catalysts—oxides, is more inconsistent than for the metals [64]. Nevertheless, if one compares the available data, one can draw definite conclusions about the relation of φ to position in the periodic table of the metal forming the oxide.[3] In Fig. 7 are

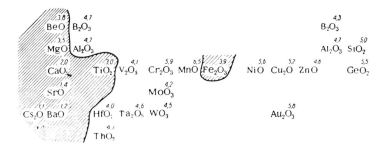

FIG. 7. The relation between work function of oxides with the position in the periodic table of the metal forming the oxide. Crosshatched oxides have a value of $\varphi < 4$ eV.

presented values of φ for stable oxides set up in a framework of the entire periodic table. To some extent, the diagram resembles Fig. 3.

Figure 8 shows the change of φ in a series of oxides of the first large period. The oxides that are n-semiconductors (TiO_2 and Fe_2O_3) have a significantly smaller work function than those that are p-semiconductors (MnO, NiO, and Cr_2O_3). Of the dielectrics, alkaline oxides and alkaline earth oxides have especially low values of φ (for Cs_2O—1.08 eV, for BaO—1.2 eV); acidic oxides have high values of φ (Al_2O_3, SiO_2, and GeO_2—$\sim$5 eV). The total interval of the change of φ for all oxides is

---

[3] A résumé of the data for the work function of semiconductors, as for their other properties, is given in the appendix at the end of this book.

1.3. WIDTH OF THE FORBIDDEN ZONE

greater than 5 eV. This significantly exceeds the changes of $\Delta\varphi$ which were obtained by the introduction of impurities into a specific oxide.

Therefore, there is a very interesting correlation of catalytic activity with work function of various catalysts, although such a correlation, as Volkenshtein indicated, does not result from existing electronic ideas in catalysis.

Boreskov and Popovskii [65] studied the oxidation of hydrogen on oxides of metals of the fourth period and found the following order of activity: $Co_3O_4 > CuO > MnO_2 > Fe_2O_3 > ZnO > Cr_2O_3 > V_2O_5 > TiO_2$.

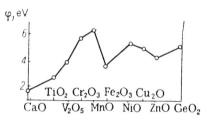

FIG. 8. The change of the work function in a series of oxides of metals of the fourth period.

On the basis of literature data, the authors cite the sequence of oxides in order of decreasing φ: $CuO > NiO > V_2O_5 > ZnO > TiO_2$, which, in principle, coincides with the sequence for activity. In the opinion of the authors, the higher the catalytic activity, the higher the work function φ, i.e., the lower the level of chemical potential of the electrons. This arrangement of oxides, in general, is verified by data from the table presented in the appendix of this book. However, Morozova and Popovskii [66] obtained in their work the following sequence according to φ: $TiO_2 > V_2O_5 > Fe_2O_3 > Cr_2O_3 > Co_3O_4 > CuO > ZnO > NiO$, almost the opposite of the first. For drawing of similar comparisons between φ and catalytic activity, values of φ at conditions and temperatures of catalysis can be made use of expeditiously. These then, can be related to data in accordance with the relationships of φ and catalytic activity of pure and modified specimens of specific semiconductors.

1.3. WIDTH OF THE FORBIDDEN ZONE

With respect to the phenomenon of intrinsic conduction of a semiconductor, the Fermi level, assuming equal statistical weights of the valence zone and the conduction zone, is found in the middle of the forbidden zone: $\epsilon_v \approx U/2$. In this case [67, 68], doping will have no influence on the location of the Fermi level; and, therefore, catalytic activity must only weakly depend on amount and character of doping and past-history of the sample of catalyst. For intrinsic conduction, a relationship must be shown between the catalytic properties of a semiconductor and its substantive properties (in contrast to doped properties), i.e., a relationship with width of the forbidden zone of a semiconductor and with position in the periodic table of the elements forming the catalyst.

In regard to the correlation between catalytic activity in oxidation-reduction reactions and the width of the forbidden zone, it has long been recognized that few insulators (which have a large width of forbidden zone U) are catalytically active; semiconductors (with a small value of U) are more active; and metals (which have $U = 0$) possess the greatest catalytic activity.[4] Nevertheless, the majority of investigators have paid attention to the relationship between catalytic activity and type of conduction, but not with U. In reality, many oxidation-reduction catalytic reactions which are catalyzed by semiconductors proceed at temperatures that correspond to the region of intrinsic conduction of the catalyst. Thus, e.g., Cr_2O_3 is used as a commercial dehydrogenation catalyst at 500–600°C [69], i.e., in the region of its intrinsic conduction [70]. There is a very probable predominance of intrinsic conduction when Cu_2O is used as a catalyst for the oxidation of propylene to acrolein near 300°C. Natural conduction predominates in the majority of cases of high temperature catalysis, in particular for complete oxidation of hydrocarbons, conversion of methane, etc.

Krylov et al. [67] studied the dehydrogenation of isopropyl alcohol on chalcogens of zinc and found that, in the sequence ZnO, ZnS, ZnSe, and ZnTe, catalytic activity increases. The activation energy of dehydrogenation of iso-C_3H_7OH drops from 25 to 46 for ZnO to 7–11 kcal for ZnTe; temperature for reaction initiation decreases from 120 to 0°C. Correspondingly, the width of the forbidden zone drops: 3.3 eV for ZnO, 3.7 eV for ZnS, 2.7 eV for ZnSe, and 2.1 eV for ZnTe. ZnO is an exception (the width of the forbidden zone for ZnO is less than that for ZnS); however, this is perhaps related to the fact that ZnO crystallizes into the wurtzite lattice instead of the sphalerite lattice (as do the remaining three compounds). In Fig. 9, the energy of activation of dehydrogenation of an alcohol, E, in the sequence ZnO → ZnTe, is compared with the half-width of the forbidden zone, $U/2$.

The data are the result of catalysis in the area of intrinsic conduction.

---

[4] Note that at present there is documented information regarding the catalytic activity of transition metals, which as a rule, are more active than oxides. At the same time, we know very little about the catalytic properties of such metals as Sn, Pb, Hg, K, Na, and others. It is entirely possible that some rules will be found for these metals.

1.3. WIDTH OF THE FORBIDDEN ZONE

Actually, even for ZnS with a very large width of forbidden zone (3.7 eV), data are available [71] that show that temperatures of catalysis (100–300°C) correspond to the area of intrinsic conduction for this semiconductor. In regard to ZnO ($U = 3.3$ eV), it has been shown [72] that when dispersion of ZnO is increased to 10^{-5} cm, its electroconductivity falls, and ZnO approaches intrinsic conduction. (Crystals of ZnO are in this size range in catalysts.) However, in the series of chalcogens of zinc, other properties of the catalyst change simultaneously with width of the forbidden zone, in particular, the type of lattice and the lattice parameter. Therefore, it would be interesting to verify the conclusion about the correlation of catalytic activity with width of the forbidden zone on a larger number of samples, selected so that changes of other parameters are minimized.

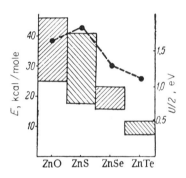

FIG. 9. The change of activation energy of dehydrogenation of isopropyl alcohol (cross-hatched columns) versus the half-width of the forbidden zones (dotted line) in the sequence ZnO → ZnTe.

The semiconductors of the so-called isoelectronic series appear to be convenient subjects for comparisons of such a nature [73, 74]. For example, an element of Group IV (A^{IV}) forms an isoelectronic series of binary compounds by combining with elements found at a distance from Group IV: $A^{III}B^{V}$, $A^{II}B^{VI}$, $A^{I}B^{VII}$. Semiconductors of the isoelectronic series have like crystalline structures and the same coordination numbers. The average number of electrons present in the outer orbital of one atom is the same (in crystals with the structure of sphalerite it is equal to 4), and the total number of inner electrons is the same. The isoelectronic germanium series formed by elements equidistant from Ge in the same period is: Ge (A^{IV}), GaAs ($A^{III}B^{V}$), ZnSe ($A^{II}B^{VI}$), and CuBr ($A^{I}B^{VII}$). All of these substances have the crystalline structure of sphalerite, equal crystalline lattice spacing of 5.65 Å, and are characterized by a width of the forbidden zone, which increases from 0.74 eV for Ge to 2.9 eV for CuBr. In the next period following Ge, one can evolve an analogous isoelectronic series: α-Sn, InSb, CdTe, and AgI. It is interesting to note that substances of the type $A_2^{II}B_3^{VI}$, i.e., of structure intermediate between $A^{III}B^V$ and $A^{II}B^{VI}$, occupy an intermediate position, according to width of the forbidden zone U. For example, the value of U for GaAs is equal to 1.45 eV, Ga_2Se_3—1.81 eV, ZnSe—2.66 eV; in the other isoelectronic series for InSb—0.23 eV, In_2Te_3—1.12 eV, and CdTe—1.5 eV. Compounds of the form $A_2^{III}B_3^{VI}$ have the same structure as sphalerite, but $\frac{1}{3}$ of their cationic nodes are statistically free, i.e., they contain gaps in the lattice (defects). Because of this, some shrinking of the lattice is observed: The lattice parameter of Ga_2Se becomes equal to 5.42 Å instead of 5.65 Å—the average value for the isoelectronic series of Ge; lack of sensitivity to doping is exhibited. Even down to low temperatures these semiconductors appear to be natural [75]. Compounds of the form $A_2^{III}B_3^{VI}$ form solid solutions at all concentrations with $A^{III}B^V$, e.g., $GaAs \cdot Ga_2Se_3$. These solid solutions are intermediate in electrical properties [76].

The study of dehydrogenation of iso-C_3H_7OH and the decomposition of hydrazine N_2H_4 to $N_2 + NH_3$ on Ge, GaAs, Ga_2Se_3, ZnSe, and CuBr [77] showed that catalytic activity, k, per square meter of surface of these zero-order reactions decreases with increase of U approximately according to the law:

$$\log k = a - bU \tag{8}$$

where a and b are constants.

The dependence of catalytic activity on U is shown in Fig. 10. According to catalytic and electrical properties, the imperfect semiconductor Ga_2Se_3 occupies an intermediate position between the semiconductors ZnSe and GaAs which lie in the isoelectronic series of

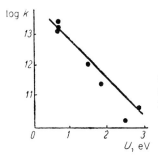

FIG. 10. The dependence of the logarithm of initial rate of decomposition of iso-C_3H_7OH at 200°C on width of the forbidden zone for semiconductors of the isoelectronic series of germanium.

germanium. Further investigation [78] showed that one can develop a correlation of rate of an oxidation–reduction reaction with the width of the forbidden zone for semiconductors of various isoelectronic series. The dependence of the dehydrogenation rate of iso-C_3H_7OH in an adsorbed layer at 200°C on the width of the forbidden zone of semi-

1.3. WIDTH OF THE FORBIDDEN ZONE

conductors (both with the structure of sphalerite and with the structure of wurtzite, e.g., ZnO) is presented in Fig. 11. In spite of the different origin of the specimens, the dependence represented by Eq. (8) is preserved. The width of the zone in Fig. 11 ranges from approximately 1 to 1.5 orders of magnitude. When U changes by 3.5 eV, catalytic activity is increased simultaneously by five orders of magnitude.

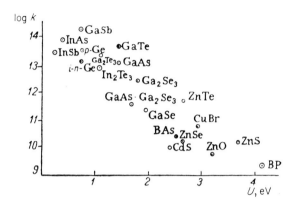

FIG. 11. The dependence of the logarithm of the rate of dehydrogenation of iso-C_3H_7OH in the adsorbed layer at 200 °C on width of the forbidden zone of a semiconductor.

At 200°C, a change of activation energy of 11 kcal/mole corresponds to the change of the reaction rate by a factor of 10^5—at constant pre-exponential factor k_0. Even though the observed change of activation energy for dehydrogenation of alcohol, E_{H_2}, was found to be close to this value, a method for relating the change of E with log k_0 could not be established. More than that, both on a less active catalyst (BAs) and on a more highly active one INAs), the values of E have nearly the same magnitudes: 16.4 and 14.8 kcal/mole. In Fig. 12 the values of E and log k_0 [27, 77] are plotted on one graph. In some cases, e.g., for three specimens of Ge, the points lie on the same line. In the majority of cases, however, there is no "compensation effect," i.e., no linear relationship (p. 12) of E and log k_0. The less active catalysts BAs, CdS, and ZnS had values of E of the same order as did active catalysts, and consequently, markedly lower values of log k_0.

If, for the semiconductors that were studied, catalysis proceeded in the region of natural conduction, the Fermi level in them would be found approximately in the middle of the forbidden zone of the semiconductor: $\epsilon_s \approx U/2$. Then, in accordance with Eqs. (1) or other equations of a similar nature, one would anticipate that at equivalent conditions, as one proceeds from one semiconductor to another, E will change by $0.5(U_2 - U_1)$. For the total studied interval it will change by $0.5 \times 3.5 \times 23 = 40$ kcal. The change, ΔE, that was found experimentally, is significantly less. Therefore, it is possible that in the present

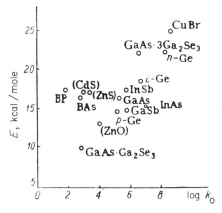

Fig. 12. The value of E and log k_0 for the dehydrogenation of iso-C_3H_7OH on various semiconductors.

sequence catalysis proceeds in the region of doped and hybrid (the transition from doped to intrinsic) conduction. The dependence of the rate of reaction on U during this transition will be less abrupt. A schematic diagram of the dependence of the rate constant on reciprocal temperature (Fig. 13) helps to explain data that have been obtained.

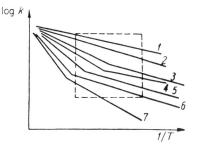

Fig. 13. Schematic representation of the dependence of log k on $1/T$ for various semiconductors. The region of measurement is enclosed by a dotted line. The lines are numbered in the order of increasing U.

It is apparent that if the most active catalysts are found in the region of intrinsic conduction (the upper lines), if the catalysts of intermediate activity are found in the area of hybrid conduction (the middle lines) and the less active in the area of doped conduction (the lower lines), a result of the small (approximately 100°C) interval of temperature, the apparent change of E for various catalysts will not be great. The concentration of current carriers (in the present case, apparently, the

1.3. WIDTH OF THE FORBIDDEN ZONE

concentration of free holes) is changed with temperature approximately according to the exponential law

$$n^+ \approx n_0^+ \exp(-\epsilon_s^+/kT)$$

In the region of doped conduction both constants n_0^+ and ϵ_s^+ have significantly lower values than in the intrinsic conduction region. If the rate constant of the catalytic reaction is proportional to n^+, then in selected temperature intervals the catalysts will possess maximum values of k_0 in the region of intrinsic conduction (Fig. 13). The high value of log k_0 for n-Ge can also be explained by the fact that, in this case, a layer of natural conduction is formed close to the surface. In p-Ge, the Fermi level lies very near to the valence zone; therefore, the experimentally found difference of activation energy for n- and p-Ge, $\Delta E_{n,p}$, is equal to half of the width of the forbidden zone, $U/2$ (see Fig. 2). However, this is most probably the value $\Delta E_{i,p}$ (where i is an index for a semiconductor with intrinsic conduction).

The straight lines in the region of intrinsic conduction in Fig. 13 can be expressed by one value. In reality, the value of n_0^+ is determined by the structure of the energy levels, which vary for different semiconductors. For the natural semiconductor Ge [79, 80] at 500°K we can obtain

$$n_0^+ = n_0^i \cong \frac{(2\pi mkT)^{3/2}}{h^2} T^3 \approx 10^{20} \text{ cm}^{-3}$$

where m is the mass of the current carrier and h is Planck's constant.

For the surface zone, by raising this value to the $\frac{2}{3}$ power, one obtains $n_0^{is} = 10^{14}$ cm^{-2}. In the doped region, the pre-exponential factor k_0, as well as the energy of activation, will be smaller than in the intrinsic region. If it is assumed that all the dopants are ionized, then in the surface zone $n_0^+ = N^{2/3}$ (where N equals the concentration of dopants). If $N = 10^{14}$ cm^{-3} (the minimal value for semiconductors [78]), then $N^{2/3} = 10^9 - 10^{10}$. However, note that log $n_0^i \approx$ log n_0^+ for intrinsic and doped semiconductors. The difference, a value equal to 4-5, approximately corresponds to the interval of Δ log k_0 [78].

The range of temperatures at which catalysis proceeds often falls within the region of hybrid conduction. Consequently, for specific semiconductors with various percentages of doping, the compensation effect (the relationship of log k_0 to E) for relatively small changes of the rate constant k will be observed. The compensation effect in the purely doped region will be observed for large changes of E and k_0 [34]. On the other hand, for many semiconductors in the intrinsic conduction region this effect is encountered less often for other, possibly chance reasons.

In Fig. 12 there is no such effect. The apparent "stretching out" of the picture in the diagonal directions arises because the regions in the upper left and lower right corners are difficult to reach by experiment (too large or too small reaction rates). The indicated explanation of the relationship between $\log k_0$ and E is analogous to the explanation given by Cremer [81], according to which the relationship of $\log k_0$ to E results from the presence on the surface of the catalyst of two or three types of centers each with sharply different values of $\log k_0$ and E.

The width of the forbidden zone, and not the type of conduction, is likely to be the determining factor in the case of catalysis and adsorption on semiconductors with small width of the forbidden zone. In a study of the adsorption of O_2 on Ge, Dell [82] did not detect differences in the adsorptive capacity of n- and p-Ge. He explained the result that he found, for adsorption on semiconductors with a small value of U, on the basis that electrons participate in interzone transfer. The magnitude of chemisorption of O_2 on Ge, according to Dell, is not related to the initial concentration of electrons in the conduction zone. This is in contrast to its initial rate.

Note that the absence of a difference of chemisorptive properties of n- and p-Ge is easier to explain by the presence on Ge of a quasi-insulated surface [16] (Section 1.2).

A number of authors [24, 83–86] have paid attention to a relationship of some kind between activation energy of a catalytic reaction and width of the forbidden zone. According to their data on the catalytic decomposition of HCOOH, H_2O_2, and C_2H_5OH and the hydrogenation of CO, C_2H_2, C_2H_4, and C_3H_6 on Ge, Si, InSb, InAs, and AlSb, in some cases a difference in the activation energies of the reaction on n- and p-semiconductors, $\Delta E_{n,p}$, is observed. For some reactions in the studied systems, the values of $\Delta E_{n,p}$ were usually equal to $U/2$ (decomposition of HCOOH on Ge and AlSb); for others $\Delta E_{n,p} = U$ (for hydrogenation of C_2H_4 on Ge and InSb, decomposition of HCOOH on InSb, and decomposition of C_2H_5OH on InSb and InAs); but also $\Delta E_{n,p} = 0$ (for decomposition of HCOOH on Si and decomposition of C_2H_5OH on AlSb). In the work of Frolov et al. [39] on dehydrogenation of C_2H_5OH and iso-C_3H_7OH on Ge, a value of $\Delta E_{n,p} = U/2$ was obtained. The changes of rate constant during this study were small and at the experimental temperatures used they did not exceed one order of magnitude. During hydrogenation of C_2H_4 on Ge and CO to CH_4 on InSb and InAs, a sharp increase of activation energy was observed with transition from low to high temperatures. The activity of catalysts according to rate constants of the decomposition of HCOOH are as follows: AlSb > InSb > InAs > Ge > Si, i.e., an order opposite to

1.3. WIDTH OF THE FORBIDDEN ZONE

that which would be expected from the values of U. One must also exclude AlSb, which has a high value of U and also high activity. However, as the authors themselves show, in the course of the reaction AlSb is decomposed into Al_2O_3 and Sb. The results are easy to explain, if one assumes that in these works, catalysis proceeds near the point of transition from doped conduction to intrinsic conduction. It is possible that when $\Delta E_{n,p} - U$ the Fermi level of the n-semiconductors lies very close to the conduction zone, but the Fermi level of p-semiconductors lies close to the valence zone, i.e., true doped conduction is obtained. When $\Delta E_{n,p} = U/2$ we have in actuality $\Delta E_{i,p}$ or $\Delta E_{i,n}$, but when $\Delta E_{n,p} \approx 0$ we are in the region of intrinsic conduction.

These facts that have been discovered concerning the regular relationship between catalytic activity and width of the forbidden zone of a semiconductor allow one, in a number of cases, to transfer the question of rules of catalyst selection into the sphere of rules of semiconductor selection.

The fact that catalysis occurs in the areas of both intrinsic and transitional conduction of a semiconductor is one of the possible causes for the existence of a relationship between catalytic activity and width of the forbidden zone U. But, in addition, in the region of doped conduction, one can obtain a relationship between activation energy of a catalytic reaction and U. Consider two individual constants k_1 and k_2, the first of which will be proportional to the concentration of electrons ($k_1 = k_1'n^-$) and the second to the concentration of holes ($k_2 = k_2'n^+$). If the product of these two constants comprises the overall rate constant of the reaction, then the width of the forbidden zone appears to be a component of the activation energy, because the product of n^+n^- is determined by the width of the forbidden zone:

$$k = k_1 k_2 = k_1' k_2' n^- n^+ = k_1' k_2' n_i^{\ i} = k_1' k_2' n_0^{ia} e^{-U/kT}$$

One might assume that such an explanation of the relationship between activation energy and width of the forbidden zone would be correct for chain surface reactions, where the steps of initiation, propagation, and termination of the chain are accelerated by the current carriers. In spite of the many theories of chain surface catalysis, there are almost no proven cases of catalysis proceeding by such a mechanism. One of the possible examples is examined in Section 5.3. Still another possible explanation of the influence of U on catalysis is that behind this influence is concealed a relationship of U to some other value which directly determines catalytic activity. It is expedient, therefore, to examine the properties of a semiconductor on which the width of the forbidden zone depends.

A regular relationship is found between the width of the forbidden zone and the chemical nature of a substance. This is determined by the location of the elements in the periodic table of Mendeleev. Ioffe [87] has surmised that the basic properties of a semiconductor are not specified on the basis of the periodic structure of a crystal but by "neighboring order," i.e., the type of chemical bond and the position of the atoms in space. The discreteness of the energy spectrum of a condensed solid is the result of discrete levels of individual atoms, and not the periodicity of the lattice as such. The interaction of levels in a dense substance leads to their diffuseness and to the emergence of a zone.

These opinions were of use in later studies regarding the rules for change of U in semiconductors and they have withstood the test of time. At present, the clearest rules are had for sp-semiconductors that do not contain transition elements—in particular for semiconductors with the structure of diamond, sphalerite, and wurtzite. These have tetrahedral coordination and sp-hybridization. Regarding semiconductors with d-bonds, see Section 1.6.

According to current ideas [76, 88–92], the value of U depends on three factors: (1) the degree of ionization of the bond, (2) the atomic weight of the elements producing the semiconductor, and (3) their polarizability. We shall examine the role of each of these factors.

The significance of the degree of ionization of the bond is shown in many works by quantum-mechanical computations, the first of which were those of Seraphin [93], Gubanov [94], and Adavi [95], and also in more recent publications. Gubanov, for example, examined the lattice of sphalerite AB (which is like the diamond lattice) in a potential field. This imposes a disturbing potential that produces opposite signs on the nuclei of both kinds of atoms A and B. Using various methods of computation, these authors proceeded to the identical result: U increases with an increase in the difference of potential of atoms A and B. In the isoelectronic series, e.g., in the series Ge, GaAs, ZnSe, and CuBr, proceeding from elementary semiconductor A^{IV} to $A^{I}B^{VII}$, the asymmetry of potential (or difference of charges) on neighboring atoms increases, consequently increasing the ionization of the bond and the width of the forbidden zone.

The degree of ionization of the bonds is mostly characterized by the difference in electronegativities Δx (Section 1.5). For semiconductors of an isoelectronic series, in which the structure is not changed, an almost linear connection between U and Δx results. One can find in the literature a series of empirical and semiempirical methods, relating U and Δx [74, 80, 88, 90, 96]. In Fig. 14 the values of the width of the forbidden zone U are compared with Δx for all of the semiconductors, for which

1.3. WIDTH OF THE FORBIDDEN ZONE

suitable information is found (see the table in the Appendix). In Fig. 14 it is evident that, even for semiconductors having different crystalline structure and type of hybridization, a good correlation between U and Δx is obtained. Oxides of transition metals do not fit the correlation.

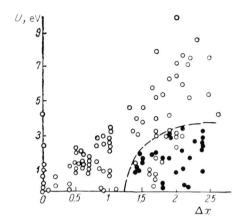

FIG. 14. The dependence of the width of the forbidden zone on electronegativity difference. The filled points pertain to oxides of transition metals.

For semiconductors, the width of the forbidden zone U, as a rule, decreases with increasing atomic weight of a specific group of the periodic table. This is true for elements (e.g., in the series C, Si, Ge, and Sn), and also for one of the atoms entering into a semiconductor (ZnO, ZnS, ZnSe, and ZnTe), or both of the atoms (BeO, MgO, ZnSe, and CdTe). To relate U to the characteristics of the valence shells of atoms, Pirson [88] suggests the use of the mean value of the principal quantum numbers of the atoms entering into a compound: $\bar{N} = \sum c_i N_i / \sum c_i$, where N_i is the principal quantum number of the ith atom; c_i is the number of i atoms in the chemical formula of the compound. Upon an increase of the atomic weight, the bonding of electrons with nuclei in valence orbitals is weakened or, as is often said, it is "metallized" [89]. In several series, the forbidden zone, in general, fades away. For example, in the series of elements of Group IV, the last of these, Pb, is a typical metal. Sometimes a linear relation between the width of the forbidden zone of a semiconductor and the total atomic number of the electropositive and electronegative elements is observed [97].

$$U = A - B(z_1 + z_2) \qquad (9)$$

Syushe [90] and Goodman [98] assume that the width of the forbidden zone, i.e., the activation energy of electrons of a semiconductor, can be broken down into homopolar U_H and ionic U_I components: $U = U_H + U_I$. Whereas U_H was found [98] to have a linear dependence on $1/d^2$ (where d is the interatomic distance), U_I depends on the difference of electronegativities Δx and the total atomic number $z_1 + z_2$:

$$U_I = \Delta x + \beta - \gamma \log(z_1 + z_2) \tag{10}$$

β and γ are constants depending on kind of hybridization. Byub [80] obtained for binary compounds of the type $Me_a X_b$ the formula:

$$U = c \frac{n_X - n_{Me}}{z_{Me} + z_X} \tag{11}$$

where n_X and n_{Me} are the number of valence electrons of the anion and cation, respectively; z_X and z_{Me} are the atomic numbers of the anion and cation.

For evaluation of the polarization of the bonds, i.e., the movement of electrons in the direction of the bonds, that results from the difference of charges of the atoms, Folberts [92] suggested the use (for polarization of the substance $A^{III}B^V$) of the value

$$J = (E_{Ion}/E_{cov})(z_{III} + z_V) \tag{12}$$

where E_{Ion} is the excess ionic energy (according to Pauling [99]); E_{cov} is the covalent energy of the bond; z_{III} and z_V are the atomic numbers of elements A^{III} and B^V.

In the series AlSb, GaAs, and InP which have the same mean principal quantum number, $\bar{N}$, the values of J and U decrease from AlSb to InP. Pirson [88], to evaluate the influence of polarization on U, makes use of the difference of electronegativities, Δx, and the ratio of the radii of the cation and anion r_{Me}/r_X. The lower the value of Δx, the stronger the polarization and the lower the effective charge of the atoms. The difference in values of r_{Me}/r_X allows us to explain why, for example, such compounds as InAs and ZnTe, which have the same values of $\bar{N}$ and Δx, have sharply different values of U (0.36 and 2.12 eV).

There is a widely known [52] empirical relation between U and the dielectric permeability ϵ:

$$U\epsilon^2 = \text{constant} \tag{13}$$

For semiconductors with the structure of diamond, sphalerite, as well as several others [100], the value of the constant happens to be 159 eV, which was substantiated theoretically.

1.3. WIDTH OF THE FORBIDDEN ZONE

In view of the fact that in binary compounds, in particular in ionic crystals A^IB^{VII} and $A^{II}B^{VI}$, the zone of conduction corresponds to the s-levels of the atom of metal A, and the valence zone to the p-levels of the atom of the nonmetal B (in oxides this corresponds to the $2p$-zone of oxygen), for computation of U one can draw upon simple formulas [101] of the type:

$$U = Me^2/r + L - I \qquad (14)$$

where M is the constant of Madelung; e is the charge of an electron; r is the distance between ions of opposite sign; L is the affinity of an atom of the nonmetal for an electron; I is the ionization potential of an atom of the metal.

Many authors investigated the different dependencies that relate U to thermodynamic characteristics. For example, Ormont [102] assumes that the width of the forbidden zone U of the semiconductors A^{IV}, $A^{III}B^V$, and $A^{II}B^{VI}$ is related to energy of atomization Ω (i.e., the heat of reaction of $A^+B^-_{cryst} = A + B$) and to the specific surface energy ω_{hkl} of various faces:

$$U = (n_B/n_A)^a (C - M + P) \omega_{hkl} \qquad U = (n_B/n_A)^a (c - m + p) \Omega \qquad (15)$$

where n_A and n_B are the number of valence electrons, respectively, for atoms A and B; M and m are functions of the total atomic number of the atom-partners; P and p are functions of the difference of electronegativities; C and c are constants.

Ruppel et al. [103] found that U increases with the heat of formation of ionic crystals, Q_{form}. For a large number of crystals, the value of U lies in the interval between Q_{form} and $2Q_{form}$. In the same work, a relationship between U and the heat of hydration was discovered. Manca [104] obtained a relationship between U and the energy of a single bond, E_s, in a semiconductor;

$$U = a(E_s - b) \qquad (16)$$

where a and b are constants, characteristic for each type of substance A^{IV}, $A^{III}B^V$, and $A^{II}B^{VI}$. This relationship is valid for small Δx.

A large number of empirical methods relating the width of the forbidden zone U to other properties of a semiconductor, [Eqs. (9)–(16)], are merely minor modifications of current theoretical ideas.

The enumerated rules for the change of U can be carried over to rules for catalyst selection, and one can seek a relationship between catalytic activity and such values as $I, L, Q_{form}, \Omega, \omega_{hkl}, r_{Me}/r_X, J, z_{III} + z_V, \Delta x$ (See p. 41), etc. This, of course, does not constitute the complete list of properties which we might like to use to seek a correlation on the basis of a relationship between catalytic activity and U. For example, it is known in a series of cases [74, 96], that U determines such values as mobility of an electron μ^- or a hole μ^+, temperature of fusion of a semiconductor, its hardness, etc.

One can observe graphically the correlation between U and the position in the periodic table of the elements that form semiconductors. Figure 15 is the example for oxides of metals. In Fig. 15b, regions are drawn that correspond to values of U in the ranges 1–3, 3–6, and above 6 eV. The graph obtained is quite regular. The width of the forbidden zone decreases from top to bottom and from the edge to the middle in the printed form of the periodic table. Oxides with maximum values of U

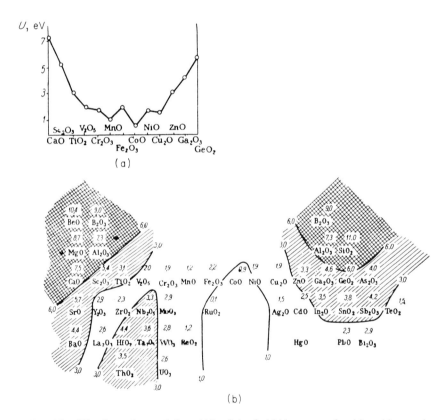

FIG. 15. The dependence of the width of the forbidden zone of stable oxides on the position in the periodic table of the metal forming them: (a) is the change of U in the series of oxides of metals of the fourth period; (b) is the values of U in stable oxides of all of the periodic system (the crosshatched area is for values of $U > 6$ and $U > 3$).

1.3. WIDTH OF THE FORBIDDEN ZONE

are located in the left and right upper corners. The growth of U (however small) in Group IV from TiO_2 to ThO_2 and in Group V from V_2O_5 to Ta_2O_5 appears to be an exception. It is possible that such variations arise from an error of presentation of the value of U in the literature.

The width of the forbidden zone is experimentally measured by various methods. The most prevalent are the methods to determine U according to temperature dependence of electroconductivity σ (i.e., the slope of the line $\log \sigma$ versus $1/T$) and according to the long wave limit of optical absorption. The obtained thermal U_T and optical U_O widths of the forbidden zone generally are not equal to each other. Mostly in ionic crystals $U_O > U_T$, whereupon the difference $U_O - U_T$ is specified as the ionic polarization of the lattice [105]. Also, there can be cases when $U_T > U_O$ [52].

A direct parallel between the values of U and catalytic activity will hardly be observed even in the case of natural conduction for semiconductors with a specific structure. For example, for diamond and its vertical analogs, A^{IV}, the face of the octahedron will be a face of cleavage (111) because the distances between atomic layers in this direction are maximal. For substances with the structure of sphalerite, the layers in this direction are arranged alternately as different type of atoms. Between these, the strength of the bond will be larger, than between layers constructed from one and the same type or between layers that are each of both types of atoms. It is understandable that for compounds with the structure of sphalerite ($A^{III}B^V$, $A^{II}B^{VI}$, and $A^{I}B^{VII}$) the faces of the rhombic dodecahedron will be faces of cleavage (110), which contain atoms of both types—e.g., for ZnS—atoms of zinc and sulfur [106, 107] (Fig. 16). Consequently, according to some data, crystalline structure limits do not strongly influence U. For example, U for ZnS, crystallized

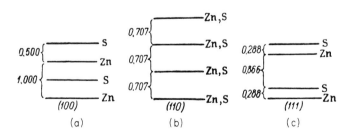

FIG. 16. The relative position of the layers of atoms Zn and S on the faces of a cube (a), rhombic dodecahedron (b), and octahedron (c), in crystaline ZnS with the sphalerite structure [106, 107]. The distances between layers are shown as the lengths of the edges of an elementary cube.

in the structure of sphalerite, has a value of 3.66 eV, in the structure of wurtzite—3.62 eV. For SiC a more noticeable difference is found [108]: U in the structure of sphalerite is 3.12 eV, in the structure of wurtzite— 2.62 eV.

In the same semiconductor the value of U can be changed depending on the crystallographic direction. For example, in rutile TiO_2 [109], the value of U in the direction parallel to the c axis is equal to 3.0 eV, but for U in the direction perpendicular to this axis it is 3.05 eV. These figures are related to the transition from the $2p$-zone of oxygen to the narrow $3d$-zone. Besides this, in TiO_2 there exists a forbidden zone for the transition from the $2p$-zone to the wide $4s$-zone with $U = 3.35$ eV.

The width of the forbidden zone depends also on sizes of the crystalline particle [110], doping, pressure, temperature; however, these changes, as a rule, are small, especially if one examines them from the point of view of the influence of U on catalytic activity.

1.4. Doped Levels

Not only the width of the forbidden zone U, but also the dope concentration in a semiconductor, and the distance from the doped levels either to the valence zone or to the conduction zone are related to properties of the principal lattice (though more indirectly than U). The maximum possible number of atoms of dope is determined by the type of crystalline lattice and the chemical correspondence between the atoms of the dope and the principal substance. We shall denote the depth of the doped levels by ΔU_d. For donor doping, ΔU_d is the distance from a local donor level to the conduction zone; for acceptor doping, it is the distance from the local acceptor level to the valence zone. The value of ΔU_d (or the energy of ionization of an atom of the dope is determined by the ionization potential of the doping atom and the dielectric constant ϵ of the substance.

For evaluation of ΔU_d one assumes the environment of the semiconductor to be undisturbed, disregarding the periodic potential field of the lattice [52]. Then ΔU_d will be equal to the energy of ionization of an atom of the dope in a vacuum, divided by ϵ^2 and multiplied by the ratio of the effective mass of a current carrier (electron or hole) m^* to its normal mass m. For a univalent "hydrogen-like" atom of dope:

$$\Delta U_d = (13.5/\epsilon^2) \times m^*/m \tag{17}$$

where 13.5 is the energy of ionization of a hydrogen atom in a vacuum in electron volts.

1.4. DOPED LEVELS

One can compute the effective mass m^* from a measurement of the Hall effect and thermoelectromotive force. Utilization of the "hydrogen-like" formula (17) for Si gave $\varDelta U_d = 0.087$ eV. Experimental data [52] show that at a concentration of donors less than 10^{16} cm^{-3} the value of $\varDelta U_d = 0.075$ eV, i.e., close to the calculated value and almost independent of the chemical nature of the atoms of the doping material. Formula (17) gives numbers corresponding to experimental values of $\varDelta U_d$ not only for Si and Ge but also for semiconductors of the type A^{III}B^V and other semiconductors with small effective mass and large dielectric constant [80]. Singly ionized atoms of dope form both local donor levels, which lie near the conduction zone, and acceptor levels which lie near the valence zone. Formula (17) is applicable only at small concentrations of dope. In the above mentioned example for Si, with increase of concentration, N, of atoms of dope, the value of $\varDelta U_d$ (in electron volts) decreases, according to the expression:

$$\varDelta U_d = (13.5/\epsilon^2) \times (m^*/m) - \beta N^{1/3} \tag{18}$$

where β is a constant.

At $N = 5.3 \times 10^{19}$, the value of $\varDelta U_d$ becomes zero.

Thus, at small dope concentrations, the depth of the doped level is inversely proportional to ϵ^2, i.e., it is determined by the chemical properties of the fundamental substance (insofar as the value of ϵ is related to chemical properties, see Section 2.3).

The model considered is applicable for singly ionized dopes; however, in this case, the "hydrogen-like" theory of local levels in semiconductors appears to be only a first approximation. It is based on an assumption concerning a purely coulombic interaction of an electron (or hole) with a charged center. Boich-Bruevich [111] showed that, in the majority of cases, this model does not hold. Therefore, corrections are necessary, taking into account the departure from Coulomb's law: At small distances because of charge error; at large distances because of charge shielding by free electrons. The consequence is that there are dependencies of $\varDelta U_d$ on the number of free electrons in a semiconductor, on temperature, and on other parameters, i.e., dependencies on properties of the basic substance become veiled.

Nevertheless, for the computation of $\varDelta U_d$, formula (17) is still used for many cases. As the example with silica shows, the values of $\varDelta U_d$ that are obtained for crystals with large dielectric constant ($\epsilon > 10$) are small—tenths and hundredths of an electron volt. Therefore, at catalytic temperatures, altering the depth of a level of singly ionized doping material, $\varDelta U_d$ (e.g., by going from the semiconductor Ge with doping by As to the same semiconductor doped by Sb of identical charge) will not strongly influence the position of the Fermi level and the catalytic activity. This is because atoms of the dope are ionized.

If however, the possibility of a doubly charged ion is taken into account, (i.e., the breaking away or joining of a second electron from an atom of dope) changes in the values of ΔU_d are obtained which, during the transition from one crystal to another, will substantially influence its catalytic activity. In the case of multicharged dope, several types of centers are formed—donor and acceptor. The addition of Au to Ge, for example, will create donor and acceptor levels, corresponding to the charge states: $[Au]^+$, $[Au]^-$, $[Au]^{2-}$, $[Au]^{3-}$, several of them are located deep in the forbidden zone. The level, lying at a distance of 0.05 eV above the edge of the valence zone, appears to be not an acceptor level but a donor level [80]. In this case formula (17) was applied for computation of ΔU_d of the levels formed upon extraction of the first electron. For example, the levels of Zn^+ in ZnO are found so close to the conduction zone that it is easily explained by the "hydrogen-like" model. For TiO_2 the doped level is located at 0.7 eV below the conduction zone, which is found to be in accordance with the ionization energy of the outer electron of the oxygen atom in an atmosphere with the dielectric constant of rutile [52]. With increase in dope concentration, ΔU_d decreases; and at concentrations greater than 0.1 %, it becomes equal to zero.

Vorobev and Zavadovskaya [112] found that, in crystals of alkaline halides, the position of the F-band in the absorption spectrum, i.e., the value of ΔU_d, is regularly changed during the transition from one substance to another:

$$h\nu_{max} = \Delta U_d = 20.65/a_0^2 \tag{19}$$

where a_0 is a constant of the lattice expressed in angstroms.

McIrvine [113] showed that, for a large number of semiconductors, the critical concentration of doping material, N_{cr}, above which doped conduction is observed, is independent of temperature; it is determined by the simple formula

$$N_{cr} = 1.6 \times 10^{24} e^{-\epsilon} \tag{20}$$

where ϵ is the dielectric constant.

Thus, this value is determined by the substantive and not by the structure-sensitive properties of the substance.

The "hydrogen-like" formula (17) is outwardly analogous to formula (13). Both say that the width of the forbidden zone U and the depth of the doped levels ΔU_d are changed in inverse proportion to the square of the

1.4. DOPED LEVELS

dielectric permeability. Thus, analogous doping in various substances will create local levels, that form a constant fraction of the width of the forbidden zone: $\Delta U_d = \alpha U$ (where α is a constant for a specific dope). In the region of doped conduction of a semiconductor, the position of ΔU_d can determine the position of the Fermi level. In turn, we can obtain a relationship between catalytic activity and the width of the forbidden zone. This alternative third way to explain the influence of U on catalysis is examined in Section 1.3.

The catalytic activity of semiconductors (see Section 1.1) is dependent not only on the Fermi level, but also on the position of the local levels v^- and w^+ of the adsorbed particles relative to the Fermi level [8, 16]. The distances $v^+ = U - v^-$ of the local donor level from the conduction zone and $w^- = U - w^+$ of the local acceptor level from the valence zone (see Fig. 1) are also determined by the nature of the lattice, i.e., they appear to be substantive properties.

Boich-Bruevich quantum-mechanically calculated the depth of the local levels v^+ of adsorbed atoms for a crystal of the type MgO [114]. This value decreases with increase of the dielectric constant of the crystal (more exactly the value $\epsilon = (\epsilon_1 + \epsilon_2)/2$, where ϵ_1 and ϵ_2 are the dielectric permeabilities of the lattice of the adsorbent and of the gas or liquid phase, respectively. Owing to the interaction of electrons, v^+ falls with increased concentration of dope in the crystal. The decrease of v^+ with increasing temperature appears to be the result of the interaction of electrons with phonons. The calculation shows that the value of $v^+ < \Delta U_d$, i.e., the depth of the local levels of adsorbed atoms is significantly less than that of the local levels specified by analogous doping materials or by structural defects.

These considerations show that both factors influencing catalytic activity—the position of the Fermi level and the position of the local levels of adsorbed particles in the energy spectrum of the crystal—are determined (although not in all cases) by the properties of the lattice. Unfortunately, at present the amount of experimental data is not sufficient to draw a comparison of the values of v and w with catalytic activity. Putseiko and Terenin [115] have determined, by optical methods, the location of local doping by a series of dyes. For example, upon adsorption of methylene blue, the local level on TlBr is located at a distance of 0.2 eV and on HgI_2 at 0.4 eV above the valence zone. Recently, the levels of adsorbed atoms and molecules have been studied in detail for Ge and Si [116].

Recently, an interesting fact has been discovered [117]: For a large number of covalent semiconductors, the position of the Fermi level on the surface ϵ_s^+ does not depend on the quantity and character of alloy dope nor even on the type of conduction (n or p) nor on the material that comprises the bulk of the semiconductor, it depends only on the width of the forbidden zone: $\epsilon_s{}^+ = \alpha' U$ (where α' is a constant value close to 0.3). This fact is still not sufficiently explained theoretically, but apparently in all of these semiconductors (A^{IV}, $A^{III}B^{V}$, and several of the type $A^{II}B^{VI}$) the value $\epsilon_s{}^+$ is determined by surface levels of nonadsorptive origin, formed by electrons of a specific type of hybridization. It is reasonable that, in this case, a dependence of catalytic activity on the width of the forbidden zone will be observed.

1.5. The Difference of Electronegativities and the Effective Charges of Atoms

Of all of the values that have been used to try to relate the methods for changing the width of the forbidden zone, the electronegativity x appears to be the most suitable. This is due to the fact that, to calculate difference of electronegativities Δx of an as yet unknown binary compound, it is not necessary to make an additional measurement. It is sufficient to refer to appropriate tables [99, 118, 119]. It is expedient, therefore, to dwell on the concept of electronegativity in somewhat more detail, in spite of the fact that recently works have appeared subjecting this concept to serious criticism [120].

Pauling [99] defines the electronegativity x as the capability of atoms in a molecule to attract electrons to themselves. According to Pauling, the difference of electronegativities, Δx, between atoms A and B is determined from the difference Δ between the experimental energy of the bond $Q(A-B)$ and the average energy of the covalent bonds $A-A$ and $B-B$: $\Delta x_{AB} = 0.208 \sqrt{\Delta}$ [where $0.208 = 1/(23.06)^{1/2}$ the conversion factor from kilocalories to electron volts]. In this equation

$$\Delta = Q(A-B) - \tfrac{1}{2}[Q(A-A) + Q(B-B)] \tag{21}$$

It seems that the values of x calculated in this way are additive. We have succeeded in attributing to each chemical element its electronegativity. Mulliken [121] preferred to define x as the arithmetic mean of the first ionization potential I and the electron affinity. Haissinsky [118] extended the idea of electronegativity to the entire periodic table and showed that this value did not always appear to be characteristic of an element since it also depends on other properties of elements in the molecule. The electronegativity of an element, in particular, depends on its valence state. The values of electronegativities have been refined by a series of modifications and corrections on the basis of thermochemical,

1.5. THE DIFFERENCE OF ELECTRONEGATIVITIES

spectroscopic, structural, and other data; however, the values of x presented in a recent survey [119] differ very little in nature from the original table of Pauling [99].

It has been shown above that there is a dependence between the difference of electronegativities and the width of the forbidden zone: the larger Δx, the smaller U. In Fig. 14, one can see good correlation between U and Δx for a large number of different semiconductors. This correlation is achieved especially well in the example of a series of compounds with the structure of sphalerite [122]. According to Pirson [88], the dependence of U on Δx for compounds of this type can be presented in the form of an equation:

$$U = U_0 + c \Delta x^b \qquad (22)$$

where c and b are constants depending on the average principal quantum number $\bar{N}$; U_0 is the width of the forbidden zone for the appropriate element A^{IV}.

The dependence corresponding to Eq. (22) for three different isoelectronic series of compounds with the structure of sphalerite is shown in Fig. 17. The points for each of the isoelectronic series lie on straight lines.

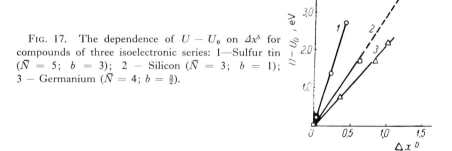

FIG. 17. The dependence of $U - U_0$ on Δx^b for compounds of three isoelectronic series: 1—Sulfur tin ($\bar{N} = 5$; $b = 3$); 2 — Silicon ($\bar{N} = 3$; $b = 1$); 3 — Germanium ($\bar{N} = 4$; $b = \frac{3}{2}$).

The extension of the correlation of U with Δx to all binary compounds, independently of their crystalline structure, coordination, type of bond and hybridization, encounters difficulty. It was shown above that the oxides of transition metals are not included in this rule. Therefore, there were assumptions [123–125] concerning the introduction of a separate range of electronegativities for each type of bond. Actually, to say that the degree of ionization of a bond is determined by the difference of electronegativities is incorrect. It depends also on the type of orbitals

(s, p, or d) and the degree of hybridization. However, the ranges of electronegativities obtained, allowing for separate types of bonds, were not different from the values of Pauling, within an accuracy of ± 0.5 units of x. Owing to the empirical character of the concept of electronegativity, it is not appropriate to attempt to refine it, but it is of use in developing general correlations.

Zvonkova [126, 127] criticized the assumptions of Roginskii and this author [67] (Section 1.3) concerning the fact that a decrease of the width of the forbidden zone U in a series of compounds ZnX during the transition from ZnS to ZnO was caused by the fact that ZnO is crystallized in the wurtzite lattice, and ZnS in the sphalerite lattice. There must be a uniform increase of U with Δx during the transition from ZnTe, ZnSe, and ZnS to ZnO. In the opinion of Zvonkova, the maximum for U on ZnS in this series can be explained by taking into account the value of the potential of the nuclei e^*/r (where e^* is the effective charge of the atom, acting on the valence layer at a distance r from the nucleus). The values of e^* depend on the shielding constant of the nucleus by the electrons of the atom and are calculated theoretically [128]. The maximum value of U in ZnS is due to the lower e^*/r of an atom of sulfur compared with those for atoms of O and Se in ZnO and ZnSe.

In every case, the use of formulas of the type (22) to compute U from Δx for compounds with different structure is unsuitable. Here one cannot speak of a functional relationship, but only about a correlational relationship. Syrkin [120] goes further and assumes that the search for a similar connection between Δx and U in general, is devoid of a basis. However, it is impossible to agree with this assumption. We shall note that for the proof of this thesis Syrkin cites incorrect values of U (without a standard source). According to Syrkin, for CuBr $\Delta x = 1$, but $U = 5$ eV; in CdS $U = 2.1$ eV at the same at the same $\Delta x = 1$, but in PbS $U = 1.17$ eV at $\Delta x = 0.9$. Actually, for CuBr $\Delta x = 1.1$ [118] and $U = 2.94$ eV[5]; for CdS $\Delta x = 0.9$ but $U = 2.38$ eV, i.e., the ratios of the value of Δx and U for CdS and CuBr almost exactly coincide. PbS actually deviates from the rule ($\Delta x = 1$, $U = 1.04$ eV), but it has a different type of lattice (NaCl, instead of sphalerite for CdS and CuBr) and a different type of orbital (p^3, instead of sp^3 for CdS and CuBr). According to Syrkin, for CdS and ZnO, the values of U are very close (2.1 and 2.2 eV), but Δx differs by a factor of 2 (1 and 2). Actually, for CdS $\Delta x = 0.9$, but $U = 2.38$ eV; for ZnO $\Delta x = 1.9$ and $U = 3.3$ eV, i.e., the ratio of U and Δx are closer to each other than Syrkin shows, especially if one

---

[5] See the table in the Appendix compiled on the basis of a critical examination of a large amount of experimental material and literature data [70, 74, 80, 129–131].

1.5. THE DIFFERENCE OF ELECTRONEGATIVITIES

takes into account the differences of their structures (sphalerite and wurtzite). It is curious that on the basis of the tabulated values of Δx and the average principal quantum number $\bar{N}$ one can draw several conclusions about the structure of the crystal. Pirson [130] showed that with increase of Δx and $\bar{N}$, instead of sphalerite, the structures of wurtzite and then of NaCl and CsCl appear.

For calculation of the degree of ionization of a bond, instead of Δx, Syrkin, as did other authors [119, 122], proposed to make use of the effective charges e^* on the separate atoms of a compound determined from experimental data. The effective charge of an atom is a real charge found to be within the volume surrounding the atom. This definition does not appear to be rigid, as it is not clear (especially in the case of covalent crystals) what volume to choose for calculation of this charge. If, as a basis for computation of the volume, one takes a specific atomic or ionic radius, one can arrive at contradictory results. Consequently, in a series of cases, the values of the effective charges are noticeably different from one another, when they are determined by different methods.

At present, several experimental methods are used to calculate the effective charges. Szigetti [132] proposed, for the determination of e^*, to calculate the atomic polarization of ionic crystals from the difference between the low frequency dielectric constant ϵ_0 and the square of the index of refraction n^2. These values are related in the following manner to the effective charge of an electron e^*:

$$\epsilon_0 - n^2 = \left(\frac{n^2 + 2}{3}\right)^2 \frac{2\pi e^{*2}}{\omega_0 r^3 m^*} \tag{23}$$

where r is the interionic distance; ω_0 is the natural frequency of the lateral vibrations of the lattice; and m^* is the reduced mass of the vibrating ions.

The value of ω_0 is usually determined from an infrared spectrum. With these data, one can calculate the value of e^*. The values of e^*/e (where e is the elementary charge of an electron) determined by this method [105, 132–134] are found (e.g., for substances of the isoelectronic series of Ge) to be equal to 0.43 for GaAs, 0.70 for ZnSe, and 1.00 for CuBr. In this case, the change of e^*/e is in the same direction as the change of Δx and U. For compounds of the type $A^{III}B^V$ we have been unsuccessful in determining a direct correspondence between the values of e^*/e and U determined in this way. An explicable linear relationship between e^*/e and the difference of polarizabilities $\alpha_B - \alpha_A$ of atoms A and B was difficult to obtain theoretically [134].

Another method of determining the effective charges is through Roentgen K-spectra. After the extraction of an electron from the K shell, the atom becomes an ion with a formal charge of $+1$, but its resultant charge, owing to the action of the chemical bond, becomes equal to $\eta = 1 + e^*/e$ or $\eta = 1 - e^*/e$. The fine structure of the K absorption band allows us to calculate the value of η and consequently e^* [135, 136]. In the same way, the method of calculation of e^* from K-spectra of absorption was used to study the position of the valence bonds in semiconductors. It was proven, for example, that the effective charge of zinc in ZnO is not equal to two (as this follows from the ionic formula $Zn^{2+}O^{2-}$), but significantly less than unity.

Kimmel [138] showed that in binary crystalline substances containing a small addition of manganese ions, the extra-fine splitting in the electronic paramagnetic resonance spectrum of the ions of manganese is linearly related to the value of e^*. This gives still another method to evaluate e^*. The effective charge can also be determined from the constant of nuclear quadrupole interaction [139]. Other methods also exist.

The values of the effective charges are very useful for the determination of the degree of the ionic character of the bond. It is expedient to try to compare in a like manner the values of e^* with the width of the forbidden zone and with catalytic activity. At present, however, there is still too little data for such a comparison; and the data we have are both insufficiently accurate and contradictory.

Suchet [140] gave these semiempirical equations for the calculation of e^*.

$$e^* = m[1 - (n/r' + n'/r)\,c] \quad \text{(for a cation)}$$
$$e^{*\prime} = m'[1 - (n/r' + n'/r)\,c] \quad \text{(for an anion)} \tag{24}$$

where c is a constant, n and n' are the number of electrons in the filled levels of the anion and cation; r and r' are their radii; and m and m' are the formal charges of the ions.

Equation (24) is based on polarization concepts. More rigorous analogous theoretical approaches [141, 142] that were based on theories of deformation of ions in crystals gave, in some cases (e.g., for alkaline halides), values of e^* close to experimental values. However, an exact theoretical calculation of e^*, according to Syrkin [120], is a matter for the distant future. Because of the absence of both theoretically proven and experimental values of e^*, the method of effective charges as yet cannot supplant the method of electronegativities.

Primarily, the effective surface charges of atoms and not the bulk effective charges influence the catalytic activity of a solid. In principle, one can determine them by any of the above indicated methods

1.5. THE DIFFERENCE OF ELECTRONEGATIVITIES

studying the dependence of e^* on the degree of dispersion, i.e., the values of the surface of the catalyst. However, such data are almost nonexistent. O'Reilly [143] and O'Reilly and Poole [144] studied nuclear magnetic resonance of aluminum oxide, pure and containing transition metal ions: Cr^{3+}, Co^{3+}, and Ni^{2+}. According to his data, on a highly dispersed sample of Al_2O_3, a strong electric field near the surface interacts with the large quadrupole moment of Al^{27} and amplifies the signal of Al^{27} that is below the limits of its detection. Adsorbed ions of the transition metals play the same role. Thus, in comparison with the bulk, surface atoms of Al carry increased effective charges. A study (by low energy electron diffraction) of the adsorption of oxygen and iodine on silicon [145, 146] showed that the distances of Si—O and Si—I in the surface layer are substantially more than the sum of the covalent radii $r_{Si} + r_O$ and $r_{Si} + r_I$ and close to the sum of the corresponding ionic radii. Obviously, in this case, e^* of the surface atoms is more than that for the bulk. A more general consideration concerning the large asymmetry of the field acting on an atom during the absence of other atoms from one of its sides leads to a conclusion about the high effective charges of surface atoms.

In a series of works on the theory of catalysis, the direct influence of effective charge on rate and activation energy of a catalytic reaction was studied. For example, Sokolov [147] studied the donor–acceptor interaction of molecules with an active center (cation) X^+ bearing a positive charge. In his opinion, not only during the interactions of molecules having a multiple π-bond, but also during the interactions of molecules with a σ-bond (e.g., H_2 with H^+) strong donor–acceptor complexes can be formed. On the surface, a donor–acceptor interaction of a lattice cation with molecules can also be brought about in the situation when, for example, two molecules with σ-bonds A—B and A'—B' form an activated complex

for the reaction $AB + A'B' \rightarrow AA' + BB'$. The energy of the donor–acceptor interaction of the cation X^+ with the activated complex exceeds that with the initial stable molecules AB and A'B'; but the energy of activation E of the reaction $AB + A'B'$ in the presence of the cation X^+ will be lowered. During this, the decrease of E will be larger, the larger the effective charge e^* of the cation. This type of interaction of reacting molecules with the cation of a catalyst is close to the question of coordination—ionic catalysis or catalysis by Lewis acids—which will be examined in more detail in Chapter 2.

Nagaev [148, 149] quantum-mechanically studied the chemisorption of a molecule on a cation of an incompletely polar crystal. Such a crystal, according to this model, resonates between ideally polar and homopolar states. In the first of these, a natural electron of the crystal is not in the adsorption site; and the particle bond is realized by means of localization into the crystal of electrons from a molecule. This produces a valence bond, i.e., it appears to be the analog of a donor σ-bond in complex compounds. In the second state, at the cationic adsorption site, we have both the natural electron of the crystal and the electron of a reactive bond of the molecule (as in the case of $H_2 + H$). On one of the anions close to the adsorption site an electron is absent. It is a fact that on the surface, in this case, we have not ions, but atoms of the catalyst. During adsorption on the surface of an incompletely polar crystal, there is a superposition of these two types of bonds (dielectronic and trielectronic according to the terminology of Nagaev). Their competition leads to a dependence of chemisorption on the effective charge of the adsorption site. The chemisorption of a molecule without rupture of the bond between its atoms is more favorable, the more polar the crystal. On the ideal polar crystal, the energy of the bond during such chemisorption is maximized, but the energy of activation E of chemisorption is equal to zero. The more the crystal differs from being ideally polar, i.e., the less Δx and e^*, the larger E of such chemisorption, and the less the energy of the bond of a chemisorbed molecule with the crystal. Physically, the origin of the energy of activation is related to repulsion by a molecule of electrons from the adsorption site. This results in an increase of the effective charge of the site. In the case of crystals with a large degree of homopolarity, chemisorption of molecules without breaking into atoms is impossible. In chemisorption with dissociation into atoms, a dielectronic bond is stronger than a monoelectronic bond. During such an adsorption of atoms, the effective charge of the adsorption site decreases. The bond of an atom with a crystal consequently appears to be stronger, the lower the effective charge e^*.

Thus, from the work of Nagaev, a new explanation of the relationship between catalytic activity and width of the forbidden zone U has resulted. It was shown above that the value of U increases with an increase in Δx and effective charge e^*. Thus, the energy of activation E for the chemisorption of molecules without bond rupture must decrease with increasing U; but E of chemisorption with breaking up into atoms (radicals) increases with increasing U. In view of the fact that in the

1.6. THE ROLE OF d-ELECTRONS

majority of catalytic processes, especially in homolytic reactions, one of the steps appears to be the chemisorption of a molecule with dissociation, it follows from the considerations presented that there is an increase of activation energy of a catalytic reaction with increase of the effective charge e^* or width of the forbidden zone U. Nagaev compares the results he obtained with previously mentioned data of Krylov and Fokina [77, 78] concerning the rules for change of the catalytic activity of semiconductors with change of the value of U (see Figs. 9–11).

This theory gives an explanation of the dependence of catalytic properties both on the width of the forbidden zone and on the amount of modifying doping material in the crystal, even when the conduction electrons and holes do not take part in bond formation of molecules or atoms with the adsorbent.

With the help of electronegativities and effective charges, we can predict, in some cases, the acid–base properties of binary compounds [119] (Section 2.1), and consequently the acid–base properties of their surfaces. This is as yet only one way to relate Δx and e^* with catalysis, but it is already useful for acid–base reactions.

1.6. The Role of d-Electrons in the Catalytic Properties of a Solid

The high catalytic activity of the transition metals appears to be one of the most significant facts of heterogeneous catalysis. A large number of investigators working in the field of catalysis paid attention to this fact, even from the beginning of the twentieth century. Not only do the transition metals themselves possess high catalytic activity, but also their alloys and compounds with nonmetals: oxides, sulfides, etc. Roginskii [6], in 1933, considered the fact that compounds of the transition metals (i.e., substances having in their structure cations of metals with unfilled d-shells) in the series Ti → Cu, Zr → Ag, and Ta → Au, are especially active in reactions of the oxidation–reduction class. Other investigators [9, 44] substantiated this observation. Roginskii explained the high catalytic activity on the basis of the fact that the cations exert an abnormally strong deforming action and that there is a gradual decrease in the potential of the chemical forces in bonds that are formed by d-electrons, in comparison with bonds formed by s- and p-electrons [150]. In recent years, the rules for selecting catalysts for oxidation–reduction reactions were studied by many investigators. It was found [39, 77, 78] that, in these reactions, substances not having d-electrons (such as Ge and compounds of the type $A^{III}B^{V}$) could also be highly active. Not only oxides in which the cations have unfilled d-shells: Cr_2O_3, FeO, Co_3O_4, NiO, etc., but also oxides of Groups I and II adjoining them in position in the periodic table: Cu_2O, Ag_2O, ZnO, and CdO (the d-shells of which contain 10 electrons, i.e., the d-shells are filled) have large catalytic activity. However, the general rule is followed: The most active catalysts in oxidation–reduction reactions are the transition metals, next come their compounds (e.g., oxides and sulfides), then follow the sp-semiconductors, containing unfilled d-shells, and, finally, dielectrics.

At present there does not exist a standard ordered scheme of the electronic structure of the oxides of the transition metals. Owing to the unfilled d-levels, metallic conduction is observed. However, for oxides of the first large period, this property is observed only for TiO, Ti_2O_3, and V_2O_3 [151]. The remaining oxides show typical semiconductor properties, i.e., a strong increase of electroconductivity with increase of temperature. Many oxides of transition metals are antiferromagnetic at low temperatures [152, 153]. In ferromagnetism, an energy of activation is not required for the movement of electrons; in antiferromagnetism, the spins of neighboring ions are oriented to be antiparallel; and the transfer of electrons proceeds with a reversal of spin, for which an energy of activation is required. In some cases, e.g., in Ti_2O_3, a discontinuity in the experimental curve of $\log \sigma$ versus $1/T$ was actually observed during the course of the transition from antiferromagnetism to paramagnetism at the temperature of Neal. However the experimental studies of the change of electrical properties during magnetic changes (inversions) in such substances as Fe_2O_3, NiO, and ferrite, do not confirm this [154]. DeBoer and Verwey [155] were the first to point out the fact that, in the oxides of transition metals, a d-zone is not formed when cations are separated by anions and their wave functions do not overlap. In such oxides, as Cr_2O_3, Fe_2O_3, CoO, NiO, and others, $3d$-electrons occupy a level localized on the cations; and so, in stoichiometric composition at room temperature, they appear to be insulators. Conduction arises in them only upon the introduction of dopants. For example, by introducing the ions Li^+ instead of Ni^{2+} into NiO, one can obtain, in order to conserve electrical neutrality, some ions Ni^{3+}. The transfer of ions Ni^{3+} within the lattice by exchange of electrons with ions of Ni^{2+} requires an activation energy.

The high value of the energy of activation of electroconductivity E_σ at large concentrations of doping material inspired in Heikes and Johnston [152] the idea that this is caused by the energy of activation of the mobility of the migration of a current carrier from one ion to its neighbor. In the example presented above, for NiO, a hole on the ion Ni^{2+} appears to be the current carrier. The creation of the hole is auto-

1.6. THE ROLE OF d-ELECTRONS

catalyzed by the deformation of the lattice around it. The activation energy of mobility is related to the action of this lattice deformation against elastic forces that arise upon the transfer of a hole from one ion to another, i.e., it has nothing in common with the usual semiconductor activation energy. The number of current carriers in oxides of transition metals are specified only by the dope, and, in contrast to electroconductivity, do not depend on temperature.

The study of complex systems on the basis of oxides of transition metals showed, that within wide ranges, the change of the dope percentage often does not influence the temperature behavior of electroconductivity. For example, the introduction of up to 10% of dope as oxides of Ni, Cu, and Mn into Co_3O_4 and the introduction of the same amounts of foreign oxides into MnO does not influence the activation energy of electroconductivity in the interval from -100 to $+800°C$ [156]. These results say that, in the use of the mechanism of Heikes–Johnston, E_a in these oxides is the activation energy of mobility.

Morin [151], not objecting to these ideas, indicates that the pure oxides of the transition metals have significant electroconductivity. To explain this, it is necessary to use zone concepts. In the oxides of the transition elements of the fourth period (Sc_2O_3, TiO_2, and V_2O_5), because of a strong cation–cation interaction, a 3d-zone exists. During the transitions to the oxides that follow (Fig. 18) there is even a cation–

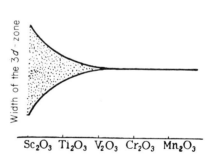

FIG. 18. Schematic diagram of the change of width of the 3d-zone for a series of oxides of transition metals [151].

anion–cation interaction; therefore, the 3d-orbitals are not overlapped and do not form a d-zone; an electron on them has a very small mobility. Due to the high mobility of holes of the 2p-zone of oxygen, conduction of the majority of oxides of the transition metals is carried out by holes. According to the works of Goodenough [157] and Geld and Tskhai [158], cation–cation interaction exists even in that case, when not more than three electrons are present on the 3d-levels. Morin [159] computed the overlap integrals of the d-orbitals of transition element cations with themselves and with sp-orbitals of anions and showed that the overlap increases with an increasing radius of the cation and anion and decreases with an increasing charge of the cation. Consequently, among the oxides of the transition metals, in the upper right corner of the periodic table (in its printed form), are found insulators and semiconductors; in the lower left corner are found metals, i.e., substances with conduction not dependent on temperature.

There is an exception to this rule. For example, WO_3 and TiO_2, according to calculation, showed a significant overlap of the $d-d-$ and $d-p$-orbitals, although they do not possess metallic conduction. According to Morin, this is a consequence of the fact that at low temperatures and in the absence of dopants, the d-zone, which has a significant width, is not filled. Almost all fluorides of transition metals appear to be dielectrics; the majority of oxides are semiconductors; however, among the sulfides of these metals, substances with metallic conduction are frequently encountered.

In NiO, pure and with doping by Li_2O, the current carriers appear on account of the following reactions [151].

$$O^{2-}(2p^6) + Ni^{2+}(3d^8) \rightarrow O^-(2p^5) + Ni^+(3d^9) \tag{25}$$

$$Ni^{2+}(3d^8) + Ni^{2+}(3d^8) \rightarrow Ni^{3+}(3d^7) + Ni^+(3d^9) \tag{26}$$

$$Ni^{2+}(3d^8) + Ni^+(3d^74s) \rightarrow Ni^{3+}(3d^7) + 4s\text{-electrons} \tag{27}$$

$$Li^+Ni^{3+}(3d^7) + Ni^{2+}(3d^8) \rightarrow Li^+Ni^{2+}(3d^8) + Ni^{3+}(3d^7) \tag{28}$$

In a pure oxide, electroconductivity will be realized as a result of processes (25)–(27) by $2p$-holes with high mobility and by $3d$-holes and $3d$-electrons with low mobility. The $4s$-electrons do not exhibit a large influence on the conductivity of oxides. The majority of pure oxides of this type (see Fig. 3), therefore, appear to be p-semiconductors. In NiO containing added Li^+, $3d$-holes arise according to Eq. (28). If the doping material is found to be in a higher valence state than the basic oxide, $3d$-electrons emerge. In Fig. 19, a diagram (from Morin) of the energy levels of NiO with addition of Li_2O is presented. At low temperatures and high concentrations of dopant, the Fermi level FF lies in the middle between the $3d$-levels of Ni^{2+} and the acceptor levels of Li^+Ni^{3+}. At high temperatures, in the region of intrinsic conduction, FF lies almost in the middle between the $3d$-level of Ni^{2+} (which lies near the $2p$-zone) and the $3d^9$-level of Ni^+.

Thus, low mobility of current carriers is one of the properties of compounds of transition metals that distinguishes them from other semiconductors. As was previously mentioned (Sections 1.1 and 1.3), there are no data in the literature about the existence of any relationship

1.6. THE ROLE OF d-ELECTRONS

between catalytic activity and the mobility of current carriers. According to the assumption of Ioffe et al. [160] (see also [161]), the localized electrons of a semiconductor-catalyst will undergo a "σ-activation" in catalytic reactions of oxidation, i.e., the breaking of a C—H bond. The delocalized electrons of a semiconductor that comprise a zone interact with π-electrons and so oxidize olefinic and acetylenic hydrocarbons through their double and triple bonds. To verify this assumption, it would be desirable to analyze data on the selection of catalysts for selective oxidation of hydrocarbons.

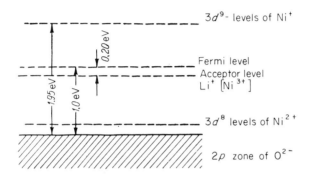

FIG. 19. Diagram of the energy levels in NiO with doping by Li_2O according to Morin [151]. The Fermi level is shown for the case of intrinsic conduction.

According to some recent investigators [162, 163], in reactions on single crystals of NiO and TiO_2, 3d-electrons in these oxides will move not along local levels, but in a narrow zone (close to 0.05 eV wide). However, their increase of electroconductivity with temperature is caused by an increase of the number of current carriers and not by an increase of their mobility. If we introduce this correction into the zone diagram in Fig. 19, the latter in general will not be principally different from the zone diagram of the usual semiconductors.

The absence of a mature theory for electronic structure of semiconductor compounds of transition elements led to the necessity of using other ideas for explanation of their catalytic properties. Up to the present time, in this area, the so-called crystal field theory and its modern modification, the ligand field theory, have been widely used. We shall examine the basic propositions of this theory [164–170].

Crystal field theory was proposed by Bethe [171]. This theory is based on the assumption that complexes of transition metals with ligands[6] are

---
[6] An anion or neutral molecule in the immediate vicinity of the central ion of a metal and interacting with it is called a ligand.

stable as a result of electrostatic interaction between the central ion and the ligands. The latter are considered to be present in an exterior electric field; no account is taken of the structures of their electronic orbitals. For an ion of a metal, the filling up of its d-orbitals by electrons is examined. The possibility of the formation of a covalent bond is completely neglected.

An isolated ion of a transition metal has a field of spherical symmetry. In it there exist five d-orbitals with the same energy. A diagram of the angular dependence of the wave function for the d-orbitals is depicted in Fig. 20. In a solution, within a cage of ligands—ions or neutral molecules,

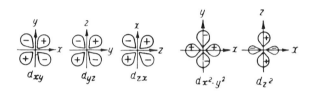

FIG. 20. A schematic representation of the angular dependence of the wave functions for d-orbitals.

or in stable oxides within a cage of ligands—ions of oxygen, a splitting of the originally nondegenerate d-level of the central ion of a transition metal into levels of different energy occurs. The magnitude and character of the split depends on the number of electrons in the d-orbital, the geometry of the distribution of the ligands, and the strength of the electric field created by them. Orbitals d_{z^2} and $d_{x^2-y^2}$ have maximum density along the axes of the coordinates x, y, and z, directed along the axes of an octahedron; the orbitals d_{xy}, d_{yz}, and d_{xz} are directed along the bisectors between these axes. If the central ion is enclosed by six ligands distributed along the axes of an octahedron, then there occurs a repulsion of their electrons from the electronic shells of the orbitals d_{z^2} and $d_{x^2-y^2}$. As a result of the repulsion, the energies of these orbitals rise in comparison with the energy of the original five-fold degenerate d-level; and the orbitals d_{xy}, d_{yz}, and d_{xz} become energetically more efficient. A splitting of the original d-level into two groups of sublevels occurs (Fig. 21): The lower group consists of three levels t_{2g}, on which six electrons can be distributed (on the basis of two with opposite spin on each level); the upper group consists of two with opposite spin on each level); the upper group consists of two e_g levels with four electrons. The orbitals t_{2g} often are designated in the literature as d_ϵ and the orbitals e_g as d_γ. The magnitude of the total split, i.e., the distance between the levels t_{2g} and e_g is designated by Δ or $10\ Dq$. The average energy of the

1.6. THE ROLE OF d-ELECTRONS

d-orbitals is conserved, and therefore the higher levels are located at $\frac{3}{5}\varDelta$ above the original d-level and the lower levels at $\frac{2}{5}\varDelta$ below it. For a tetrahedral distribution of ligands, by analogous considerations, the d-level is split into three upper levels t_2 and two lower levels e_2.

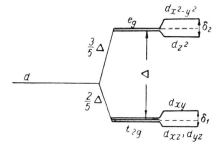

FIG. 21. The splitting of the d-level of an isolated ion in an octahedral ligand field.

When electrons are distributed in the d-levels in complexes, two trends compete: (1) the tendency of an electron to occupy more stable orbitals and (2) the tendency of electrons to be distributed in different orbitals without spin pairing. The first of these is specified by the ligand field, i.e., by electrostatic repulsion that displaces the electrons in an octahedral complex to lower levels t_{2g}. The second tendency is related to an exchange interaction: the stabilization of an atom during an interaction increases with an increase in the number of unpaired electrons that have parallel spins. In a strong ligand field ($\varDelta$ is great) the first tendency prevails with the formation of low-spin configurations; in a weak field ($\varDelta$ is small) the second tendency prevails. For each given ion of a metal there exists some critical value $\varDelta$, above which low-spin complexes are formed and below which high-spin complexes are formed. The so-called spectrochemical series of ligands[7] was found in which the ligands are empirically located according to increasing $\varDelta$ [167]: I$^-$, Br$^-$, Cl$^-$ $\sim$SCN$^-$, diethylthiophosphate, F$^-$, urea $\sim$OH$^-$, C$_2$H$_5$OH, $\underline{N}$O$_2^-$, HCOO$^-$, oxalate ion (2O), H$_2$O, malonate ion (2O), S$\underline{C}$N$^-$, aminoacetate $\sim$ ethylene diamine tetraacetate ion (2N, 4O), pyridine $\sim$ NH$_3$, ethylene diamine (2N) $\sim$ diethylene triamine (3N) $\sim$ triethylene tetraamine (4N), $\underline{S}$O$_3^{2-}$, phenanthroline, $\underline{N}$O$_2^-$, CN$^-$.

The order of ligands in this series is almost independent of the nature of the metal. The value of $\varDelta$ is usually determined experimentally from an analysis of molecular spectra. For oxides and hydroxides of divalent metals (such as for complexes of Me^{2+} with oxygen-containing ligands

---

[7] The atoms of an anion that coordinate with an atom of the metal are underlined or set in parentheses.

in solutions) it amounts to 7500–12,500 cm^{-1} (0.9–1.6 eV or 20–35 kcal); for corresponding complexes of trivalent metals it amounts to 13,500–21,000 cm^{-1}, but for quadrivalent (Pt^{4+}) it is ~30,000 cm^{-1}. In a tetrahedral field the value of Δ is less than in an octahedral field.

During filling by the first d-electrons of the lower orbitals t_{2g} (see Fig. 21) in a crystal field, a gain of energy occurs (in comparison with an isolated ion of a transition metal or the so-called stabilization by a crystal field[8]). If, in an octahedral field, an ion of a transition metal has a coordination of d-orbitals: $(t_{2g})^m(e_g)^n$ (where m and n are, respectively, the number of electrons in the t_{2g} and e_g orbitals), then the energy of stabilization Δ by a crystal field is equal to $(4m - 6n)/10$; for a tetrahedral configuration $(e)^p(t_2)^q$ it is equal to $(6p - 4q)/10$. The distribution of d-electrons in orbitals of complexes of an octahedral configuration and the energy of stabilization by a crystal field in units of Δ is shown in Table I.

TABLE I

DISTRIBUTION OF THE 3d-ELECTRONS IN THE OCTAHEDRAL CONFIGURATION[a]

Number of 3d-electrons	Ion	High spin (weak field)		Gain of energy in units of Δ	Low spin (strong field)		Gain of energy in units of Δ
		t_{2g}	e_g		t_{2g}	e_g	
1	Ti^{3+}, V^{4+}	↑		0.4	↑		0.4
2	Ti^{2+}, V^{3+}	↑ ↑		0.8	↑ ↑		0.8
3	V^{2+}, Cr^{3+}, Mn^{4+}	↑ ↑ ↑		1.2	↑ ↑ ↑		1.2
4	Cr^{2+}, Mn^{3+}, Fe^{4+}	↑ ↑ ↑	↑	0.6	↑↓ ↑ ↑		1.6
5	Mn^{2+}, Fe^{3+}, Co^{4+}	↑ ↑ ↑	↑ ↑	0.0	↑↓ ↑↓ ↑		2.0
6	Fe^{2+}, Co^{3+}	↑↓ ↑ ↑	↑ ↑	0.4	↑↓ ↑↓ ↑↓		2.4
7	Co^{2+}, Ni^{3+}	↑↓ ↑↓ ↑	↑ ↑	0.8	↑↓ ↑↓ ↑↓	↑	1.8
8	Ni^{2+}	↑↓ ↑↓ ↑↓	↑ ↑	1.2	↑↓ ↑↓ ↑↓	↑ ↑	1.2
9	Cu^{2+}	↑↓ ↑↓ ↑↓	↑↓ ↑	0.6	↑↓ ↑↓ ↑↓	↑↓ ↑	0.6
10	Zn^{2+}, Cu$^+$	↑↓ ↑↓ ↑↓	↑↓ ↑↓	0.0	↑↓ ↑↓ ↑↓	↑↓ ↑↓	0.0

[a] According to Dunitz and Orgel [168].

It is evident from Table I that, in weak octahedral (also in weak tetrahedral) fields there occurs a stabilization of energy of all ions of transition metals, except for ions with the configurations d^0(Sc^{3+}, Ti^{4+}, V^{5+});

[8] By "crystal field" in this case is meant the usual electrostatic field of ligands in the matrix and not, certainly, in the field of anions in the crystal lattice.

1.6. THE ROLE OF d-ELECTRONS

d^5(Mn^{2+}, Fe^{3+}, Co^{4+}); d^{10}(Cu^+, Zn^{2+}). There is analogous behavior for ions of transition metals whose values of charge and radius are of similar magnitudes. For these, there are no energy stabilization effects in a crystal field. Ions of the configurations d^0, d^5, d^{10} have spherical symmetry In a strong field, ions of d^0 and d^{10} exhibit such properties, but an ion of d^5 is stabilized by a crystal field.

Experiment shows that for ions of transition metals of the first large period, a weak ligand field (i.e., high-spin complexes) occurs more frequently. In the spectrochemical series, the critical value of Δ is usually found somewhere between H_2O and NH_3; oxygen-containing ligands more often form a weak field; nitrogen-containing, a strong field. The structures d^6 represent an exception, e.g., the ion Co^{3+}, which almost always gives complexes with coupled spins, characteristic for strong fields. In the second and third large periods low-spin complexes form more often. The values of Δ for these are increased.

The crystal field theory explained magnetic properties and optical spectra of complexes of transition metals in solution. Dunitz and Orgel [169] applied this theory to explain the crystalline structure of stable ions of compounds of transition metals, especially oxides. For example, for the ions Cr^{3+} and Ni^{2+}, an octahedral configuratoin gives a larger energy stabilization than does tetrahedral. Therefore the latter configuration for these elements in solids is almost unobserved. For ions of d^0, d^5, d^{10}(Ti^{4+}, V^{5+}, Cr^{6+}, Mn^{2+}, Fe^{3+}, Cu^+), both tetrahedral and octahedral structures occur. The theory allows us to explain the distribution of cations along octahedral and tetrahedral locations in spinels.

In fields with lower symmetry than octahedral, additional degeneracy and splitting of levels t_{2g} and e_g occur (or in a tetrahedral field e and t_2). This is the so-called Jahn–Teller effect. For example, in an elongated octahedral structure with two long and four short bonds, the levels are split as shown in Fig. 21.

Beginning with the works of Van Vleck [172] a more complete theory for an explanation of the properties of complex compounds began to develop based on the method of molecular orbitals—the ligand field theory. In contrast to the crystal field theory, it considers the electronic structure not only of the central ion but also of the ligands.

The molecular orbital of a complex is formed by means of a combination of orbitals of the central ion and of the ligands, which possess the same properties of symmetry (relative to the general system of coordinates). A detailed description of this theory is contained in a series of monographs [165–167]. We shall present here a picture for formation of energy levels (molecular orbitals) of an octahedral complex from atomic orbitals of the metal and ligands (Fig. 22). Twelve electrons from ligands take part in the interaction (two in each of the σ-orbitals of the ligands) and n electrons of the central metal ion. The ligand electrons are distributed to the molecular bonding orbitals a_{1g}, t_{1u}, and e_g, formed, respectively, from the atomic 4s-, 4p-, and 3d-orbitals of the metal and a_{1g}, t_{1u}, and e_g-orbitals of the ligand. The antibonding orbitals are designated in Fig. 22 a_{1g}^*, t_{1u}^*, and $e_g{}^*$. The degeneracy multiplicity is in parentheses. The remaining d-electrons of the central metal atom are distributed in the unbonded orbital t_{2g} and the antibonding orbital $e_g{}^*$ with capacities, respectively, of six and four electrons.

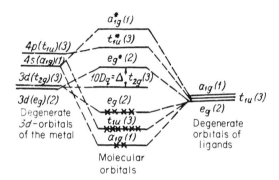

FIG. 22. A molecular orbital diagram of the energy levels for an octahedral complex of an ion of a transition metal of the fourth period.

The distance between them, $\varDelta = 10\,Dq$, is determined from spectral data and is completely analogous to the corresponding value in the crystal field theory. In this case, however, the orbital t_{2g} appears to be nonbonding and is found at the same level as the original degenerate d-level. On the other hand, the orbital $e_g{}^*$ is located above the orbital t_{2g} because of its antibonding character. Therefore, during filling of the t_{2g}-orbital by electrons, there is no effect of stabilization of energy, while the filling of the $e_g{}^*$-orbital is energetically disadvantageous. Nevertheless, the relative energy of these orbitals and the order of filling them with electrons in a strong and weak field of ligands remain the same as in the crystal field theory.

We shall note that stabilization in a crystal field (in a ligand field) is only one of the factors providing stability for transition metal complexes. In Fig. 23 are represented experimental data on heats of hydration, Q_H, of aqueous complexes of divalent and trivalent ions of the first large period [164, 166, 170]. Points (crosses) are located on a curve with two maxima and a minimum at Mn^{2+} and Fe^{3+}. If we subtract the crystal field stabilization energy, the "adjusted" values of $Q-$ (the blackened

1.6. THE ROLE OF d-ELECTRONS

points) are located on a smooth curve, rising to the end of the period. This fundamental contribution for complex stability is determined by the attraction of ligands to the shell of a spherically symmetrical metal ion. It increases with increasing charge and decreasing ionic radius. The order of stability of complexes of divalent ions of metals of the first large period with almost all ligands [164] increases in the series $Mn^{2+} < Fe^{2+} < Co^{2+} < Ni^{2+} < Cu^{2+} >$ (sic) Zn^{2+}.

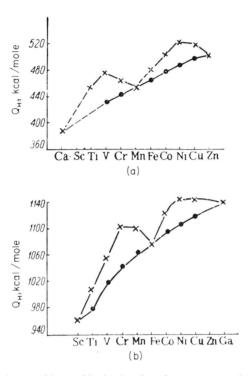

FIG. 23. The change of heat of hydration Q_H of aqueous complexes of divalent (a) and trivalent (b) ions of metals of the fourth period [164, 166, 170] × —experimental data; ●—precise (sic) data.

The splitting of the d-levels in crystalline compounds of transition metals creates an additional complication of the level structure in a solid in comparison with the diagram represented in Fig. 19 (without accounting for the effect of splitting). In nonstoichiometric MnO or CuO a hole is formed in the e_g-levels: During introduction, e.g., of Li_2O into MnO, the ion Mn^{3+} is formed with the structure of d^4, instead of d^5, as in pure MnO; five electrons leave level e_g (see Fig. 21). In FeO, CoO, and NiO a hole is formed in the t_{2g}-levels [151].

The transitions of electrons within the limits of the d-shell among the levels schematically shown in Figs. 21 and 22, i.e., the d–d transitions, give low intensity lines in the visible or nearly visible portion of the optical spectrum. At shorter wavelengths, often in the ultraviolet region, compounds of transition metals show significantly more intensive spectra for charge transfer. These lines correspond to, for example, the excitation of an electron of a ligand t_{1u} to level e_g of the central atom (see Fig. 22). In the zone diagrams for oxides of transition metals, the transfer of an electron across the forbidden zone corresponds to the charge transfer spectra: from the $2p$-zone of oxygen to the d-level of the metal (see Fig. 19). The order for frequency change of absorption bands of charge transfer spectra [167] in complexes corresponds to the rules listed for the width of the forbidden zone (Section 1.3). For example, in the series of halide ions of certain metals $[Me(NH_3)_5X]^{n+}$ in solution, the absorption of long wave maxima increases but the energy of transition falls from Cl^- to Br^- and I^-. In a series of crystalline halides of certain metals, the width of the forbidden zone decreases during the transition from chloride to bromide and then to iodine (see the table in the Appendix). In complexes of various ions with Cl^- or Br^- in solution, the order of frequency lowering corresponding to charge transfer is the same as the order for decrease of the width of the forbidden zone of the corresponding oxides. There is a little in the literature on charge transfer spectra for complexes of various metals:

$$Ti^{4+} \sim Sn^{4+} > Pb^{4+} > Fe^{3+} > Cu^{2+}$$

Recently Jorgensen [167] showed that the frequency hv (in cm^{-1}) corresponding to the transfer of charge between the metal Me and the ligand L in the complex MeL_n, is approximately proportional to the difference of electronegativities of Me and L. If we express this statement mathematically $hv = \{x_{opt}[L] - x_{opt}[Me]\}\,30{,}000$, the "optical electronegativity" x_{opt} (according to the terminology of Jorgensen) that will be obtained will be close to the electronegativity value of Pauling (see Section 1.5). The relationship between hv and Δx_{opt} is completely analogous to the relationship between width of the forbidden zone U and Δx.

In a number of cases, the energy corresponding to crystal field stabilization and to charge transfer spectra is fairly close to the activation energy of a catalytic reaction. A question arises concerning the correlation of these values.

Dowden and Wells [173] applied crystal field theory to explain the dependence of catalytic activity of oxides of transition metals on the position of the metal in the periodic table. In their opinion, a simple electronic theory cannot explain the changes of catalytic activity observed

1.6. THE ROLE OF d-ELECTRONS

experimentally during transition from one oxide to another. The electronic levels of metal cations of the fourth period in oxides (referred to a vacuum) are approximately equal. The catalytic activity of metal oxides of the fourth period often changes, but not monotonically—rather, proceeding through minima and maxima. Most often, minima are observed in systems with the configuration of cations d^0, d^5, d^{10}, and the maxima are between these. Thus, change of catalytic activity implies change of other properties of compounds of transition metals (see, e.g., Fig. 23).

Dowden et al. [174] studied the H_2–D_2 exchange on oxides of the fourth period and found that their catalytic activity (Fig. 24) begins

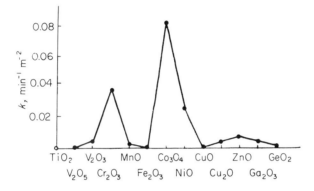

FIG. 24. The change of catalytic activity of oxides of metals of the fourth period in the H_2–D_2 exchange reaction [174].

almost with zero at TiO_2 (the configuration of Ti^{4+} is d^0), passes across a maximum at Cr_2O_3 (d^3) to a minimum at Fe_2O_3 (d^5), then to a new high maximum at Co_3O_4 (d^6–d^7) and is lowered to CuO and ZnO (d^9, d^{10}). The two-spiked diagram of the catalytic activity of metal oxides of the fourth period is also observed for other reactions, e.g., dehydrogenation and disproportionation of cyclohexene [175]. The rules for change of catalytic activity of metal oxides of the fifth and sixth periods, in general, reiterate the fourth period with complications produced by the ease of reduction of the metal oxides of Group VIII to the metal. In the fifth period, for the H_2–D_2 exchange reaction, ZrO_2 appears to be inactive and Nb_2O_5 and MoO_3 are very slightly active; the high activity of ReO_2 and Rh_2O_3 is possibly due to the presence of metallic Re and Rh in them [43]. The two-spiked diagram for catalytic activity of metal sulfides of the fourth period was found for the hydrogenolysis reaction of carbon bisulfide [173].

According to Dowden and Wells [173], one can liken chemisorption on a cation of a transition metal to a reaction of a complex in which there is an increase in the number of ligands. In this process, one of the ligands will be an adsorbed molecule; and the others will be anions of the crystalline lattice. During adsorption, an increase of coordination occurs. For example, during adsorption on a cation of a lattice of the type NaCl on the (100) face, there will be a change of the number of ligands from 5 to 6 and a change of coordination: tetragonal pyramid → octahedron; on face (110)—tetrahedron → tetragonal pyramid → octahedron; on face (111)—triangle → tetrahedron → tetragonal pyramid → octahedron. Analogous changes occur during adsorption on cations of other types of lattices that have octahedral coordination around the cations (of the types NiAs, pyrite, CdI_2, corundum, spinel, etc.). Changes in crystal field stabilization energy during such coordination changes are presented in Table II. The calculation is made using experimental values of Δ, measured directly on oxides, or in some cases, from Δ for corresponding aqueous solutions or hydrated salts. A minus sign corresponds to an

TABLE II

The Changes of the Crystal Field Stabilization Energy (kcal/mole) for Changes of Coordination[a]

Ion	Change of coordination					
	Triang. ↓ Tetrahed.	Tetrahed. ↓ Tetragon. pyramid	Tetragon. pyramid ↓ Octahed.	Tetrahed. ↓ Octahed.	Triang. ↓ Tetragon. pyramid	Triang. ↓ Octahed.
d^0 (Ca^{2+}, Sc^{3+})	0	0	0	0	0	0
d^1 (Ti^{3+})	+6	−9.5	+2.9	−6.7	−3.6	−0.7
d^2 (V^{3+})	+11	−18	+5.5	−13	−6.8	−1.3
d^3 (V^{2+})	+27	−23	−7.0	−30	+3.2	−3.8
d^3 (Cr^{3+})	+34	−30	−9.2	−39	+4.2	−5
d^4 (Cr^{2+})	+14	−29	+12	−16	−14	−2.1
d^4 (Mn^{3+})	+20	−39	+17	−22	−20	−2.9
d^5 (Mn^{2+}, Fe^{3+})	0	0	0	0	0	0
d^6 (Fe^{2+})	+3.6	−5.7	+1.7	−4	−2.1	−0.42
d^6 (Co^{3+})	+6.4	−10	+3.0 (−21)[b]	−7	−3.8	−0.74
d^7 (Co^{2+})	+6.7	−11	+3.2	−7.4	−4.0	−0.78
d^8 (Ni^{2+})	+17	−15	−4.6	−19	+2.1	−2.5
d^9 Cu^{2+})	+14	−27	+12	−16	−14	−2
d^{10} (Cu^+, Zn^{2+})	0	0	0	0	0	0

[a] According to Dowden and Wells [173]. (An approximation of a "weak" field.)
[b] Approximation of a "strong" field.

1.6. THE ROLE OF d-ELECTRONS

increase of stabilization energy. It is evident from Table II that, in almost all cases (except the second column and several points of the fourth and sixth columns), an increase of coordination number leads to an exothermic effect. The structures of cations d^0, d^5, and d^{10} which in a weak crystal field of ligands do not have any additional energy stabilization caused by this field (and the configurations d^0 and d^{10} in a strong field) are exceptions.

The general picture suggests the two-spiked diagram for the change of catalytic activity of oxides of transition metals in H_2–D_2 exchange and other reactions (Fig. 24). If we hypothesize that adsorption or some step carried out with an increase of coordination number appears to be the limiting step of the reaction, then the minimal exothermic effect will be observed for the systems d^0, d^5, and d^{10}, which have also the minimal catalytic activity. The authors [173] also explain several exceptions to this simple rule. For example, according to the data of Table II, the transitions tetragonal pyramid–octahedron (face 100) on Co_2O_3 (configuration of cation d^6) and CuO (d^9) are carried out with an endothermic effect. However, Co^{3+} forms only complexes having a strong field; in the case of CuO the face (100) is almost always absent; and the mentioned configurational transitions are not observed during adsorption. The most catalytically active systems are those with the electronic configuration d^3, d^6, and d^8. These show the maximum crystal field stabilization during an increase of coordination number.

Changes of coordination during catalysis can occur either without electronic transitions or in conjunction with them. For example, the adsorption of H_2, according to Dowden, can proceed without electronic transitions by strong polarization of the molecule:

$$\begin{array}{cc} H^- & \cdots & H^+ \\ \vdots & & \vdots \\ Me^{n+} & \cdots & O^{2-} \end{array}$$

this is especially favorable on lattices containing V^{2+}, Cr^{3+}, Co^{3+}, and Ni^{2+}, or other cations of the configurations d^3, d^6 and d^8. During desorption, crystal field stabilization is observed on face (100), which contains V^{3+}, Cr^{2+}, and Ni^+ (see the structures d^2, d^4, and d^7 in the fourth column of Table II). Therefore, the conversion of ions V^{2+} into V^{3+}, Cr^{3+} into Cr^{2+}, etc. will favor desorption. Hence, it follows that semiconductor lattices containing pairs of ions, V^{2+}/V^{3+}, Cr^{2+}/Cr^{3+}, Co^{2+}/Co^{3+} and Ni^+/Ni^{2+}, will favor H_2–D_2 exchange. The configurations d^0, d^5, and d^{10} will be the least active.

Another example: Decomposition of N_2O proceeds with an

electronic transition, during which the limiting step appears to be desorption of oxygen $2Me^{(n+1)+} \cdots O^- \to Me^{n+} + Me^{(n+1)+} \cdots O_2^- \to 2Me^{n+} + O_{2(gas)}$. During the transition, a change of coordination occurs: octahedron → tetragonal pyramid, with transfer of an electron to the catalyst. The minimum endothermic effect, from the point of view of the crystal field, will be for the structures d^0, d^1, d^2, d^5, d^6, d^7, d^9, and d^{10}. The most active p-semiconductors Cu_2O, MnO, CoO, and others are derived from ions with this configuration. Thus, it is evident that for oxidation–reduction reactions, the crystal field theory in conjunction with the electronic theory of catalysis on semiconductors (Section 1.1) gives the best predictions.

The transition from oxides to sulfides should, according to the crystal field theory, give the same type of dependence. However, during this transition, the electronic structure of the solid is substantially changed: Usually, the width of the forbidden zone is reduced.

The crystal field theory is convenient for chemists in that it relates catalytic activity directly to chemical properties of catalysts. This theory adequately explains the similarity of rules for change of catalytic activity of compounds of transition metals that have been observed in a series of cases in homogeneous and heterogeneous catalysis. For example, in oxidation–reduction reactions in solutions, the relative catalytic activities of ions Co^{2+}/Co^{3+}, Ni^{2+}/Ni^{3+}, and Mn^{2+}/Mn^{3+} are approximately the same as for C_2H_4 oxidation on the solid oxides of Co, Ni, and Mn. The ions Mn^{2+} and Fe^{3+}, free or in the form of complexes, usually behave in the same way as ions of nontransition metals. Their activity is significantly higher or lower (but not equal) than the activity of ions of the other transition metals, which is explained by the absence of crystal field stabilization for these ions [176].

Complex ions of transition metals (except systems of d^5) most often of all are only slightly active catalytically. This is particularly so for metals of the platinic group, the coordination sphere of which is filled. It is interesting to note that in natural complex biocatalysts (e.g., hemoglobin, chlorophyll), and in complex producing ion-activators for fermentations are found ions of nontransition metals: Zn^{2+}, Ca^{2+}, Mg^{2+}, and others, the ions of d^5—Fe^{3+} and Mn^{2+}, and also Cu^{2+}, Fe^{2+}, Co^{2+}, Ni^{2+}, Cr^{2+}; the ions of heavy transition metals are almost never encountered [177]. Mo represents an exception, whose actual valence state in biocatalysts is as yet unknown. Concerning the catalytic action of ions, see Section 2.2.

In a strong ligand field there can be other rules; and the configuration d^5 does not always have a similarity with nontransition metals. It was already said above that nitrogen-containing ligands, e.g., NH_3, form a strong field in contrast to H_2O [178]. Therefore, the type of catalytic

1.6. THE ROLE OF d-ELECTRONS

activity had by ions of transition metals with ligands of NH_3 will be different than that had with ligands of H_2O. The maximum crystal field stabilization in a strong field corresponds to the d^6 cation configuration. It is possible that by d^6 structure stabilization one can explain the high catalytic activity of Co_3O_4, compared to other oxides, for oxidation of NH_3 [179].

Following the work of Dowden and Wells [173], critical comments were expressed concerning their proposed explanation of the nature of the catalytic activity of transition metal oxides. Doubts regarding the accuracy of the very two-spiked dependence of activity of these oxides in H_2–D_2 exchange were also expressed (see Fig. 24). Stone [48] assumed that NiO in this reaction is catalytically inactive, the activity appearing upon partial reduction of NiO to metallic nickel. It was also shown [180] that CoO (d^7) is significantly less active than Co^{3+} (d^6) in spite of the data of work [174]. In the opinion of De et al. [181] and De and Stone [182], it is expedient to compare the influence of the crystal field on catalytic activity even for oxides with a specific crystalline structure. In their work, the H_2–D_2 exchange was studied on oxides with the structure of corundum: Ti_2O_3 (cation configuration d^1), V_2O_3 (d^2), and Cr_2O_3 (d^3). In the first two, the site has a significant overlap of d-orbitals down to the zone layer (See Fig. 18) as well as a pairing of spins of neighboring cations; in Cr_2O_3 below 45°C, a cation–anion–cation interaction is observed, characteristic for antiferromagnetism; but above this temperature there is paramagnetism. At low temperature (100–300°C), Cr_2O_3 is more active, which is in accordance with the data of Dowden and other authors (See Fig. 24). At higher temperatures, Ti_2O_3 and V_2O_3 are highly active. By correlating results for magnetic changes, De et al. [181] showed that, with increase of temperature, the activity of Ti_2O_3 and V_2O_3 increases because of both spin pairing and excitation of d-electrons on the e_g-level (See Fig. 21). It is possible that at high temperatures the configurations d^1 and d^2 in H_2–D_2 exchange are more active than d^3, although there is no direct data for Cr_2O_3 at a temperature above 300°C for this case. The molecules of H_2, according to the data of these authors, are adsorbed on two atoms of the metal and not on the metal and oxygen, as was shown above.

Nevertheless, in the overwhelming majority of recent works, the two-spiked diagram for change of catalytic activity of metal oxides of the fourth period in oxidation–reduction reactions is verified. The non-occurrence of this relationship in a given concrete case does not justify consideration of arguments against the expedience of applying crystal field theory in catalysis. In each case, it is necessary to seek a concrete explanation of a specific result. During this, it can be especially advantageous to obtain, during adsorption and catalysis, measurements of optical spectra that give an indication of the actual position of the d-levels of the surface cations and their changes. Endeavors in this direction are, as yet, few and have still not resulted in significant successes [183].

Up to now, efforts to compare different types of catalytic activity of oxides of transition metals of the fourth period have not been conducted incorporating charge transfer spectra of both interzone transitions and of transitions from local doped levels to the d-level. The expedience of similar comparisons can be seen by comparing Figs. 15a and 24. It is evident that the maxima of catalytic activity in H_2–D_2 exchange correspond to minima of U and, on the other hand, on Fe_2O_3 (d^5), a minimum of catalytic activity and a maximum U are observed. Also, one can find, in this case, a parallelism between homogeneous and heterogeneous catalysis. It is known that charge transfer in homogeneous reactions in solutions (for instance, $Fe^{3+} \to Fe^{2+}$) often initiates reactions in which free radicals participate.

Calculation of the changes of crystal field stabilization energy in the absence of charge transfer enables us to explain the two-spiked diagram for change of catalytic activity of the oxides of the first large period. However, this does not explain why this diagram is observed only in oxidation–reduction reactions.

In the examples presented, application of crystal field theory and ligand field theory leads to the same results. However, if the ligands are able to form a π-bond (O_2, NO, CO, C_2H_4, benzene, and others), a purely electrostatic approach to crystal field theory cannot explain the properties of complexes. According to ligand field theory [166], two e_g-orbitals (d_{z^2} and $d_{x^2-y^2}$), the s and p-orbitals of the metal atom, and the corresponding ligand orbitals participate in the formation of octahedral σ-bonds (See Fig. 22). But, if the ligands contain π-bonds, they interact with the t_{2g}-orbitals of the metal (d_{xy}, d_{yz}, d_{xz}). During this, in contrast to the usual donor–acceptor bond, a reverse movement of electron density from the central atom to the ligand occurs. Such a bond is often called dative. Thus, for a dative bond, not empty, but full d-orbitals are necessary. It is understandable that the best complex producers for olefinic and acetylenic hydrocarbons (to form π-complexes with metal ions) appear to be ions with the configuration d^{10}: Hg^{2+}, Cd^{2+}, Cu^+, Ag^+, Pt^0.

In catalytic reactions in which olefins and acetylene participate, compounds of metals inclined to dative interaction are the most active. For example, in vapor phase reactions of acetylene (hydration, hydrochlorination, hydrocyanidation, and synthesis of vinyl acetate) the

1.6. THE ROLE OF d-ELECTRONS

following sequence of activity of cations of salts was obtained [184, 185]:

$$Hg^{2+} > Bi^{2+} > Cd^{2+} > Zn^{2+} > Ni^{2+} > Fe^{3+} > Mg^{2+} > Ca^{2+} > Ba^{2+}$$

Among the ions of metals with configuration d^{10}, catalytic activity increases as one proceeds downward in the periodic table in accordance with the increase of capability for dative interaction. Compounds of Hg^{2+} possess an especially high catalytic activity. This is due to the high mobility of its $5d$-electrons and the ready polarizability of the cation, surpassing almost all other cations [185].

Compounds most active for polymerization of olefins are those of cations with the electronic configuration d^1: $TiCl_3$ in Ziegler–Natta catalysts [186], and oxides of molybdenum and chromium where there appear to be active centers, seemingly Mo^{5+} and Cr^{5+} [187]. It is impossible to expect a large stabilization during interaction of d^1 cations with π-bonds. As was shown above, the most stable complexes of such a family are formed by cations with filled d-shells. However, Cossee [188] presented a diagram of the molecular orbitals of a complex of a transition metal with ethylene that qualitatively explains the polymerization mechanism (Fig. 25). The reactions proceed as follows:

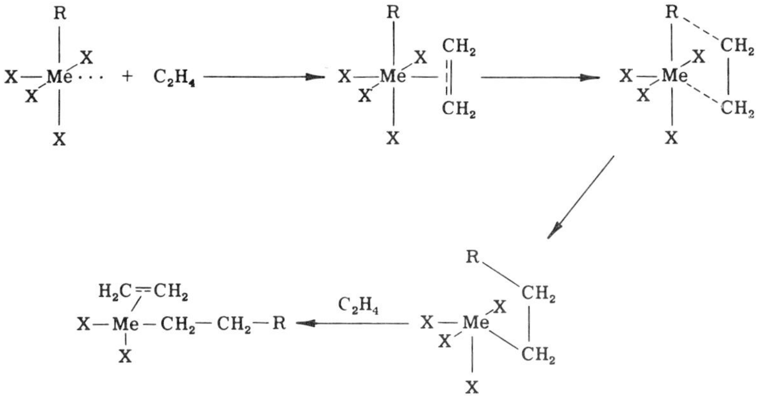

$$
\begin{array}{c}
R \\
| \\
X-Me\cdots + C_2H_4 \longrightarrow X-Me-\begin{array}{c}CH_2 \\ \| \\ CH_2\end{array} \longrightarrow X-Me\begin{array}{c}CH_2 \\ | \\ CH_2\end{array}
\end{array}
$$

$$
\begin{array}{c}
H_2C=CH_2 \\
X-Me-CH_2-CH_2-R \xleftarrow{C_2H_4} X-Me\begin{array}{c}CH_2 \\ | \\ CH_2\end{array}
\end{array}
$$

(29)

where X represents an atom of Cl, Br, or I; R is an alkyl group.

In the Ziegler–Natta catalyst, $TiCl_3 + Al(C_2H_5)_3$, aluminum appears to be the alkylating agent that supplies the radical R to the transition metal. The ion of the transition metal Me (Ti^{3+}) is found in octahedral coordination on the surface of solid $TiCl_3$. Four coordination bonds are occupied by the anions X (Cl^-), the fifth by the alkyl R, and the sixth is free for interaction with ethylene. In Fig. 25, ψ_1 and ψ_2 are bonding orbitals. The complex is stabilized because of reverse motion of d-

electrons from one of the d_{xy}, d_{xz}, d_{yz} orbitals. However, only on the latter is there an electron. Consequently, the complex cannot be very stable. But, because of the attraction of the ψ_2 orbital ($d_{yz} + \pi^*$—the ethylene orbital) for the orbital of the metal–alkyl bond, ψ_{RMe} (i.e., the decrease of the value of ΔE to $\Delta E'$, See Fig. 25), a new possibility appears: the transition of an electron from orbital ψ_{RMe} to ψ_2 (d_{yz}). The R—Me bond is broken during this process; and the olefin molecule fixes itself at the base of a short growing polymer chain. Subsequent to this,

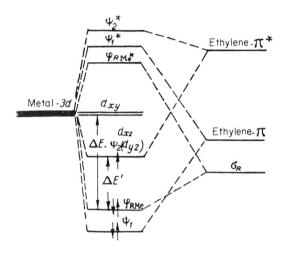

Fig. 25. Diagram of molecular orbitals of a complex of the ion Ti^{3+} with ethylene on the surface of $TiCl_3$ alkylated with the group R [188].

the process repeats itself, while the olefin and short polymer chains constantly change locations in the coordination sphere of the ion of the transition metal. Thus, to accomplish the growth process of a chain, the presence of a vacancy in the d_{xy}, d_{xz}, or d_{yz} shells is necessary. An examination of specific energetic distances between molecular orbitals [188] showed that the d-level in the better olefin polymerization catalysts is very near to the π-level of ethylene.

1.7. The Color of Solids

It is known that substances that are catalytically active in oxidation–reduction reactions are usually colored. The first to pay attention to the relationship between catalytic activity and color were Roginskii and Schul [14] in 1928. They studied the effect of various solid oxides on

1.7. THE COLOR OF SOLIDS

the decomposition of the unstable oxygen-containing compounds $KMnO_4$, $KClO_3$, HgO, and Ag_2O. They found that the most active oxides (NiO, Co_3O_4, MnO_2, CuO) are intensely colored materials in which black and brown tones predominate. Then come the less active substances, which have red, orange, and yellow colors (SnO_2, PbO, ZnO, CdO, and others). The slightly active and inactive substances are white (Al_2O_3, MgO, TiO_2, SnO_2). The relationship between the catalytic activity of various oxides and their color was also verified for *iso*-octane oxidation [189] and hydrogen peroxide decomposition [190]. In all cases, intensely colored compounds showed strong catalytic action, but white compounds were inactive or only slightly active. In the course of this work, some levels of activity were observed for uncolored compounds. The catalytically more active among the uncolored substances had absorption bands in the near ultraviolet spectral region [191]. Other investigators working in the area of catalysis have also paid attention to the relationship between catalytic activity and color [9].

In inorganic chemistry, a qualitative relationship [99, 192–194] is recognized between the color and the electronegativity differences Δx of atoms which enter into the formation of a substance: The smaller Δx, the more intensely colored is the substance; and the absorption spectrum decreases from the ultraviolet range to the visible. The change of color can be followed, using as an example a series of analogous compounds. For example, in the series BeO, MgO, ZnO, CdO, and HgO, the first two oxides are white; ZnO is white or weakly yellow; CdO is yellowish brown; and HgO is intensely colored yellow or red. In this same direction, bond ionization and Δx decrease; corresponding to this, the proportion of covalent bonds increases. Sulfides have a lower value of Δx than oxides and are more intensely colored. In a series of binary compounds of silver: AgCl, AgBr, AgI, and Ag_2S, the first compound is colorless; the second is light yellow; the third is yellow; and the fourth is black. In a series of analogous compounds of Sb^{3+} the sequence is similar: colorless, yellow, red, orange, and black. Δx falls in this same direction. Exceptions to this rule are comparatively few: Cs_2O has a greater value of Δx (2.8) than CaO (2.5), but the former compound is orange-yellow and the latter is colorless.

The relationship between color and Δx is easy to explain if one takes into consideration (Section 1.3) the relationship between width of the forbidden zone and Δx. Actually, absorption in the visible region of the spectrum (400–760 mμ) corresponds to the transfer of electrons with energy of 1.2–2.4 eV ($\sim$30–60 kcal). Consequently, pure substances, having a width of the forbidden zone greater than 2.4 eV, will be colorless.

The color of compounds of transition metals can be explained

[167, 195] from the point of view of ligand field theory, examined above (Section 1.6). As has been already shown, the absorption spectra of these compounds consist of both intensive charge transfer bands and the less intensive bands that correspond to transition from one d-orbital to another. If we examine the series of oxides of the fourth period: Sc_2O_3 and TiO_2 (white); V_2O_3 and VO (black); Cr_2O_3 (green); MnO_2 and Mn_2O_3 (black); MnO (green); Fe_2O_3 (brown); CoO (greyish green); NiO (green); Co_2O_3 (brownish black); Cu_2O (red); ZnO, Ga_2O_3, and others (colorless) it can be seen that structures of d^0 and d^{10} (except Cu_2O) form colorless compounds, and structures of d^5 (MnO and Fe_2O_3) form less intensely colored oxides than do the configurations d^1–d^4 and d^6–d^9. Apparently, the color of transition metal oxides is specified in principle by charge transfer spectra (See Fig. 15). In other cases, d–d transitions are more important. For example, solid cupric sulfate $CuSO_4$ is colorless, since the SO_4^{2-} ligands form a very weak field and so the d–d transitions fall in the infrared region. The hydrate $CuSO_4 \cdot 5H_2O$ is blue because the ligands (molecules of H_2O) shift the d–d spectrum into the visible region. Finally, copper–ammonia complexes have an intensely blue-violet color on account of the strong field of the NH_3 ligands.

Thus, in the case of pure semiconductors, the relationship between catalytic activity and color is the relationship between catalytic activity and the width of the forbidden zone U or the structure of the d-shell of the cation. These were examined in detail in Sections 1.3 and 1.6.

Crystals with a wide forbidden zone can be intensely colored, e.g., bright red ruby, the lattice of which is composed of corundum with $U = 7.3$ eV. The color centers in them are the so-called F-centers; they can also be found in alkaline halides. In these cases the color is caused by additions (doping) to the basic lattice. Local levels form in the forbidden zone with depths that correspond to electronic transitions in the visible region of the spectrum. According to contemporary ideas, the F-centers appear to be formations composed of an anionic vacancy and an electron bonded with it. The electron is formed as a result of the ionization of atoms of the alkaline metal that was introduced into the lattice of the halide.

The catalytic properties of compounds such as dielectrics, the colors of which are specified not by the absorption spectrum of the basic lattice, but by the doping materials, have been insufficiently studied. However, high catalytic activity for dehydrogenation was discovered recently [46] for "black" aluminum oxide, obtained by partial reduction of Al_2O_3 in hydrogen. It is possible that, in such substances, there is a favorable distance between local levels of dopants introduced into the lattice and local levels of adsorbed species (Section 1.4) for catalysis.

1.7. THE COLOR OF SOLIDS

Even though it is impossible to assume the presence of color as a necessary criterion for catalytic activity for oxidation–reduction reactions, still, in every case, the correlation between activity and color can be explained by existing theories of catalysis. The energy of electronic transfers in the visible part of the spectrum is rather close in order of magnitude to the activation energy of many oxidation–reduction catalytic reactions.

2 • Solid Properties and Catalytic Activity in Acid–Base Reactions

2.1. Proton-Acid and -Base Properties of a Surface

According to Bronsted, the definition of acids and bases is contained in the following: An acid is a substance capable of losing a proton; a base is a substance capable of adding a proton [196]. This definition apparently does not relate acid–base properties to the aggregate state of the substance (the majority of acids are liquids). Therefore, in a specific case, a solid surface can be an acid and a base. Bernardskii [197] and, after him, Pauling [99], introduced ideas about acidic properties of solid aluminosilicates.

Acids and bases in the liquid phase appear to be catalysts for heterolytic type reactions: polymerization, condensation, hydrolysis, hydration, and dehydration. Roginskii and Ioffe [198] showed that if, in the liquid phase, a reaction is accelerated by acids or bases, then it will be accelerated by a solid possessing acidic or basic functions.

There exist three principal methods to determine the acidity or basicity of a surface. One is ion exchange adsorption, which is the titration of aqueous suspensions of a catalyst or an adsorbent by either a base (for definition of acidity) or an acid (for definition of basicity). During cation-exchange adsorption of alkali on silica–alumina catalysts [199–201] acidic hydrogen of the surface is substituted for the alkali metal ion: Li^+, Na^+, and K^+. Catalytic cracking activity decreases linearly as a function of the quantity of Li or Na ions (Fig. 26) introduced into the catalyst. This is the result of the linear decrease in the

2.1. ACID AND BASE PROPERTIES OF A SURFACE

number of acidic hydrogen atoms, which are apparently the active sites for catalysis.

The number of acid sites is determined in this manner. One can also use this method to determine the strength of the acid sites by changing the pH of the ion exchange solution. For example, in a solution of the salts NaCl or CH_3COONa, only strongly acidic hydrogen—the relatively mobile H^+ of silica–alumina—is exchanged for sodium. Under the action of the alkali NaOH, not only such hydrogen, but also the weakly

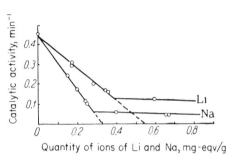

FIG. 26. Dependence of catalytic activity on amount of Li and Na ions introduced into silica–alumina [199, 200].

acidic hydrogen of surface OH-groups is exposed to exchange. It was successfully shown [199] that the cracking reaction on silica–alumina proceeds on strongly acidic sites and is poisoned upon exchange of H^+ of the surface with the alkali NaOH and salts NaCl and CH_3COONa. Isomerization of cyclohexene to methyl cyclopentene proceeds on weaker acidic centers and is poisoned only at high pH (through the action of the alkali NaOH). The disadvantage of the method is that there is the possibility of changing the catalyst surface as a result of its interaction with the aqueous medium.

The second method is the adsorption of acids or bases from the gas phase. From the amount of adsorbed base or acid, one can estimate the number of correspondingly acidic or basic sites on the surface. The strength depends on the stability of the bonds with the surface. A linear relationship between catalytic activity and the number of acidic sites was found for a series of cases. Figure 27 shows data [201] illustrating, for gas oil cracking, the relationship between the amount of quinoline chemisorbed at 316°C and gasoline yield, thus characterizing the catalytic activity of various catalysts.

Krylov and Fokina [202–204] studied the adsorption of a weak base—pyridine—from the gas phase for determination of the number of acidic centers, and the adsorption of a weak acid—phenol—for determination of the number of basic centers. At low temperatures (20°C), on the majority of catalysts, the amount of adsorbed phenol and pyridine is the

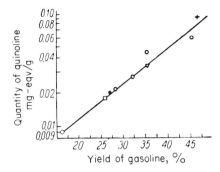

FIG. 27. The relationship between catalytic activity as yield of gasoline for various gas oil cracking catalysts and the number of acid sites determined by chemisorption of quinoline [201].

same and amounts to 4–7 μmoles/m^2, i.e., almost all the surface is filled Hence, the conclusion was drawn that a large fraction of the adsorbed pyridine and phenol is affixed by hydrogen bonding, e.g.,

$$\begin{array}{c} \text{pyridine} \cdots \text{H} \\ | \\ \text{O} \\ | \\ -\text{O}-\text{Al}-\text{O}- \end{array} \tag{30}$$

$$\begin{array}{c} \text{phenol} \\ | \\ \text{O} \\ | \\ \text{H} \quad \text{H} \quad \text{H} \\ \text{O} \quad \text{O} \quad \text{O} \\ \text{Si} \quad \text{Si} \end{array} \tag{31}$$

Multilayers are not formed during this. It is known [205] that, even at low temperatures, the adsorption of amines on common adsorbents occurs by a monolayer. The activation energy of such adsorption is close to zero. Consequently, surface heterogeneity is not found during a study of adsorption of weak acids and bases from the gas phase by the differential isotope method [204]. The adsorption of phenol on strong bases [CaO, Ca(OH)$_2$] and pyridine on strong acids (silica–aluminas) leads to rather large adsorption values—up to 11 μmoles/m^2.

One could theorize that the bond stability of adsorbed pyridine and phenol—i.e., the strength of acidic and basic centers—varies according to the decrease of adsorption with increasing temperature. Figure 28 shows the temperature dependence of the amount of phenol adsorption on oxides. It is evident that the stability of the phenol bond, i.e., the

2.1. ACID AND BASE PROPERTIES OF A SURFACE

basicity of the surface, is highest for CaO and lowest for silica–alumina. Among the catalysts studied, the bond stability of phenol decreases in the order CaO, CdO, MgO, Ca(OH)$_2$, ZnO, BeO, Al$_2$O$_3$, CdS, ZnS, and silica–alumina. One cannot succeed in desorbing phenol from the surface of CaO and CdO, even at 400°C. Almost opposite correlations were obtained for the adsorption of pyridine on these same catalysts: The more stable bond of adsorbed pyridine is observed with the surface of acid (silica–alumina) and amphoteric (Al$_2$O$_3$ and BeO) compounds. It was also shown [203] that the acid catalysts that most stably adsorb pyridine are most catalytically active for alcohol dehydration; basic catalysts that stably adsorb phenol are more active for alcohol dehydrogenation.

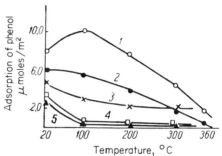

FIG. 28. Dependence of phenol adsorption on temperature. Adsorption isobars are computed from the saturation points of kinetic phenol adsorption isotherms: 1—CaO; 2—BaO; 3—Al$_2$O$_3$; 4—Silica gel; 5—silica–alumina.

The third method to study acidity and basicity is to titrate the surface of catalysts with weak bases and acids in nonaqueous media, with adsorption of the indicators on the surface. This method conveys, for solid surfaces, the concept of acidity function that was introduced by Hammett [206] as a characteristic of aqueous solutions.

To determine acidity by the indicator method in a homogeneous medium [207], it is necessary that the color of the base-indicator B differ from the color of its acid form BH$^+$. The equilibrium constant of the reaction BH$^+ \leftrightarrows$ B + H$^+$ is equal to:

$$K_{BH^+} = \frac{a_{H^+} a_B}{a_{BH^+}} = \frac{a_{H^+} f_B}{f_{BH^+}} \cdot \frac{c_B}{c_{BH^+}} \tag{32}$$

where a_{H^+} is the thermodynamic activity of a proton; c_B and c_{BH^+} are the concentrations of the base B and its protonized form; f_B and f_{BH^+} are their activity coefficients.

Designating $-\log K_{BH^+} = pK$ we obtain from Eq. (32)

$$-\log a_{H^+} \frac{f_B}{f_{BH^+}} = pK - \log \frac{c_{BH^+}}{c} = H_0 \tag{33}$$

where H_0 is the so-called acidity function characterizing the proton-donor properties of a homogeneous medium.

By assembling the indicators with different pK's, one can measure acidity functions on various media. In dilute solutions, $H_0 = $ pH. In concentrated solutions of acids, H_0 can achieve large negative values. For example, in a 100% solution of H_2SO_4, $H_0 = -10.3$.

It happens that an indicator that is adsorbed on a surface also changes color. The appearance of the color of the acid form of the indicator shows that the value of the function H_0 on the catalyst surface is lower than the pK of the appropriate indicator. The lower H_0, the more acidic properties the surface possesses. As an approximation [from Eq. (33) the second term $\log(c_{BH^+}/c_B)$ is neglected], one can assume that, at the instant of color change, $H_0 \approx$ pK. This method was applied for the first time by Walling [208] while studying the color of indicators: n-dimethylamino-azobenzene, 1,4-di-isopropyl-amino-anthraquinone, benzene-azodiphenyl amine, and azoparanitro-diphenylamine adsorbed from a solution of *iso*-octane on various substances. The following sequence of decreasing H_0 (increasing acid strength) was found by him: CaF_2, TiO_2, $CaSO_4$, Al_2O_3, $CaCl_2$, $AgCl$, SiO_2, Sb_2O_5, H_2WO_4, $Mg(ClO_4)_2$, $AgClO_4$, $Fe_2(SO_4)_3 \cdot xH_2O$, $Al_2O_3 \cdot SiO_2$, $MgO \cdot SiO_2$, $CuSO_4$.

Other investigators [209–216] also applied this method. Benesi [210] found that the value of H_0 for silica gel ranges between $+4.0$ and $+6.8$, i.e., its acidity is very slight; but, for silica–alumina, $H_0 < -8.2$, which is close to the acidity of sulfuric acid applied on silica gel. By utilizing a greater number of indicators, one can encompass a range of catalysts with large differences in acidity.

In the work of Dzisko and Borisova [212], these indicators were used: neutral red (pK of the transition $+6.8$), bromocresol purple ($+6.1$), bromophenol blue ($+3.86$), benzene-azodiphenylamine ($+1.5$), paranitrobenzene-azoparanitro-diphenylamine ($+0.43$), dicinnamyl acetone (-3), benzal-acetophenone (-5.3), and anthraquinone (-8.2). It was found that Al_2O_3, SiO_2, and ZrO_2 have a weakly acid surface ($H_0 \sim +4$); TiO_2, hydrosilicate of magnesium and aluminum borate have medium strength (H_0 from -3.0 to -5.6); and silica–aluminas and zirconium-silicates have high acidity ($H_0 \sim -8.2$).

The number of acid centers in the above mentioned work [212] (see also [213–216]) is determined by titration with butylamine from a solution of heptane. For some reactions studied, a linear relationship was found between catalytic activity and the number of acid centers, as determined by the complete adsorption of butylamine. The dependence of catalytic activity for dehydration reactions of ethyl and isopropyl alcohols on the

2.1. ACID AND BASE PROPERTIES OF A SURFACE

number of acidic groups on zirconium-silicate catalysts, according to Dzisko [216], is shown in Fig. 29. Thus, catalytic activity, expressed as milliequivalents of acid centers, is a constant value. On the other hand, a simple relationship between acid strength, i.e., the value of H_0, and catalytic activity, was not obtained. It is known [217] that in homogeneous catalysis the Hammett relationship is often observed between acidity and catalytic activity:

$$\log k + nH_0 = \text{const.} \qquad (34)$$

where k is the rate constant of the reaction; $n = 1$ or 2.

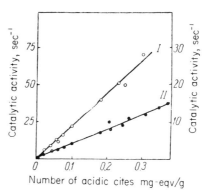

FIG. 29. The dependence of catalytic activity for isopropyl (I) and ethyl (II) alcohol dehydration on the number of acid centers on zirconium-silicate catalysts [216].

In works [213, 216], however, Eq. (34) is not verified for butylene polymerization, isobutylene dimerization, and alcohol dehydration. There is some correlation between the values of k and H_0. For example, in butylene polymerization, when H_0 increases by five units, k is changed not by an order of five but by a factor of five. The dimensionless coefficient n in Eq. (34) is equal to 0.15–0.6. This result is rather difficult to explain, assuming, as in homogeneous catalysis, that the reactive form is obtained by the addition of a proton to the adsorbed molecule.

In contrast to [212, 213], a relationship between acidity and activity of alcohol dehydration catalysts was not found by Topchieva.

The above three methods allow us to estimate the number and strengths of acid (or basic) centers on a surface with a specific degree of precision. They do not give direct indications of their nature. There is a lively discussion in the literature about whether these centers appear to be acids of the Bronsted type (p. 70) or acids of the Lewis type (p. 81). Many authors [199, 200, 218–221] relate the surface acidity measured by such methods to the presence of a proton or mobile hydrogen of an acidic group on the surface of stable acids, i.e., acids of the Bronsted type.

For the well studied silica–alumina catalyst, they assume [218] that Al replaces Si in quaternary coordination:

$$\begin{array}{c} | \quad \quad | \quad \quad | \\ O \quad \quad O \quad \quad O \\ | \quad \quad | \;\; H^+ \;\; | \\ -O-Si-O-{}^-Al-O-Si-O- \\ | \quad \quad | \quad \quad | \\ O \quad \quad O \quad \quad O \\ | \quad \quad | \quad \quad | \end{array} \qquad (35)$$

Because of the fact that the valence of aluminum is a unit less than that of Si, a proton must exist to neutralize the resulting charge. This proton apparently does not enter into the composition of the OH group; and it possesses high mobility on the surface. The existence of aluminum in a tetrahedral configuration in silica–aluminas is demonstrated in a series of spectroscopic studies [222, 223]. According to other ideas [224], the quaternary coordination of Al is optional; and the mobility of a proton can arise as the result of electron transfer from the surface silanol group to aluminum.

$$\begin{array}{c} H^+ \\ | \\ O \\ | \\ -O-Al-O-Si-O- \\ | \quad \quad | \\ O \quad \quad O \\ | \quad \quad | \end{array} \qquad (36)$$

In catalytic reactions, the proton combines with the reactant molecules, forming reactive carbonium ions. For example, ethylene is converted into the ion $C_2H_5^+$, capable of further conversions. The formation of a carbonium ion on acid centers of the Bronsted type probably precedes the catalytic reactions of olefin polymerization, hydrocarbon cracking, toluene disproportionation to benzene and xylene; and, according to some viewpoints, alcohol dehydration [224–228]. Ohta [229] even proposes a measure for the capability of a solid to catalyze a heterolytic reaction (e.g., alcohol dehydration) to be called the proton-donor intensity of a solid.

By studying the infrared spectra of adsorbed amines, one can distinguish centers of type (35) and type (36) or those like them. In the first case, ions of the ammonium type must be formed from NH_3, pyridinium type from pyridine, etc. If hydrogen bonding arises, e.g., according to the diagram (30), the displaced absorption bands of the original amines and the excited surface OH groups must be observed in the spectra. The more excited the OH group, the more acidic properties it has. According to Terenin [230, 231], on the surface of silica–alumina catalysts with no

2.1. ACID AND BASE PROPERTIES OF A SURFACE

adsorbed water, free protons are not observed; but adsorbed amines form a hydrogen bond with surface OH groups (see also [232–234]). The study of hydrogen atom structure by nuclear magnetic resonance [235, 236] also showed that the majority of protons in silica–alumina is found in the surface Si—OH groups, which, by their properties, are almost indistinguishable from the same groups of silica gel.

However, in work [236], upon adsorption of pyridine on silica–alumina (in contrast to adsorption on Al_2O_3 and SiO_2) the spectrum of the pyridinium ion $C_5H_5NH^+$ was discovered. Apparently, on the surface of catalysts, there actually exists an assembly of acid centers of different activity. The heterogeneity of the surface according to acid–base properties was demonstrated experimentally [237–240]. By maximum adsorption of ammonia, pyridine, and some other substances, one can even measure the total acidity, which includes OH groups, mobile protons, and also (p. 81) acid centers of the Lewis type. Of these, the OH groups, in truth, are catalytically inactive or slightly active. According to nuclear magnetic resonance data [235], the upper limit of concentration of mobile protons in silica–alumina that one can evaluate is 3.10^{13} H^+ on 1 cm^2. This figure is close to the number of active centers for catalytic reactions that proceed by an acid mechanism. The addition of a proton to hydrocarbon molecules has been demonstrated spectroscopically. For example, in work [223] on silica-aluminas and zirconium-silicates, the existence of the anthracene ion AH^+ (A is a molecule of anthracene) was discovered. In the opinion of Levy and Folman [241], ions of H^+ can appear at low temperatures not only from OH groups of the surface but also as a result of the autoprotolysis of adsorbed molecules. The latter is a consequence of the presence of the high dielectric constant of the adsorbed layer, e.g.,

$$2CH_3OH \leftrightarrows CH_3OH_2^+ + CH_3O^-$$

The existence of ions was demonstrated in this case by measurement of electroconductivity of the adsorbed layer.

Basic centers of the surface are those centers possessing proton-acceptor properties. Spectroscopic data [242] show that, during adsorption of phenol on the surface of silica gel, the OH groups of the surface are hardly disturbed; but a strong disturbance of the OH group of phenol occurs. This is really result of the interaction of acidic OH groups of phenol with the oxygen atoms of silica gel, as shown in the diagram (31). On the basis of adsorption studies of phenol [202–204] and CO_2, which utilized infrared spectroscopic and differential-isotope methods [243], one can draw conclusions about the presence of various types of alkaline centers on the surface of stable oxides and hydroxides: (a) strongly basic centers—ions of O^{2-}, (b) basic centers of average strength—atoms of oxygen neighboring with OH groups, and (c) weakly basic centers—OH groups of the surface. The latter possess alkaline properties, even on the surface of strong bases: CaO, SrO, BaO, and, to a lesser extent, on MgO. In diagram (31) are pictured basic centers of type (b). The adsorption of phenol (as well as the adsorption of amines on acidic centers) permits us to measure the total number of alkaline centers without their differentiation. A small portion of the surface is responsible for the catalytic activity of solid bases, probably centers of type (a). Low values of the preexponential factor k_0 are explained by this for reactions catalyzed by solid bases, e.g., in isopropyl alcohol dehydrogenation on CaO and BaO [244].

The majority of investigations of acidic properties of a surface refer to composite catalysts: silica–aluminas, and also magnesium silicates, zirconium-silicates, and others. A question arises concerning the comparison of the properties of simple metal oxides, which, as a rule, possess less acidity than composite oxides.

According to Pauling [99], one can explain the strength of oxygen acids on the basis of simple electrostatic considerations. If we write the formula of an acid in the form $XO_m(OH)_n$ (where X is the central atom), then the larger m, the weaker the bond of an atom of hydrogen with that of oxygen in the OH group, the stronger the acid. For example, in the series $Si(OH)_4$, $PO(OH)_3$, $SO_2(OH)_2$, ClO_3OH, acidity increases during transition from silicic to perchloric acid, because of the weakening of the electrostatic attraction of the proton.

In solid hydroxides, the less stably bonded the proton with oxygen, the easier it is to form hydrogen bonds. Therefore, for measurement of acidity and basicity in hydroxides, one can assume a length of the hydrogen bond $O-H \cdots O$ or "hydroxide radius" r_{OH} as well as a displacement of the absorption maximum of the valence band of the OH group relative to the undisturbed OH group [245–247]. Near basic hydroxides, the hydrogen bond $O-H \cdots O$ is practically not formed; its length (3.2–3.3 Å) and radius r_{OH} (1.6–1.7 Å) are great; the valence band of the OH group is not disturbed; and the absorption maximum lies at wavelength $\lambda = 2.7$–$2.8\,\mu$. For acidic hydrates, the distance $OH \cdots O$ is shortened to 2.9 Å but r_{OH} is shortened to 1.3 Å because of the formation of a hydrogen bond; the absorption maximum of the valence band of the OH group is displaced into the long-wave region (3.4–3.5 μ). The relationship between wavelength corresponding to the maximum absorption λ and r_{OH} [245] is shown in Fig. 30. As one moves from left to right along the curve, acidity drops, but basicity increases.

One can find, in the literature, other attempts at evaluating the

2.1. ACID AND BASE PROPERTIES OF A SURFACE

acidity and basicity of simple solid oxides. For example, on the basis of a qualitative comparison between the tendency of oxygen acids to add a proton and the ease of separation of an ion of oxygen, a definition was given for acid systems: A substance is called a base if, as a result of the separation from it of an ion of oxygen, an acid is formed:

$$\text{base} \rightarrow \text{acid} + O^{2-}$$

$$CaO = Ca^{2+} + O^{2-}$$

$$SiO_3^{2-} = SiO_2 + O^{2-}$$

Proceeding from this definition, it was found that the larger the acidity, the larger the energy of the bond of the element with oxygen. Schvartzman and Tomilin [249] proposed the following sequence of decreasing acidity of oxides, coinciding with decreasing energy of the bond O—Me: P, V, Si, Ge, B, As, Sb, Zr, Ti, Al, Th, Be, Sc, La, Y, Sn, Ga, In, Pb, Mg, Li, Zn, Ba, Ca, Sr, Cd, Na, K, Rb, Hg, and Cs.

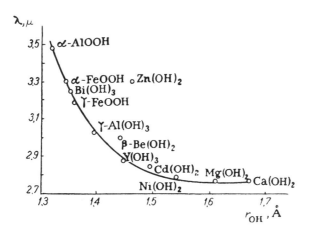

FIG. 30. The relationship between position of the absorption maximum of the valence OH-band in hydroxides and radius of the OH-group [245].

Consequently, it is possible to indicate, by some general rules, the change of acidity and basicity of simple oxides and their hydrates. With increase of charge of a transition metal, the acidity of the corresponding oxide increases, and the basicity falls; e.g., $HMnO_4$ is an acid, Mn_2O_7 is an acidic oxide, $Mn(OH)_2$ is a base, and MnO is a basic oxide. In the periodic table of elements, an increase of atomic number of a metal in a specific group increases basic properties. Proceeding from left to right in a specific period, acid properties grow. Acidic oxides have mainly a covalent bond; basic oxides have an ionic bond. Therefore, for determination of the acidity and basicity of oxides, it is expedient to settle a question concerning the effective charge on the oxygen atom. In ref. [119], a series of oxides is presented arranged in order of decreasing effective charge on oxygen or difference of electronegativities: Na_2O, CaO, MgO, Ag_2O, BeO, Al_2O_3, CdO, ZnO, SnO, H_2O, B_2O_3, FeO, SiO_2, Ga_2O_3, Fe_2O_3, P_4O_6, SnO_2, P_4O_{10}, GeO_2, TiO_2, As_4O_6, CO, As_4O_{10}, SO_2, SeO_2, SO_3, SeO_3, N_2O_5, and Cl_2O_7. Water is used for calibration of amphoterism. Oxides in this series to the left of water are bases; those to the right are acids.

We notice that the order of arrangement of oxides in the above series is somewhat changed. As also in Fig. 30, the order of hydroxides according to acidity does not entirely coincide with the order according to difference of electronegativities.

It is interesting to note that solid bases, in many cases, catalyze not only heterolytic but also homolytic reactions. For example, calcium oxide is an active dehydrogenation catalyst for alcohol [250]. In this case, one can formally write the heterolytic mechanism, which involves transfer of a proton from alcohol to the basic catalyst site and the formation of a carbonium anion:

$$C_2H_5OH + O^{2-}_{cat} \to C_2H_5O^-_{ads} + H^+O^{2-}_{cat}$$

which is further decomposed according to the reaction:

$$C_2H_5O^-_{ads} + H^+_{cat} \to CH_3CHO + H_2$$

The comparison of kinetic data for alcohol dehydrogenation on solid bases and on semiconductors causes us to assume that the removal of hydrogen is a homolytic process:

$$\begin{array}{c} \text{H} \\ CH_3\text{—}C\text{—}O \\ \text{--|---|--} \\ \text{H} \quad \text{H} \\ \vdots \quad \vdots \\ Ca \quad O \quad Ca \quad O \end{array} \qquad (37)$$

During this, for all practical purposes, on every surface atom a surplus charge ($+1$ on Ca, -1 on O) will be found, analogous to a "free valence," and favored by homolytic bond rupture. According to the terminology of Volkenshtein and Kogan [56], this corresponds to a quasi-insulated surface.

2.2. Coordination Capacity of Metals

In Section 1.6, one of the most important properties of transition metals was examined: the capability of their cations to form coordination complexes with anions or neutral molecules. However, coordination complexes are also formed by ions of metals that do not contain d-electrons.

A large number of investigators [201, 209, 231, 251–257] have concluded that it appears that the proton is not the catalytically active center of silica–alumina; rather, the active center is aluminum with a deficiency of an electron pair in its shell. Such Al (found on the surface, replacing Si) has a coordination number of 3 and not 4, as is diagram (35). When reactive molecules that form a carbonium ion, e.g. [201], interact with Al, the coordination number of 4 is restored:

$$\text{RH} + \underset{\text{paraffin}}{\begin{bmatrix} \overset{|}{O} \\ Al-O-Si- \\ \underset{|}{O} \end{bmatrix}} \rightarrow R^+ \begin{bmatrix} \overset{|}{O} \\ H : Al-O-Si- \\ \underset{|}{O} \end{bmatrix}^- \quad (38)$$

It is logical to apply the Lewis definition of an acid to such Al: A Lewis acid is a substance capable of receiving a pair of electrons (a Lewis base is a substance having a free pair of electrons) [196]. A Lewis acid and base form between themselves a so-called donor–acceptor bond; the acid is the acceptor and the base the donor of an electron pair. Examples of donor–acceptor complexes have been examined earlier in Section 1.6.

The coordination properties of surface atoms of nontransition metals will evidently be determined by the same factors as the coordination properties of metal ions in solution. The stability of complexes of metal ions in solution having the same charge e increases with decrease of its radius r. For ions of approximately equal radius it increases with increasing charge of the metal ion, e.g.: $Na^+ < Ca^{2+} < Y^{3+} < Th^{4+}$ [170] Therefore, the polarizing action of the cation is employed as a qualitative measurement of stability of a complex. This is determined by the ratio e/r. The stability of some complexes (the logarithm of the stability constant) is linearly related not to the value e/r but to e/r^2 or e^2/r or $e/(r_{Me} + r_L)$ (where $r_{Me} + r_L$ is the sum of the radii of the cation and the ligand). The ionization potential of a metal atom and the electron affinity of a metal ion change symbately with polarizing action. These values are also used as qualitative measures to compare the stability of complexes.

Nevertheless, a simple correlation with e/r (or the other values noted) is often not found because the bond in complexes does not appear to be purely ionic. As a rule, the donor–acceptor bond is weakly covalent but strongly electrostatic. Mulliken [258–260] quantum-mechanically solved the problem concerning the bond in "charge transfer complexes" and showed that such complexes could exist in certain donor–acceptor complexes of metal ions having electronegative ligands. A charge transfer complex can exist in two energetic states. In the ground state, both components—donor and acceptor—are connected by ion-dipole, hydrogen, and London dispersion forces, and also, to a small extent, by electrostatic and covalent forces resulting from charge transfer from the donor (ligand) to the acceptor (metal). In the excited state, practically complete transfer of an electron from the donor to the acceptor occurs. The extent of the covalent bond increases [164]: (a) with charge of the metal ion, (b) with electronegativity of the metal ion (high electron affinity), and (c) with polarizability of the ligand (e.g., during transition from H_2O to NH_3 and from F^- to Cl^-).

In the case of transition metal ions, it is also reasonable to consider (Section 1.6) crystal field energy stabilization. In general, one can say that, under otherwise identical conditions, systems with large crystal field stabilization must be weaker acids. So the ion $Cr^{3+} \cdot aq$ (of structure d^2) is a weaker acid than $Fe^{3+} \cdot aq$ (of structure d^5). The order of acidity increase for metal ions in complexes is presented below. Acidity is determined [176] from ionization constants in aqueous medium for the reaction

$$Me(H_2O)^{n+} \leftrightharpoons Me(OH)(H_2O)^{(n-1)+} + H^+:$$

$Tl < Ba^{2+} < Ca^{2+} < Mg^{2+} < Ni^{2+} < Fe^{2+} < Zn^{2+} < Cd^{2+} < Cu^{2+} < Pb^{2+} < Al^{3+} < Sc^{3+} < Hg^{2+} < In^{3+} < Cr^{3+} < Fe^{3+}$.

Metal ions are stronger acids in tetrahedral complexes than in octahedral because of reduced charge shielding.

Besides the value e/r and the crystal field stabilization, when studying the catalytic action of metal ions one should also consider their purely geometric dimensions. For example, in complexes of metal ions of Group II with EDTA (ethylenediamine-tetra acetic acid), the complexes of Ca^{2+} are more stable than Mg^{2+}, Sr^{2+}, and Ba^{2+}. In the opinion of Martell and Calvin [261], this can be explained by the fact that the ion Mg^{2+} is too small for all the donor groups to be distributed without steric hindrance. The ions Sr^{2+} and Ba^{2+} are less stable than Ca^{2+} because of the small value of e/r. Therefore Ca^{2+} is found as the optimum. Small ions form more stable complexes with small ligands; large ions with large ligands.

The rules for change of catalytic activity of Lewis acids have been studied principally for homogeneous catalysis [176, 262]. For example,

2.2. COORDINATION CAPACITY OF METALS

for the reaction of decarbonylation the following sequence of catalytic activity of ions was found: $Al^{3+} > Fe^{3+} > Cu^{2+} > Fe^{2+} > Zn^{2+} > Mg^{2+} > Mn^{2+} > Ca^{2+}$, coinciding, in general, with the order of decreasing e/r, Lewis acidity, or complex-forming activity. From this point of view, the proton H^+ would have the maximum catalytic activity. This, however, is not always so. As a good example, one can cite the decarboxylation reaction of dimethyloxalacetic acid [263]. Its dianion gives off CO_2, forming enolate:

$$^-O_2C-\overset{O}{\underset{\|}{C}}-C(CH_3)_2-CO_2^- \rightarrow {^-O_2C}-\overset{O^-}{\underset{|}{C}}=C(CH_3)_2 + CO_2$$

The addition of H^+ to the carboxyl group gives a monoanion, which is decarboxylated more rapidly:

$$HO_2C-\overset{O}{\underset{\|}{C}}-C(CH_3)_2-CO_2^- \rightarrow HO_2C-\overset{O^-}{\underset{|}{C}}=C(CH_3)_2 + CO_2$$

The decrease of negative charge accelerates the transition of a pair of electrons from the separating CO_2 to the remaining molecule. A multicharged cation connecting all the negative charge fulfills this function even better:

$$\underset{\underset{Cu}{O\,\overset{+}{}\,O}}{O=C-C}-C(CH_3)_2-CO_2^- \longrightarrow \underset{\underset{Cu}{OO}}{O=C-C}=C(CH_3)_2 + CO_2$$

In general, one can say [176] that metal ions will catalyze a reaction under the condition that the possibility of forming an intracomplex bond exists. Then, the metal ion appears to be more effective than H^+. One must also consider that not mobile H^+ but, rather, the hydronium ion H_3O^+, exists in aqueous solutions.

The following order of activities was found [264] for bromination of cyclic ethers: $Cu^{2+} > Ni^{2+} > La^{3+} > Zn^{2+} > Pb^{2+} > Mn^{2+} > Cd^{2+} > Ca^{2+} > Ba^{2+}$. The order of activities here coincides with the expected order of stabilities of the complexes. If the limiting step of the reaction consists not of the formation but of the decomposition of the cyclic complex, the order of activities cannot coincide with the order of e/r.

Apparently, one can transpose, to a certain degree, the rules found for ions in solution to ions on the surface of a solid. For example, the order of acidity of ions in solution (see above) in general coincides with the order of acidity of solid oxides (see Section 2.1). The adsorptive and catalytic activities of these ions are determined by the same factors:

their sizes, the value of e/r, and the crystal field stabilization energy. Consequently, for a solid surface, one can expect different catalytic activity of ions on different crystal faces, specified by unequal degrees of shielding of charge and a different coordination number.

Probably, the ions of small dimensions (Si^{4+} and Al^{3+}) that are strongly shielded by large atoms and groups (O^{2-}, OH^-, and others) will be available to interact with adsorbed ligands. This will be so, even on faces of higher orders or in the presence of a large number of structural defects. Actually, different methods for determining the number of Lewis acid centers show that, on silica–alumina, Al_2O_3, and SiO_2, the number of centers amounts to 10^{12}–10^{14} cm^{-2}, i.e., significantly less than the total number of surface atoms.

Experimental methods for determining the Lewis acid–base properties of a surface are analogous to the methods for determining the proton acidity or basicity (i.e., the Bronsted acidity). In a series of cases, these two types of acidity are difficult to distinguish. For example, the method of ion-exchange adsorption of strong bases from solutions can lead to hydrolysis of the bonds Si—O—Al in silica–alumina and to their hydration [200]. As a result, the total (proton and nonproton) acidity is determined. According Miesserov [251], the exchange properties of silica–alumina are caused not by hydrogen, but by Al, which is hydrated in an aqueous medium.

The most common method to determine the number of active centers is adsorption of Lewis bases.

Often, without additional information, it is not clear after adsorption of a weak base or a weak acid whether the determined acid (base) is a Lewis or Bronsted acid (base). Ammonia and organic amines are adsorbed not only on OH groups according to diagram (30) or on weakly bonded protons H^+, but also on nonproton centers, i.e., they show total acidity. Spectroscopic data [230, 231, 265–267] show that ammonia and organic amines can form a very stable bond with a Lewis acid, e.g., with the surface ion Al. During the study of ultraviolet absorption spectra of aromatic amines during adsorption on silica–alumina, it was found [230, 267] that, in contrast to SiO_2, the first adsorbed molecules of the amines develop a displacement of the adsorption maximum to the short wavelength region. This is characteristic for attachment of an electron acceptor agent to an amino group. On the surface, this agent cannot be proton acid centers, since the spectral shift is preserved during neutralization by base. Apparently these centers are surface coordination-unsaturated atoms of Al. On iron oxide, the initial quantities of ammonia are adsorbed not on the OH groups but on the ion Fe^{3+}. The heat of adsorption Q_{ads} of such ammonia is not great: 11.5 kcal/mole. In

2.2. COORDINATION CAPACITY OF METALS

ref. [268], it is indicated that such a low value of Q_{ads} can hardly be explained by the formation of a coordination bond of the d-electrons of Fe^{3+} with a pair of electrons of NH_3. More probably, this is the energy of an ion–dipole interaction. During the adsorption of NH_3 on Al_2O_3 significantly large heats of adsorption were found [269]: up to 20–25 kcal/mole.

In accordance with the Lewis principle, oxygen-containing ligands such as dioxane and ethers are often used for evaluation of the number of nonproton centers. The measurement of the infrared absorption spectra of molecular compounds of metal halides with NO, acetonitrile, pyridine, diethyl ether, methanol, and cyclohexane showed [270] that these halides possess strong acceptor properties, stronger than HCOOH and CH_3COOH. These properties ("Lewis acidity") decrease in the order: $AlBr_3 > AlCl_3 > SnCl_4 > TiCl_4 > SnI_4 > FeCl_3 > BiCl_3$. With respect to ether-halides $R_2O \cdot MeX_n$, the most stable compounds with ether R_2O are those of the halides of aluminum and beryllium [271].

Water also can form a stable coordination bond with surface metal atoms that are not shielded by OH groups: $>Me:OH_2$. According to IR data from refs. [234, 272], the initial quantities of water adsorbed on porous glass do not completely shift the OH group absorption bands. They are adsorbed on ions of Si possessing strong acceptor properties. Subsequent quantities of water are connected by weaker hydrogen bonds with the OH groups. This result was verified in ref. [273] by a nuclear magnetic resonance study of the adsorption of water on silica gel. The initial quantities of water (0.17 μmole/m^2) did not change either the width or the second moment of the NMR signal of the OH groups. The heat of adsorption of these portions was near 20 kcal/mole, i.e., somewhat greater than the energy of hydrogen bonding. The authors explain their results on the basis that coordination unsaturated ions of Si exist on the surface. This corresponds to a theory of Weyl [274].

The analog of water, H_2S, at low surface coverages forms a coordination bond with the Al of γ-Al_2O_3. The heat of adsorption is as high as 38 kcal/mole [275].

The adsorption of triphenylmethane and condensed aromatic hydrocarbons (anthracene, perylene, and tetracene [256, 257, 276]) for the determination of the number of Lewis acid centers is still used. During the adsorption of anthracene on a Lewis center, a separation of the hydride ion H$^-$ occurs, with the formation of an ion of anthracene A^+, which, in the ultraviolet spectrum gives an absorption band which is different from the band of the protonized ion AH$^+$ formed on proton–acid centers [223]. Molecules of condensed aromatic hydrocarbons have large dimensions and can shield several active centers at once. Probably, therefore, the number of nonproton centers determined by this method is too low.

The indicator method is also used to evaluate nonproton acidity. For example [200, 207], one can titrate a catalyst suspension with a benzene solution of ethylacetate or of dioxane in the presence of an indicator. Analogously to the proton acidity function, one can define the Lewis acidity function [210]:

$$H_0 = -\log(a_A f_B/f_{AB}) \tag{39}$$

where a_A is the activity of the Lewis acid or the acceptor of electrons: f_B and f_{AB} are the activity coefficients of the base B and the complex AB.

After calcination *in vacuo* at 150–500°C, the acidity of Al_2O_3, SiO_2, TiO_2, CeO_2, $BaSO_4$, $CaSO_4$, $MnSO_4$, ZnS, $HgCl_2$, and $CaCl_2$ (as determined by adsorption of methyl red, dimethyl yellow, and bromothymol blue from a solution of *iso*-octane) increased [278]. Adsorbed water and partially adsorbed OH groups were removed from the surface by the calcination. Apparently, acidity determined by this method is of the nonproton type. It is interesting that $NiSO_4$ acidity in such experiments increases with an increase in temperature of calcination up to 350°C. At higher temperature it diminishes.

Nonproton acidity of silica–aluminas and Al_2O_3 was compared with their catalytic activity in heterolytic reactions. The selective poisoning of the surface by Lewis bases and its influence on catalytic activity were studied. The reactions of skeletal isomerization of butylenes and of isomerization about a double bond, of cracking of low molecular weight olefins, and, apparently, also of dehydration of alcohols, proceed [225, 255–257, 279] on Lewis acid centers, i.e., the coordination-unsaturated atoms of Al. Catalytic activity for these reactions increases with increased percentage of Al_2O_3 in the silica–alumina. The most active is pure Al_2O_3. In accordance with diagram (38), the mechanism of reaction on such atoms of Al is often terminated by separation of the hydride ion and formation of a carbonium ion that is capable of further conversions. In the work of Leftin and Herman [257], e.g., it was shown with the help of ultraviolet spectra that during isomerization of butylenes on silica–aluminas, π-allyl carbonium ions undergo a shifting of the double bond via reactive intermediates:

$$\left[\begin{array}{c} H \\ | \\ H \diagdown C \diagup CH_3 \\ C = C \\ | \quad | \\ H \quad H \end{array} \right]^+ \quad \text{and} \quad \left[\begin{array}{c} H \\ | \\ H \diagdown C \diagup H \\ C = C \\ | \quad | \\ H \quad CH_3 \end{array} \right]^+ \tag{40}$$

2.2. COORDINATION CAPACITY OF METALS

Coordination-ionic complexes play a large role in catalytic polymerization processes. Section 1.6 discussed the polymerization of olefins on Ziegler–Natta catalysts and other compounds of transition metals. The coordination mechanism was shown for polymerization of ethylene oxide and propylene oxide on oxides, hydroxides, and carbonates of metals of Group II, aluminum, and iron [280–282]. During decomposition of hydroxides and carbonates *in vacuo* and their conversion into oxides, catalytic activity increases in proportion to the number of surface metal atoms that are not shielded by OH groups. On oxides of Mg, Be, and Al calcined *in vacuo* at 300–500°C, the number of these atoms on 1 cm² is (2×10^{13})–(2×10^{14}). Infrared spectra showed that OH groups are not disturbed in the ethylene oxide polymerization process. Based on a study of the reaction mechanism, it was assumed that the polymerization reaction (growth of the chain) proceeds by adsorption of a molecule of ethylene oxide on a metal atom which simultaneously retains the growing chain of polymer. The subsequent "insertion" of this molecule then takes place at the base of the chain, e.g., on MgO;

(41)

Surface coordination-unsaturated metal atoms can also catalyze some homolytic reactions. On aluminum oxide and silica–alumina, H_2–D_2 exchange [255, 283], dehydrogenation of paraffins [227], and even oxidation of benzaldehyde to the benzoate ion [224] were effected. In the article of Dowden and Wells examined in detail earlier [173], it was shown that the H_2–D_2 exchange can proceed by a heterolytic mechanism. However, this assumption appears to be hardly probable. Optical data and EPR spectral data taken during adsorption of aromatic amines and some condensed aromatic hydrocarbons show that, in some cases, the surface metal atoms possess significant electron-acceptor properties, even down to a charge of $+1$. Owing to this, a covalent bond of the adsorbed molecule with the adsorbent can form; and homolytic bond breaking can occur in a molecule.

It was shown by the work of Sidorova and Terenin [265] that, for most strong acids of the Lewis type (CeO_2, $CuSO_4$, $BiCl_3$, and bentonite), during adsorption of diphenyl amine, complete ionization occurs, with the formation of a colored ion. Here, it is interesting to note that in investigations of Walling [208] that were mentioned above regarding determination of surface acidity by the indicator method, anhydrous $CuSO_4$ also seemed to be an acid near in strength to silica–alumina and mineral acids. On the majority of less strong Lewis acids (e.g., Al_2O_3, MgO, TiO_2, SiO_2, $BaSO_4$, ZnS, TlI, and AgI) coloring of diphenylamine is not observed. Analogous phenomena were discovered during adsorption of other amines: n-phenylene diamine, triphenylamine, aniline, and also the hydrocarbons, perylene and anthracene [225, 276, 284]. During adsorption, band formation in the optical (visible and ultraviolet) spectrum corresponding to the ionized form of the molecule is often accompanied by the appearance of EPR signals. This attests to the formation of cation radicals on electron acceptor centers. The intensity of the optical spectra of these cation radicals and the corresponding EPR signals change symbately with the Lewis acidity of the surface. For example, in ref. [284], it is shown that the intensity of the bands of the ionized form of malachite green increases in the sequence: $CsCl < RbCl < KCl < NaCl < LiCl$ and $BaSO_4 < CaSO_4 < MgSO_4 < SiO_2 < CaO < MgO$, i.e., it increases symbately with the polarizing capability of the cation.

The number of strong electron-acceptor centers is significantly less than the total number of surface metal atoms: On silica–alumina, according to ref. [285] it is less by 1.5–2 orders. The reasons for the existence of some quantity of strong electron-acceptor centers among other surface metal ions that possess the capability of accepting a pair of electrons is not entirely clear at the present time. Additionally, in insulators of the type Al_2O_3, the atoms $Al^{\delta+}$ create local acceptor levels lying deep in the forbidden zone. They occur at a cleavage of the crystal or upon dehydration of its surface. Their number can be a nonequilibrium value. Sometimes the heating of the crystal leads to the interaction of these local acceptor levels ("positive free valences") with the equivalent number of donor levels ($O^{\delta-}$ or "negative free valences") and to their mutual destruction. It is probable that homolytic oxidation–reduction reactions also proceed on these centers.

On semiconductors, an excess charge, e.g., positive, can be removed from the surface metal atom as a result of interaction with a conduction electron:

$$: \overset{\delta+}{\underset{..}{Zn}} : + e = : \underset{..}{Zn} :$$

Because of this, the metal atom (in the example given, a Zn atom of zinc oxide) receives the capability to accept a Lewis base without the formation of a covalent bond; and this contributes to the course of a heterolytic reaction. It is possible that, by this mechanism, one can explain the small acceleration of alcohol dehydration (a typical heterolytic reaction) observed by some authors when an admixture that increases electron conductivity (raising the Fermi level) is introduced into the semiconductor [286]. For example, the catalytic activity of TiO_2 in this reaction increases somewhat upon doping with WO_3.

Schwab [287] attempted to apply the ideas of the electronic theory of catalysis (Section 1.1) to acid–base catalysis (the cracking reaction) on silica–aluminas. In his opinion, the substitution of Si^{4+} for Al^{3+} in the silica–alumina lattice corresponds to the introduction of p-type doping into the semiconductor. To some degree, the p-semiconductor is actually analogous to an acid and the n-semiconductor to a base. p-Semiconductors, as well as acids, possess the properties of an acceptor of either an electron or an electron pair; n-semiconductors and bases possess the properties of being a donor of either an electron or an electron pair. However general this analogy may be, it should be noted that the mechanisms of homolytic and heterolytic reactions are different.

In the paragraph on Lewis acids it was stated that Lewis bases appear to have properties not different from the properties of Bronsted bases. For example, surface ions of oxygen in oxides possess both properties: they can combine with either a proton or a molecule (of the type BF_3) both of which lack a pair of electrons.

2.3. Dielectric Constant

The dielectric constant or dielectric permeability ϵ is a dimensionless value showing how many times the interaction force between charges in a given medium is reduced in comparison with a vacuum. The Clausius–Mosotti formula relates ϵ to the polarizability of atoms, molecules or ions, i.e., to the chemical properties of a substance:

$$\frac{\epsilon - 1}{\epsilon + 2} = \frac{4\pi}{3} \sum N_i \alpha_i \qquad (42)$$

where N_i is the number of atoms, molecules, or ions of the type i in the bulk of the substance; α_i is their polarizability.

The values of α_i are periodically changed, depending on the position of the element in the periodic table. In contrast to the polarizing action of ions, their polarizabilities α_i increase with increase of the principal quantum number and decrease with increase of charge. In solids, because of the complexity of interactions, ϵ usually cannot be calculated from the polarizabilities of the separate ions. Polarizability in a solid substance consists in principle of two parts: electronic α_{el}, specified by the displacement of electrons relative to the nucleus; and ionic α_{ion} related to the displacement of ions relative to other ions. In molecular crystals, orientation polarizability can be added to this, and, in ferroelectrics, the spontaneous polarizability of unbroken regions of the crystal.

The electronic polarizability, determined in principle by the volume of the particle, maintains its value in fields of high frequency. One can determine the dielectric permeability ϵ_∞ corresponding to high frequencies by measuring the index of refraction n: $\epsilon_\infty = n^2$. The value of α_{el} can be determined from Eq. (42) by substituting n^2 for ϵ. For the majority of ionic crystals, the value of ϵ_∞ varies in the range from 2 to 4.

The ionic polarizability is determined by the effective charge e^*, the mass of the ions m^*, and the natural frequency of optical oscillations ω_0 (Section 1.5, see Eqs. 23). The ionic part of the dielectric permeability is varied over a wider range: from 5 to 30 and more. The orientation polarizability in ionic crystals amounts to a small value. Thus, it is evident that the main contribution (50–80%) to the dielectric permeability of ionic crystals is made by the ionic polarizability of semiconductors. In crystals with covalent bonds, the values of ϵ are rather close to n^2. The theoretical calculation of ϵ for ionic crystals is difficult, because of the lack of knowledge about many of the constants. Therefore, when studying the relationships among various solid properties, one usually compares them [288] with experimental values of $\epsilon_\infty = n^2$ and $\epsilon_0 = \epsilon$, i.e., with high frequency and static dielectric permeabilities.

The values of ϵ_0 are presented in the table of the Appendix. The discrepancies in data from the literature for ϵ are sometimes very great and are caused for various reasons. For example, the value of ϵ can depend on crystallographic direction. Thus, the value of ϵ for rutile TiO_2 in the direction perpendicular to the optical axis of the crystal is equal to 86 but in the parallel direction it is 170. Sometimes, very small quantities of admixtures of metal compounds of particular valences can change ϵ of ionic crystals by many multiples [289, 290]. For example, 0.03% of U_3O_8 in TiO_2 increases the value of ϵ from 100 to 10,000; 0.1% Nb_2O_5 and Ta_2O_5 in TiO_2 increases ϵ to 40,000; and an admixture (1–5%) of Bi_2O_3 in ZnO increases ϵ from 10–20 to 1000. This is explained by the high value of the electronic polarizability of the hydrogen-like orbital of the admixed atom which results from the large radius of its orbital.

The dielectric permeability enters into the series of formulas (examined

2.3. DIELECTRIC CONSTANT

in Chapter 1) that explain its influence on catalytic activity. All things considered, it is determined by the fact that an interaction between charges (e.g., ions, electrons, dipoles) occurs in a medium decreasing this interaction ϵ times. To estimate the interaction between charges on the surface of a solid, the arithmetic mean value of ϵ between ϵ_{cryst} and ϵ_{gas} is frequently examined [114]. In reality, such a calculation appears to be a very rough approximation. It is not possible to calculate the polarization of the adsorbed layer in the same manner, by simply adding the polarizations of the adsorbate and adsorbent. Besides this, in microcracks, cracks, and in the presence of uniform doping, the value of ϵ can be higher than this mean value $(\epsilon_{cryst} + \epsilon_{gas})/2$.

In catalytic processes with an electronic mechanism, one can expect a symbate change of ϵ^2 and catalytic activity in those cases when the latter diminishes with increase of the width of the forbidden zone U or depth of the doped levels ΔU_d. As indicated above [see Eqs. (13) and (17)], both of these values are changed inversely proportionately to ϵ^2.

In the case of acid–base catalytic processes, ϵ of a medium can influence the energy for transfer of a proton from the crystal to the reagent and back, i.e., the activation energy of the process. One method to calculate this value is as follows [291]: We shall assume that the proton occupies one of the cationic nodes in the ionic crystal. According to dielectric theory [101], the work of removal of the cation from one node of the crystalline lattice is

$$W = W_{latt} - \tfrac{1}{2}e\varphi \qquad (43)$$

where W_{latt} is the energy of the crystalline lattice; $\tfrac{1}{2}e\varphi$ is the work of polarization of the surrounding medium (e is the charge; φ is the potential in the center of the vacancy after the removal of the charge specified by this polarization).

The work of polarization is approximately equal to:

$$\tfrac{1}{2}e\varphi \approx \tfrac{1}{2}[1 - (1/\epsilon)] e^2/R \qquad (44)$$

where R is the radius of the empty node remaining after the removal of the cation.

For a lattice of the type NaCl $R = 0.6\,a$, $a = 1.66$ Å (a is the lattice parameter). If the cation (proton) is removed from the surface, the energy of its bond with the lattice will be lower than W_{latt}. Simultaneously the value of $\tfrac{1}{2}e\varphi$ will also be lower because of the lower value of ϵ. We say approximately that the cation is half immersed in the crystal (which has a high value of ϵ) and half found in the gas (with $\epsilon = 1$). For NaCl $\epsilon = (5.6 + 1)/2 = 3.3$. Then $\tfrac{1}{2}e\varphi = \tfrac{1}{2}[1 - (1/\epsilon)] e^2/R = 2.9$ eV or 67 kcal/mole. For CaO and Al_2O_3 with $\epsilon = 12$ the values of polarization energy will be even higher.

It is known [292] that analogous expressions are obtained in the course of calculating the activation energy of a homogeneous acid-catalyzed reaction.

We shall assume that ϵ in the example presented above increases by unity. Then the value of $\frac{1}{2}e\varphi$ will be equal to 71 kcal/mole, i.e., it is increased by 4 kcal/mole. Therefore, a small change of ϵ of a catalyst or its surface layer is sufficient to obtain a significant change of activation energy and, consequently, rate of a catalytic reaction. The introduction of dopants into the crystal, the change of character of its surface structure, and the appearance of cracks and pores on the surface, etc., can lead to this effect.

At high values of ϵ the term $1/\epsilon \to 0$ and, correspondingly, the work of polarization according to Eq. (44) will be constant. Therefore it is hard to expect significant anomalies in the change of catalytic activity during the transition to ferroelectrics, for which ϵ attains values of 10^4–10^5. Acid–base catalytic processes on the surface of ferroelectrics have not been studied. In the case of oxidation–reduction processes, Parravano [293] studied the oxidation of CO on the ferroelectrics $NaNbO_3$, $KNbO_3$, and $LaFeO_3$. During the transition across the Curie point, in which the dielectric permeability falls sharply, comparatively small changes of catalytic activity were found. When examining the similar nature of the dependencies, it is necessary to bear in mind that the surfaces of ferroelectrics have different properties than the interiors [294]. The surface layer, which sometimes encompasses several atomic layers, cannot possess ferroelectric properties and have a low value of ϵ.

3 • Catalytic Activity and Structure of Solids

3.1. Crystalline Lattice Parameter

Crystalline lattice parameters may exert a direct as well as an indirect effect on catalytic activity. The indirect effect may be inferred from either one of the previously discussed theories. Thus, the relationship between crystalline lattice parameter and catalytic activity follows directly from the assumption that there is a connection between catalytic activity and the polarizing strength of a cation, expressed by e/r. The width of the forbidden zone (all conditions being equal) decreases with an increasing crystalline lattice parameter. The crystalline lattice parameter for the series of oxides of the transition metals of the first large period exhibits three maxima and two minima (Fig. 31). This has been explained, however, simply in terms of difference in the crystal field [295], instead of resorting to the relationship between catalytic activity and crystalline lattice parameter. Such an explanation may, however, unduly ignore the direct relationship between crystalline lattice parameter and catalysis. This direct relationship comes into play when, during a surface reaction, the various reactant molecules undergo adsorption by forming bonds with catalyst surface atoms in a geometrically ordered fashion. The principle of a definite geometrical relationship between the interatomic distances in the reactant molecules and the atomic spacing on the catalyst surface was presented originally in 1929 in terms of the Balandin's "multiplet theory of catalysis" [296].[1] The following are several examples of the application of the multiplet theory of catalysis to nonmetallic catalysts.

---

[1] The present state of the multiplet theory of catalysis is outlined in ref. [297].

According to Balandin, most catalytic reactions fall into the so-called "duplet" class:

$$
\begin{array}{ccc}
\overset{\bullet}{A} \quad \overset{\bullet}{C} & \overset{\bullet}{A}---\overset{}{C} & \overset{\bullet}{A}\text{———}\overset{}{C} \\
\bullet | \quad |\bullet \rightarrow \bullet | \quad |\bullet \rightarrow \bullet \quad\quad \bullet \\
B \quad D & B---D & B\text{———}D \\
\underset{\bullet}{} \quad \underset{\bullet}{} & \underset{\bullet}{} \quad \underset{\bullet}{} &
\end{array}
\qquad (45)
$$

A reaction can occur only if atoms of the reactant molecules, A, B, C, and D, in the schematic diagram (45), are congruent with surface atoms of the catalyst, shown as dots. Such a congruence condition calls for certain definite geometrical relationships among the atoms within the reactant molecules and among the atoms at the surface of the catalyst. If such a match is not provided, the reaction will not take place.

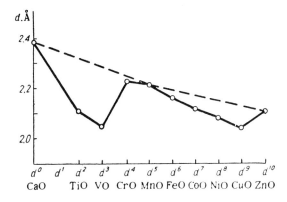

FIG. 31. Changes in metal to nonmetal spacing in a series of divalent oxides of metals of the fourth period

Dehydrogenation of alcohols falls into the duplet class [297, 298] involving adsorption via $\underset{\underset{H}{|}\underset{H}{|}}{C\text{—}O}$-group;

$$
\begin{array}{ccc}
\begin{array}{c} R_2 \\ \diagdown \\ R_1 \quad C\text{—}O \\ K| \quad |K \\ H \quad H \\ K \end{array} K \rightarrow
\begin{array}{c} R_2 \\ \diagdown \\ R_1 \quad C\text{----}O \\ K| \quad |K \\ H\text{----}H \\ K \end{array} K \rightarrow
\begin{array}{c} R_2 \\ \diagdown \\ R_1 \quad C{=}O \\ K \quad\quad K \\ H\text{——}H \\ K \end{array}
\end{array}
\qquad (46)
$$

where K are catalyst atoms.

In dehydration of alcohols, according to the multiplet theory of catalysis, the $\underset{\underset{H}{|}\underset{O}{|}}{C\text{—}C}$-group participates. The interatomic distance is

3.1. CRYSTALLINE LATTICE PARAMETER

1.43 Å in C—O and 1.54 Å in C—C. Consequently, it has been postulated that optimum value of the lattice parameter is greater for alcohol dehydration than for dehydrogenation. Thus, in the lattice parameter range limited by optimum values for these two reactions, an increase in the value of a should enhance dehydration activity, while a decrease in this value should promote dehydrogenation activity. This prediction has been proved by Balandin and Egorova [299], who showed that an increase in lattice parameter in the series of V_2O_5, Nb_2O_5, and Ta_2O_5 was paralleled by increasing yields of dehydration products obtained over these oxide catalysts. Obviously, it would be more appropriate to present such changes in catalytic activity in terms of absolute rate of dehydration per square meter of catalyst surface are instead of presenting them in terms of relative yields of the evolved water. However, catalyst surface areas were determined neither in this study nor in the subsequent work by Balandin and Sokolova [300] on decomposition of isopropyl alcohol over Ta_2O_5 and Nb_2O_5. Both the distinct dehydration activity of anatase and the distinct dehydrogenation activity of rutile have been related by Rubinshtein [298] directly to the magnitude of the Ti—O distance. A similar conclusion evolved from an attempt of Rubinshtein *et al.* [301] to rationalize the differences in catalytic activity of ZnO, ZnS, and ZnSe in decomposition of isopropyl alcohol. If, however, the catalytic activity were related to a unit surface area [301], then the dehydrogenation activity would have been found to increase with increasing Zn—X distance, which is contrary to expectation. Similar trends in activity changes in the series, ZnO, ZnS, ZnSe, and ZnTe, were reported by Krylov *et al.* [302], and Krylov and Fokina [67].

A "duplet" type activated complex and a contribution of lattice defects of the ZnO crystal lattice were proposed by Dowden *et al.* [184] and Balandin and Rozhdestvenskaya [303] for H_2–D_2 exchange over ZnO:

$$\begin{array}{c} Zn^{2+} \\ H\text{-----}D \\ Zn^{2+} \mid \quad 2e \quad \mid Zn^{2+} \\ H\text{-----}D \\ Zn^{2+} \end{array} \qquad (47)$$

Adsorption of H_2 and D_2 occurs in the vicinity of an ionic vacancy caused by substitution of an O^{2-} ion by two electrons (so-called "F center"). According to this picture an increase in electrical conductivity should result in an increased rate of hydrogen–deuterium exchange.

Initially, the multiplet theory was found to apply to metals. It accounted satisfactorily for hydrogenation of benzene and dehydrogenation of cyclohexane in terms of matching of these six-membered molecules with the metal's pinacoid configuration of the hexagonal lattice (A3) and the octahedral face-centered cubic lattice (A1). The active center on the metal involves six atoms and is called a "sextet." A pictorial presentation of sextet-type complex, according to Balandin [297], is shown in Fig. 32. The multiplet theory predicts that metals with interatomic distances ranging from 2.48 to 2.77 Å should exhibit catalytic activity in these two cases since the metal spacings match the interatomic distances in the six-membered cyclic molecules. This prediction has been verified experimentally.

FIG. 32. Schematic of the sextet complex, according to Balandin [297].

Subsequently, Balandin and Brusov [304] observed dehydrogenation of cyclohexane over oxides such as Cr_2O_3 and MoO_3, which have neither octahedral faces nor atoms with dimensions or configuration allowing a flat adherence of six-membered rings. To account for such cases, a two-point adsorption scheme, or a so-called duplet mechanism, has been proposed:

$$\begin{array}{c}\text{H}_2\\ \text{H}_2\text{C}\diagup{}^{\text{C}}\diagdown\text{CH}_2\\ |\qquad\qquad|\\ \text{H}_2\text{C}\diagdown_{\text{C}}\diagup\text{CH}_2\\ \text{H}_2\end{array} \longrightarrow \begin{array}{c}\text{H}_2\\ \text{H}_2\text{C}\diagup{}^{\text{C}}\diagdown\text{CH}_2\\ |\qquad\cdot|\cdot\\ \text{H}_2\text{C}\diagdown_{\text{C}}\diagup\text{CH}_2\\ \text{H}_2\end{array} \qquad (48)$$

where the dots represent atoms involved in adsorption centers.

The rate of reaction is much smaller in the case of a duplet-type mechanism than in the case of a sextet-type mechanism. According to Balandin, Krylov, and their co-workers [303–305], dehydrogeneation over Cr_2O_3, ZnO, and CaO calls for substantially higher activation energies (about 30 kcal/mole) than it is the case over metallic catalysts (about 10 kcal/mole). The duplet-type mechanism of cyclohexane dehydrogenation over oxides was inferred by Herington and Rideal [306] from the presence of cyclohexene in the reaction product. Balandin and Isagulyants [307] found that decalin and cyclohexane dehydrogenate over Cr_2O_3 at an equal rate, even though a flat orientation of decalin requires significantly more surface area. This fact also attests to the duplet scheme in dehydrogenation of cyclic hydrocarbons over oxide

3.1. CRYSTALLINE LATTICE PARAMETER

catalysts. Cyclopentane dehydrogenates over Cr_2O_3 under conditions similar to those required for cyclohexane [308]. According to Balandin, the duplet-type mechanism developed for cyclohexane applies also to paraffins such as butane.

A duplet mechanism was suggested also for aromatization of olefins [309, 310]. High catalytic activity of MoO_3 was explained in terms of its partial reduction to MoO_2. On the (100) faces of MoO_2, the distance between molybdenum atoms is 2.79 Å, which favors a two-point adsorption of the olefins, with a minimum angular strain for Mo−C−C (angle equal to 108°). Also, on the spinel-type lattices such as γ-Al_2O_3, γ-Cr_2O_3, or γ-Fe_2O_3, the distances between metal atoms on (100) face are suitable for two-point adsorption of hydrocarbons. Long distances separating surface metal atoms prevent adsorbed hydrocarbons from undergoing multiple transformations. The lattice γ-Al_2O_3 is rather inactive, but in the chromia–alumina system it contributes to the aromatization activity by stabilizing the active state of γ-Cr_2O_3. According to Plate [311], on the corundum-type lattice (α-Al_2O_3, α-Cr_2O_3, α-Fe_2O_3) aromatization may involve a sextet-type scheme, which might be visualized as an adherence of a six-membered ring to the octahedral face (made up of either metal or oxygen atoms). MoS_2 was found to be more active than MoO_2 as a hydrogenation–dehydrogenation catalyst despite a more strained angle in Mo−C−C (112°). This suggests a two-point adsorption of the olefin. Griffith [309] attributes the higher activity to a hydrogenation–dehydrogenation mechanism over MoS_2 that involves six surface atoms of molybdenum. NiS_3 is unsuitable for two-point adsorption of olefins [298]; and it adsorbs only those species which prefer one-point adsorption, e.g., CS_2.

The geometric fit of the reactants and the catalyst is, in itself, not a sufficient condition for the occurrence of catalytic activity. According to Balandin, catalytic activity will require also an "energetic fit," namely, an optimal bond strength between the reactants and the catalyst. Application of the principle of the energetic fit permitted an explanation [312] in terms of the duplet-type mechanism of the differences among oxides of the metals of the second group of the periodic table for dehydrogenation of hydrocarbons: inactivity of BeO and MgO, low activity of ZnO, and high activity of CdO.

In general, the principle of energetic fit seems to be correct, and several authors [313–317] have employed it as a basis for considering the problem of catalyst selection. At the present, the energy of bonding between the reactants and the catalyst is being routinely determined from the experimental data on the kinetics of a catalytic reaction. The problem associated with energetic factors in catalysis have been omitted

98　　　　　　　　　　　　　　　　　　　　　　2. ACID–BASE REACTIONS

from the present discussion, because it is the intention of this author to relate catalytic activity solely to the inherent properties of the solid state.

When considering the geometric correspondence, it should be borne in mind that the spatial arrangement in the surface layer may considerably differ from that in the bulk. Recent investigations of the structure of germanium and silicon [145, 146, 318] by low energy electron diffraction (LEED) revealed structural differences between the two-to-three uppermost layers and the bulk of the solid. In Fig. 33 a schematic of

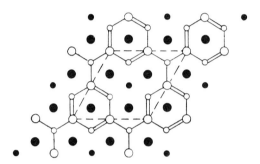

Fig. 33. Distribution of atoms in the surface layer on the Ge crystal face according to Lander and Morrison [146]: ○—the uppermost layer; ◦—the second uppermost layer; ●—fourth layer from the uppermost one; and •—the sixth layer.

the atomic distribution in the surface layers of germanium's face (111) is shown, after Lander and Morrison [146]. Instead of assuming a diamond-like structure, the surface atoms of germanium arrange themselves preferentially into benzene-like (but not flat) rings covalently bonded to the layers beneath them. In such benzene-like rings, the Ge—Ge interatomic distance is 1.16 Å, instead of being equal to the regular Ge—Ge distance of 1.22 Å. A further distortion of the surface geometric picture comes from the adsorption of foreign atoms and molecules which do not participate in catalysis.

It should be realized that geometric correspondence with the catalyst may not necessarily apply to the reactant, but rather to one of the intermediate structures [191] involved in the limiting step of a surface chemical reaction.

It was implied by Balandin [297] that in a multiplet complex, such as that shown in schematic (46), one atom of a molecule undergoing a catalytic reaction may interact with atoms of one constituent of the catalyst; while another atom of this same molecule may interact with atoms of another catalyst constituent. It has, however, been customary

3.1. CRYSTALLINE LATTICE PARAMETER

that, in studies of nonmetallic catalysts from the viewpoint of multiplet theory, one considered the adsorption on only one type of active center, usually on the metal atom.

Another group of studies of geometric factors in catalysis revolved around the possibility of adsorption on another type of active centers, namely, centers that induce bond polarization. This mode of catalytic action was first advanced by Shilov [319]. For the case of homogeneous catalysis, this author proposed a six-membered activated complex composed of a polar molecule and a polar catalyst center, the assumption here being that such a cyclic structure would have minimum angular stresses. This type of activated complex may conceivably operate in heterogeneous catalysis too. A point in case would be the following complex involved in dehydration of alcohols over alumina [319]:

$$CH_2-CH_2 + HO-Al \rightleftharpoons \begin{bmatrix} H_2 & H_2 \\ C-C \\ H & OH \\ O\cdots Al \\ H & O \end{bmatrix} \tag{49}$$

$$\swarrow \nearrow$$

$$C_2H_4 + H_2O + AlO(OH)$$

The coordinately unsaturated Al-ion polarizes the OH-group of the alcohol, while the polarized hydrogen atom of the methyl group interacts concomitantly with a surface OH-group. In essence, this process represents a proton exchange between the catalyst and the reactants. According to Eucken [320, 321] and Wicke [322], the cyclic activated complex in dehydration of alcohols over alumina has the following structure:

$$\begin{array}{c} -C-\!-\!-C- \\ | \quad\quad | \\ H \quad O \\ | \diagup \diagdown \\ O \; H \; H \; O \\ \diagdown | \diagup \\ O \quad O \quad O \\ | \quad | \quad | \\ Al \; Al \; Al \end{array} \tag{50}$$

In this case, not the OH-group of the catalyst, but the strongly electronegative ion O^{2-} serves as a donor of the electron pair; the less negative OH-group acts as an acceptor. In the course of the reaction, the O^{2-} ions and the hydroxyl groups OH^- change positions. These authors proposed that, for dehydration of alcohols on oxide catalysts, a cyclic transition state involving positive polarization of hydrogen resulted from adsorption on oxygen and negative polarization of hydrogen resulted from adsorption on metal:

$$\begin{array}{c} -\underset{|}{\text{C}}-\underset{|}{\text{C}}-\text{O} \\ \ominus\text{H} \quad \text{H}\oplus \\ \downarrow \quad\quad \downarrow \\ \text{O}^{2-}\text{Zn}^{2+} \quad \text{O}^{2-}\text{Zn}^{2+} \end{array}$$

(51)

Although schematic (37) of Section 2.1, resembles schematic (51), in the former the C—H and O—H cleave homolytically, while in the latter a proton and a hydride ion split off. However, heterolytic cleavage to hydride ion is energetically unfavorable, Interestingly, Eucken and Heuer [320] arrived at an opposite conclusion despite using schematics (50) and (51), which mainly resemble the multiplet theory duplets. They proposed that an increase in the crystalline lattice parameter should enhance dehydrogenation activity, while a decrease in this parameter should enhance dehydration activity.

Turkevich and Smith [323, 324] considered cyclic complex for isomerization and polymerization of olefins. The following schematic

of double bond isomerization calls, according to these authors, for a catalyst capable of accepting and donating hydrogen atoms at the active centers, which are separated by a distance of about 3.5 Å. This condition satisfies liquid phase catalysts such as H_3PO_4 (3.46 Å), H_2SO_4 (3.46 Å), $HClO_4$ (3.41 Å), and solid catalysts such as moist $AlCl_3$ (3.46 Å), silica gel (3.50 Å), and partially hydrogenated nickel (3.52 Å). Cyclic activated complexes, made up of substrates and catalysts and involving strongly polarized bonds and hydrogen exchange, are likely to participate in catalytic cracking over aluminosilicate [219, 325] and in catalytic elimination of hydrogen chloride from alkyl chlorides [326, 327]. In the latter case, the reaction takes place readily on salts of

3.1. CRYSTALLINE LATTICE PARAMETER

alkali metals, due to the formation of the following highly polarized complex:

$$\begin{array}{c} -\underset{|}{\overset{|}{C}}-\underset{|}{\overset{|}{C}}- \\ \underset{\cdot\cdot}{Cl}\quad \underset{\uparrow}{H} \\ \downarrow\quad\;\;\cdot\cdot \\ Na\quad Cl \end{array} \rightarrow \quad -\underset{}{\overset{|}{C}}=\underset{}{\overset{|}{C}}- \quad + HCl \atop NaCl \qquad (52)$$

The distance between H and Cl atoms in alkyl chloride is approximately 2.5 Å; and, according to the published data [327], a lattice spacing of this magnitude results in a minimum activation energy for dehydrochlorination of a majority of the alkyl halides. A later detailed study [328] revealed that the mechanism of HCl elimination from alkyl halides is rather complex and, in fact, occurs as cis- as well as trans-elimination.

Cyclic transition states have also been considered for catalytic oxidation–reduction reaction. It was proposed [329] that oxidation of carbon monoxide on oxide catalysts proceeds via polarization of CO molecules on surface metal and oxygen atoms, as shown in the following schematic:

$$\begin{array}{c} O\text{---}C \\ |\;\;\;\;\;\;| \\ |\;\;\;\;\;\;| \\ -O\text{---}Me\text{---}O\text{---}Me- \end{array}$$

According to this study [329], oxides with small metal–oxygen lattice according spacings are active catalysts for CO oxidation; these are: MnO_2 (1.84 Å), Ni_2O_3 (1.80 Å), and Co_3O_4 (1.92 Å).

The mechanistic concept, based on polarization of substrate bonds in the fields of surface donor and acceptor sites, leads directly to a simple rule for catalyst selection. Thus, solids with an optimum spacing between surface donor and acceptor ions and with either ions of maximum polarizing strength or ions with labile protons (Sections 1.1 and 1.2) will exhibit catalytic activity.

If a substrate molecule adsorbs by three or more points, then the catalyst may affect not only the rate and selectivity but also the stereochemistry of the reaction. Thus, reaction between acrylonitrile and methylcyclohexanone in the presence of basic catalysts gives a product which is optically inactive. An "asymmetric synthesis" was accomplished [330] when this reaction was conducted with an alkali impregnated on levorotary quartz. The reaction product exhibited optical activity. However, in this case, the correlation with the crystalline lattice parameter is extremely complex.

methylcyclohexanone(CH₃) + $CH_2=CHCN$ → methylcyclohexanone(CH₃, CH_2CH_2CN)

A final case is that of the effect of the crystalline lattice parameter on the stereospecificity of polymers. A special study [186] has been devoted to the mechanism of formation of stereospecific, or so-called isotactic, polymers. A solid surface may cause stereospecificity simply by exerting a specific steric hindrance. According to Arlman and Cossee [331], polymerization of propylene and polymerization of ethylene on the $TiCl_3 + Al(C_2H_5)_3$ catalyst are alike; the ethylene case is presented in Section 1.5, schematic (29). A molecule of propylene shown in Fig. 34 is adsorbed on α-$TiCl_3$ alkylated with a C_2H_5-group.

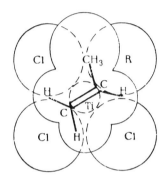

FIG. 34. Adsorption of a molecule of propylene on α-$TiCl_3$, according to Arlman and Cossee [331].

The active site here is the square formed by three Cl^- ions and an alkyl radical located in the four corners and a Ti^{3+} ion in the square's center. The fourth Cl^- ion (not shown in the figure) is oriented toward the crystal and, thus, for the case of six coordination titanium, there is an exposed surface vacancy for the attachment of an olefin via its π-electrons. The plane of the square forms a 54.7° angle with (001) face of $TiCl_3$ As shown in Fig. 34, the propylene orients itself in such a fashion that the steric abstraction from its methyl group is reduced to a minimum. It is unlikely that the bulky methyl group rests on Cl^- ions since, in the coordination sphere of titanium, two pairs of Cl^- ions are at different levels with respect to each other. Upon introduction of monomer and chain initiation [Section 1.6, schematic (29)], the growing chain and the vacant site of the catalyst exchange places, retaining, however, their mutual spatial orientation. Under such conditions, the isotactic polypropylene chain grows at the catalyst surface; and its methyl groups are distributed along only one side of the polymer chain. The configuration of atoms of the surface of α-$TiCl_3$ as well as the spatial arrangements of propylene leading to atactic and syndiotatic polypropylene and to inhibition of the polymerization reaction, are also presented in ref. [331].

Figure 34 illustrates the general principle of the multiplet theory: A rigid adhesion of a substrate molecule to the catalyst is invoked. The substrate–catalyst complex is required to be neither a duplet nor a multipoint-type attachment, but one involving only one-point linkage to the catalyst surface, such as that resulting from an interaction of π-electrons with a Ti^{3+} ion. Nevertheless, even for such a picture of the substrate–catalyst complex, there should be some indirect influence of the interatomic spacing of the surface on the catalytic activity. For example, a substitution of Cl^- ions by larger ions such as Br^- and I^- should lead to greater steric hindrance and lower reaction rate. Indeed, the rate of formation of isotactic polypropylene was found to be greater on $TiCl_3$ than on either $TiBr_3$ or TiI_3.

3.2. Type of Crystal Lattice

The type of crystal lattice, *per se*, should not directly affect catalytic activity; but it may serve as an auxiliary criterion for classifying solids.

Thus, e.g., oxidation–reduction reactions are catalyzed by semiconductors, which are prone to assume a certain type of crystal lattice. For semiconductors not involving transition elements, Mooser and Pearson [332, 333] propose that, to have chemical bonding in the crystal lattice between semiconductor components, the following conditions are required: (1) The linkages in the crystal should preferably be covalent; (2) in simple semiconductors, all s- and p-electrons should be lumped together; but in two-component semiconductors, at least one of the atoms should have its s, p-electrons fully paired; and (3) the presence of vacant "metallic" orbitals (e.g., d-orbitals) in certain atoms of a multicomponent semiconductor should not result in a conversion of the semiconductor into the metal, provided such atoms are not linked to each other, (4) the linkages in the crystal should represent a continuous chain in one, two, or three directions. Excluding transition elements, only atoms of the elements of Groups IV–VII of the periodic system may have their s- and p-orbitals completely occupied. This served as a basis for the formulation by Mooser and Pearson [334] of the octet rule, which is a criterion for a "semiconductor linkage" in the solids:

$$n_e/n_a + b = 8 \qquad (53)$$

where n_a is number of atoms of elements of Groups IV–VII in the semiconductor formula, b is number of linkages which each of such atoms forms with atoms of the same kind, n_e is number of valence electrons corresponding to the semiconductor formula (e.g., for Ge, $n_e = 4$, $n_a = 1$, and $b = 4$; for GaAs, $n_e = 8$, $n_a = 1$, and $b = 0$; for In_2Te_3, $n_e = 24$, $n_a = 3$, and $b = 0$; for CdSb, $n_e = 7$, $n_a = 1$, and $b = 1$).

As a result of sp^3 hybridization, the majority of semiconductors have a tetrahedral lattice network. The cubic structure of sphalerite (zinc blende) and the hexagonal structure of wurtzite are particularly common. As the covalency of the linkage is increased, i.e., with decreasing electronegativity difference Δx, wurtzite is formed in preference to sphalerite. The structures of sphalerite and wurtzite, depicted in Fig. 35, are the

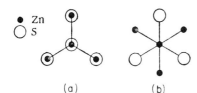

FIG. 35. Schematic presentation of atomic distribution in ZnS: (a) wurtzite; (b) sphalerite.

most common among semiconductor catalysts [335]. In very polar compounds, such as ZnO, there is an attraction between oppositely charged groups which facilitates arrangement of alternate cation and anion layers into wurtzite structure in the manner of hexagonal closest packing. In homopolar crystals (e.g., in ZnSe and ZnTe) a repulsion of the electron clouds prevails, resulting in arrangement of alternate layers into the sphalerite structure analogous to cubic closest packing. ZnS may assume either this or some other structure. Because of electrical attraction, the Zn—S distance in the wurtzite structure is somewhat shorter (2.34 Å) than in the sphalerite structure (2.36 Å).

By replacing one or more atoms in the wurtzite and sphalerite structures, one may obtain other semiconductor-like structures with tetrahedral sp^3 bonds that satisfy rule (53). If a monovalent metal in compound A^IB^{VII} (e.g., AgI) becomes replaced by a divalent one (e.g., Hg^{2+}), one obtains the tetragonal lamellar lattice of HgI_2, which can be considered to be a defect structure derived from a sphalerite-type lattice. Conversely, by replacing one bivalent metal in the $A^{II}B^{VI}$ structure by two monovalent ones (e.g., $2Cu^+$), one produces Cu_2S, which crystallizes in the cubic face-centered fluorite lattice (more precisely, antifluorite). Oxides, sulfides, selenides, and tellurides of alkaline metals (e.g., Na_2O, Na_2S) also crystallize into the influorite lattice. They are semiconductors, provided their electronegativity difference is not too large.

Mooser and Pearson have shown that large differences in electronegativity Δx and large average value of the principal quantum numbers N in binary compounds leads to an octahedral structure, while small Δx and N lead to a tetrahedral structure.

Goryunova [336] formulated the following rule for formation of

3.2. TYPE OF CRYSTAL LATTICE

semiconductor componds with sphalerite and wurtzite structures: It is necessary that the specific electroaffinity constant R/Z of each constituent element be greater than 7.5 eV:

$$R/Z > 7.5 \quad \text{eV} \qquad (54)$$

The electroaffinity constant R represents the energy required for attraction of the first electron to the valence band. This constant reflects the ion (or atom) field strength. The specific electroaffinity constant is the value of R divided by the ionic charge Z. According to Goryunova [74], rule (54) characterizes the possibility of formation of tetrahedral semiconductor compounds better than the more qualitative principles of Mooser and Pearson. Formation of ionic and metallic structures is favored when the values of R are small for either one or both constituents of the compounds. The greater the value of R, the more favored is the formation of semiconductor-like structures with covalent bonding.

Semiconductors with sp^3 bonds are obtained also from combinations of tetrahedra other than those represented in wurtzite and sphalerite. Among such structures are cuprite, GeS_2, SiS_2, and TlSe.

Semiconductors with octahedral coordination also occur. For example, PbS has a cubic lattice of the NaCl-type. It was proposed [337] that, in this case, the interaction between atoms involves not sp^3-bonding but, rather p^3-bonding, in which orbital axes from a 90° angle. This manifests itself in the greater stability of p-orbitals in heavy elements.

Interestingly, the halides and sulfides of metals whose ionic configuration are d^0, d^5, and d^{10} [i.e., the structural configuration of the ions is that of the transition metals (Section 1.6)] assume tetrahedral coordination into sphalerite and wurtzite structures. In other compounds of transition metals with different d-electron configurations, octahedral coordination predominates. For example, MnS (d^5) crystallizes into the sphalertie lattice while the neighboring sulfides (NiS, CrS, and VS) crystallize into the hexagonal lattice of NiAs.

Ionic crystals with large Δx (e.g., CaO, BaO) are solid bases and, therefore, catalyze reactions proceeding via the basic mechanism. Binary compounds with small Δx favor reactions involving acidic mechanisms; strictly speaking, they are not ionic crystals. Nevertheless, in considering structures of such oxides as SiO_2 and Al_2O_3, an ionic picture (not well founded) is frequently proposed; in many cases, assumption of ionic structures led to correct predictions of the behavior of these two compounds.

Dimensions of ions in crystals are characterized by ionic radii. It is essential that the sum of the ionic radii of the constituents of a crystalline compound be equal to the distance between two adjoining ions in the crystal lattice. In addition, the radii of the cations and anions must not be too different. The ratio of ionic radii of a cation and an anion r_{Me}/r_X is a criterion which determines the conditions for the formation of crystalline structures with various coordination numbers [99, 169]. This was originally shown in 1926 by Goldshmidt. The following are the prerequisites for the formation of ionic crystals: (1) noncontraction of the ionic shells, (2) contact between ions with opposite charges (3) maximum possible coordination member, and (4) minimum repulsion between ions of the same charge. From the first two requirements conditions for formation of structures with different coordination numbers were delineated as a function of the r_{Me}/r_X ratio. Structures with coordination number (cn) of 3 are, thus, stable at $r_{Me}/r_X > 0.155$; cn 4 at $r_{Me}/r_X > 0.225$; with cn 5 or 6 at $r_{Me}/r_X > 0.414$; and with cn 8 at $r_{Me}/r_X > 0.645$. Formation of structures with cn 5 is hardly probable since the interanionic distance in such structures would be smaller in an octahedron; and application of the third requirement indicates, at suitable r_{Me}/r_X, preferred formation of an octahedral structure.

Among the ionic crystals of the $A^I B^{VII}$-type, salts of Cs^+ and RbCl with particularly high r_{Me}/r_X ratio (greater than 0.8) assume a body-centered cubic structure of CsCl-type with cn 8; the alkaline halides assume a face-centered cubic lattice of NaCl-type with cn 6. The NaCl-type structure is also characteristic for many oxides and sulfides of the $Me^{2+}X^{2-}$ type. The above-mentioned simple requirements (derived from a crystal model built with ions having noncontractible electron shells) do not apply to MeX-type semiconductors (e.g., ZnS, CuBr, and AgI) because the actual Me—X distance in their crystals does not agree with the calculated sum of the ionic radii.

For example, an ionic model applies satisfactorily to MeX_2 type compounds. Two structures that are common among compounds of this type are: the cubic fluorite structure with cn of the metal equal to 8 and the tetragonal ring-type rutile structure with metal cn of 6. All compounds of MeX_2 type with $r_{Me}/r_X < 0.73$ assume a rutile structure (TiO_2, SnO_2, ZnF_2, and others), and those with $r_{Me}/r_X > 0.73$ assume a fluorite structure (CaF_2, ThO_2, $BaCl_2$, and others). Except for Cs_2O, the alkaline oxides crystallize in an antifluorite-type structure (in which the anion and the cation exchange their positions).

Compounds of transition elements crystallize frequently into crystal structures with reduced symmetry; this can be explained in terms of Jahn–Teller effect [164, 169] of the crystal field theory (see the discussion in Section 1.6). For example, in octahedral coordination, the Cu^{2+} ion has an electronic configuration $d^9 = t_{2g}^6 e_g^3$ (see Table I and Fig. 21).

3.2. TYPE OF CRYSTAL LATTICE

In any field the d^{10} ion assumes a spherical symmetry. The d^9 ion differs from the d^{10} ion in that it has a hole in the d-shell which may occur in either one of the e_g-orbitals: $d_{x^2-y^2}$-orbital or d_{z^2}-orbital. In the first case, the nuclear charge is screened less along the x and y axes, the ligands (anions) are more strongly attracted, and, so, the bond lengths along the x and y axes are shorter than those along the z axis (Fig. 36). In the

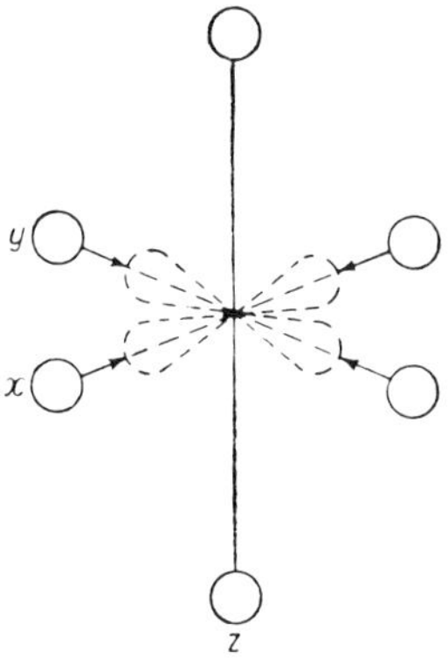

FIG. 36. Schematic presentation of bonding in the complex of an ion with d^9 electronic configuration (Cu^{2+}).

second case, conversely, the bonds along the z axis are shortest. It was established experimentally that $CuCl_2$, $CuBr_2$, and CuI_2 belong to the first case: a distorted octahedral lattice is formed with two long and two short bonds. For example, in $CuCl_2$, four of the $CuCl_2$, four of the Cu—Cl bonds are 2.30 Å in length, and two Cu—Cl bonds are 2.95 Å in length. In such lattices, splitting of the t_{2g} and the e_g orbitals occurs, (as shown in Fig. 21) with resultant additional structural crystal field stabilization. An extreme case, which actually occurs in CuO and in the monoclinic structure of tenorite will result from complete removal of the two most remote ligands. Ion Cu^{2+} in CuO has a cn of 4 and a square-type configuration. Analogous d^9-structures with four short and two long bonds occur in high-spin complex of ions with d^4 configuration (Cr^{2+}, Mn^{3+}) and in low-spin complexes with d^7 configuration (Co^{2+}, Ni^{3+}). Definitions of high- and low-spin complexes were given in Section 1.6.

In high-spin complex d^1, d^2, d^6, and d^7, and in low-spin complexes d^4 and d^5, the distortions of the crystal structures from regular octahedron are significantly smaller as a result of the Jahn–Teller effect. This stems from the fact that the symmetry axes of the t_{2g}-orbitals (d_{xy}, d_{xz}, and d_{yz}) are not oriented toward the ligands (Fig. 20). Therefore, the unsymmetrical electron occupancy of the t_{2g}-orbitals affects the bond lengths to a considerably smaller degree than such occupancy of the e_g-orbitals ($d_{x^2-y^2}$ and d_{z^2}). There will be no distortions in structures with completely filled (or half filled) t_{2g} level and empty e_g level, as is the case in complexes, d^3, d^5, and d^8, and in low-spin complexes d^6 [169].

In absence of the Jahn–Teller dislocation effect, the rules for formation of oxides of transition elements (VO, MnO, and NiO) agree well with the structure of ionic molecules, as described above. Higher oxides with low r_{Me}/r_X values (V^5, Cr^{6+}, Mn^{7+}, Re^{7+}), all have cn 4 and those with higher r_{Me}/r_X values (Ti^{4+}, Zr^{4+}, Hf^{4+}, Nb^{5+}, Ta^{5+}) have cn 6. The largest ions from the first group (V^{5+}) and the smallest ions from the second group (Ti^{4+}, Nb^{5+}) occur in strongly distorted as well as in intermediate structures. For example, in V_2O_5 ion, V^{5+} is associated with five oxygen atoms by bonds of varying length: 1.58, 1.78, 1.88, 2.02, and 2.79 Å; and the sixth O^{2-} ion is even more remote [338]. According to Orgel [294], V_2O_5 exhibits high activity as an oxidation catalyst because of its peculiar structure which permits the occurrence of the V^{5+} ion in tetrahedral and octahedral configuration and also because of the changeable valence of vanadium.

Stone [48] drew on the peculiarity of the crystal structure in explaining high activity of Cu_2O in oxidation of CO and in oxygen exchange. On (001) face, the O^{2-} ions located in "convexed" sites are those that participate in oxygen exchange as well as in catalytic reaction. Adsorption of CO on such surface O^{2-} ions creates a favorable situation for the reaction of CO with such O^{2-} ions or with adsorbed oxygen. Also, on other principal faces of Cu_2O, the O^{2-} ion exhibits high reactivity. According to Stone, from a reactivity viewpoint no such favorable location of O^{2-} ions on the surface occurs in oxides with a regular cubic face-centered lattice (e.g., in NiO).

According to the rules of coordination, the presence of free sites (vacancies) in the coordination shell around the cation is required for a catalytic reaction to occur. In chlorides and sulfides, the fact that radii of Cl^- and S^{2-} are greater than the radius of O^{-2} may lead to a complete shielding of the cation, and, consequently, to reduction in catalytic activity. In such cases, crystals frequently possess a layer-like structure. In this connection it should be mentioned that Brennan [339] failed to observe catalytic activity in halides such as AlF_3, TiF_3, $CrCl_3$, MnF_2, CoF_2, $CoBr_2$, $CuCl_2$, and ZnF_2 for H_2–D_2 exchange. He attributed the low activity of NiF_2, $CoCl_2$, and $NiCl_2$ to the presence of metallic admixtures.

As discussed in Section 1.6 and Section 3.1, the occurrence of a vacancy in the layer of Cl ions on the $TiCl_3$ surface is mandatory for polymerization of olefins by the Ziegler–Natta mechanism. Arlman [340]

3.2. TYPE OF CRYSTAL LATTICE

produced evidence that such vacancies form in the corners and on edges of a crystal.

Finally, it should be borne in mind that the multiplet theory of catalysis assigns great importance to the lattice type in the case of formation of multipoint attached complexes. For example, hexagonal and cubic face-centered lattices are well suited for accommodating benzene molecules in their faces (Fig. 32).

Part II • PRINCIPLES OF SELECTION OF CATALYSTS FOR VARIOUS REACTIONS

Introduction

Part I of this monograph was concerned with various properties of the solid state which, according to the several theories of catalysis, may either determine catalytic activity or correlate with it in some way. Many of these properties were shown to be not entirely independent but contributing; for example, width of the forbidden zone is related to the difference in electronegativity, and this correlates with the effective ionic charge; lattice parameter and type of crystal lattice were also shown to be connected, etc. In the final analysis all such properties are a function of the electronic structure of the atoms and ions involved in the structure of a catalyst.

Obviously, not all the properties which may affect catalytic activity have already been considered. Many of the neglected properties, however, are somehow related to those that were considered. For example, ionization potential correlates with the work function and ionic and atomic radii correlate with lattice parameter. Another problem that has been neglected was the question relating to catalyst selection based on bond energy between adsorbed reactants and catalyst surface atoms. Most of the emphasis was placed on those properties of the solid state which can be measured experimentally independently of catalysis. In considering various catalytic concepts, primary attention was given to those conclusions which, at the present state of theory, permit catalytic activity to be correlated directly with the measured parameters. This, however, is not meant to imply that any theory precludes a possible correlation between catalytic activity and properties of solids which lie outside its framework.

Part II of this monograph considers practical catalytic reactions; it encompasses data published during the past 12 years relating to the selection of binary-compound-type catalysts ($A_n B_m$).

Those studies which have presented results on the catalytic properties of at least two to three catalysts of different chemical nature used in a single reaction system are analyzed in depth. No regard has been given to problems relating to admixtures (promoters, poisons, modifiers) as well as to mixed or complex catalysts. Also, all information from patent literature has been neglected as its credibility is somewhat questionable.

An attempt has been made to correlate catalytic data with the properties of the solid state which were covered in Part I. The feasibility of such a correlation has emerged from recent advances in the methods of studying catalysts and adsorbents. Primarily, this refers to the development of simple and precise methods of determining surface areas of solids [341, 342], progress in vacuum technique, and availability of a series of nondifferential methods for determining rates of catalytic reactions [343]. Despite that, even in many very recently published studies, there is a pitiful lack of data on specific surface area; and catalysis is investigated in either static or dynamic systems without due attention to the process macrokinetics; another shortcoming is that measurements were frequently taken over too narrow a range of temperature, pressure, etc.

The rules of selection of oxidation–reduction catalysts can be derived from more credible data than those available for catalysts for acid–base reactions. But, even in the former case, most data refer to a rather limited number of reactions, such as oxidation of carbon monoxide and hydrocarbons, decomposition of N_2O and ethyl- and isopropyl alcohols, and dehydrogenation of cyclohexane. In each of these case, the number of available published studies is confined to 10–15 references.

The extent of generalization and depth of analysis of data, from the point of view of selection of a catalyst for a particular reaction, has been varied in this monograph over a wide range. This, by no means, reflects the relative importance of any given reaction, but merely relates to the current level of understanding of each of the catalytic reactions considered in Part II.

4 • Decomposition of Alcohols and Acids

4.1. DECOMPOSITION OF ALCOHOLS

The two basic modes of alcohol decomposition are: (a) dehydrogenation to form an aldehyde (in the case of primary alcohols) or a ketone and hydrogen (in the case of secondary alcohols), and (b) dehydration to form an olefin and water. At high temperatures (300–600°C), severe decomposition of an alcohol may occur leading to the cleavage of carbon–carbon bonds and the resultant formation of paraffins, CO, CO_2, etc. At near ambient temperatures, ether is the major reaction product. This reaction will not be considered here since, for the purpose of catalyst selection, it is sufficient to regard only the general reactions (a) and (b). Comprehensive studies of these reactions revealed that each of them calls for a different type of a catalyst.

The mechanism of alcohol decomposition was touched upon several times in Part I of this monograph. A summary of conclusions, derived from the various theories of catalysis, for catalyst selection in alcohol decomposition are:

1. According to the widely known conclusions based on the electronic theory of catalysis [17, 24], the rate limiting step for dehydrogenation is the migration of vacancies, while for dehydration, it is the migration of free electrons in the catalyst; consequently, dehydrogenation is expected to be catalyzed by p-type semiconductors and dehydration should be catalyzed by n-type semiconductors (see p. 11). However, some authors [38, 344] have arrived at opposite conclusions.

2. The multiplet theory of catalysis (Section 3.1) implies two-point

adsorption as a condition for alcohol decomposition. Optimum interatomic spacing between surface atoms of the catalyst for adsorption leading to dehydration is greater than that leading to dehydrogenation. Therefore, an increase in crystal lattice parameters will aid dehydration activity, while a corresponding decrease will aid dehydrogenation activity.

3. According to the classification of catalytic processes (p. 5), dehydration is an acidic reaction and, therefore, should be catalyzed by solid and liquid protonic and aprotic acids; dehydrogenation, being an electronic reaction (in the general usage of this term), should be catalyzed by metals and also by semiconductors.

4. In accordance with the concept of cyclic complexes in catalysis [319, 320] (Section 3.1), adsorption of an alcohol involves a two-point attachment to the electropositive and electronegative surface atoms; an increase in metal–oxygen distance will facilitate dehydrogenation; and a decrease in such distance will facilitate dehydration.

5. Based on the premise that alcohol decomposition is a multistep reaction [345], dehydration requires acidic catalysts with n-type conductivity and dehydrogenation requires basic catalysts with p-type conductivity.

6. Allowing for the role of the width of the forbidden zone (U) (Section 1.3) and for the presumption that catalysis involves specific conductivity, the rate of alcohol dehydrogenation will increase with decreasing value of U. There should be no effect of U on dehydration, since this is an acid catalyzed reaction.

All these conclusions, which are based on the various theories of catalysis, will now be reviewed in light of empirical data.

Decomposition of isopropyl alcohol has gained a prominent place as a model reaction for studying the principles of catalyst selection. The two main paths of this decomposition are:

$$\begin{array}{c} CH_3 \\ | \\ CH_2\text{---}OH \\ | \\ CH_3 \end{array} \begin{array}{c} \nearrow CH_3COCH_3 + H_2 \quad \text{dehydrogenation} \\ \\ \searrow CH_3CH=CH_2 + H_2O \quad \text{dehydration} \end{array}$$

both of them are free of side reactions.

In Part I, some consideration was given to certain simple rules regarding the selection of catalysts for decomposition of alcohols. In order to pinpoint the general rules for catalyst selection, resort is made to mathematical and statistical methods [346]. Attention is drawn

4.1. DECOMPOSITION OF ALCOHOLS

to a group of studies on isopropyl alcohol decomposition [77, 78, 212–216, 244, 286, 299–303, 305, 320, 321, 347–370]. Whenever sufficient data were available, the catalytic activity was expressed in terms of reaction rate per square meter of catalyst surface [$k = (dc/dt)(1/s)$].

In some of these studies, the only data given were the yields of the product or the degree of decomposition. In those cases where the only information given was the temperature at which reaction commenced or the temperature level at which measurable rates of reaction were discernable, the value of $e^{1/T}$ was used as a measure of catalytic activity, where T is given in degrees Kelvin. (For constant pre-exponential factor k_0, T is inversely proportional to activation energy.)

Information derived by this approach cannot be compared reliably. This author [371] places maximum confidence in these studies which report specific reaction rates (rate per square meter of catalyst surface area) and which were performed in the kinetically controlled range. Some examples are studies of catalysis at low pressures [320], in the absorbed layer [78], or in recycle reactors [213]. For determining catalytic activity, in these cases, a statistical weight of $w = 4$ has been assumed. For those studies from which k could be computed [301, 351, 356], but which were likely to be biased by diffusion since they were conducted in a flow system, a weight of 3 has been used. A weight factor of 2 was assumed for those studies in which reaction rates could not be expressed in standard units due to the lack of any two factors, for example, when catalyst surface area was given but information was missing regarding the kinetic conditions (diffusion or kinetic region). Also, whenever specific surface area was unknown, the value of w was assumed to be unity.

Since the values of catalytic activity obtained from the various references differed by orders of magnitude, only the best known systems were compared. With this viewpoint, ZnO was taken as a base catalyst for the dehydrogenation of isopropyl alcohol and assigned a relative activity of 100 (or 2 on a logarithmic scale), and Al_2O_3 was taken as a base for the dehydration of this alcohol and given a relative activity of 10 (or 1 on a logarithmic scale). The activities of the other catalysts are placed appropriately in relation to these two points of reference. Whenever possible, a comparison of actual catalytic activity (or extrapolated values of catalytic activity) was made at 150°C.

In Table III (for various nonmetallic catalysts) are shown average values of catalytic activity on a logarithmic scale obtained in the following way, $\log k = \sum w \log k / \sum w$. The table contains the sums of the statistical weights $\sum w$, which reflect the degree of credibility of the presented information. A credibility equal to unity was assigned to data

TABLE III

RELATIVE CATALYTIC ACTIVITY OF VARIOUS BINARY COMPOUNDS FOR DECOMPOSITION OF ISOPROPYL ALCOHOL

Catalyst	Dehydrogenation		Dehydration	
	log k	Σw	log k	Σw
α-Al$_3$O$_3$	1.56	4	−2.00	4
γ-Al$_2$O$_3$	0.57	9	1.00	12
BAs	0.80	4	1.00	4
BP	0.50	4	1.10	4
BaO	1.03	7	0.40	7
BeO	0.59	13	−0.16	10
CaO	0.99	15	−0.97	10
CdO	2.13	14	0.26	12
CdS	3.10	4		
CeO$_2$	1.98	2	0.51	2
Cr$_2$O$_3$	1.44	8	0.24	6
CrS	0.47	3	0.62	3
CrSe	0.40	3	1.28	3
CuBr	1.25	5		
CuO	1.89	5		
Cu$_2$O	2.11	5	−1.50	3
FeO	1.43	4	0.40	3
Fe$_2$O$_3$	1.52	6	0.72	7
GaAs	2.86	5		
GaSb	4.76	4		
Ga$_2$Se$_3$	2.78	4		
Ga$_2$Te$_3$	3.12	4		
Ge	3.07	5		
HfO$_2$	0.55	2	0.27	2
HgO	2.81	4		
HgS	2.42	4		
InAs	3.17	4		
InSb	2.96	4		
In$_2$O$_3$	1.54	1	0.82	4
KF	−0.07	4	−1.52	4
La$_2$O$_3$	1.19	2	0.96	5
MgO	0.79	16	−0.79	14
MnO	1.84	11	−0.07	3
Mn$_2$O$_3$	1.56	5	0.03	3
MoO$_3$	1.91	2	0.50	3
MoS$_2$	2.30	1	0.74	1
Nb$_2$O$_2$	−0.14	1	1.11	3
NiO	2.67	7	0.15	3
NiS	0.15	3	2.07	3
NiSe	0.80	3	0.35	3
PbO	2.72	4		

4.1. DECOMPOSITION OF ALCOHOLS

TABLE III (continued)

Catalyst	Dehydrogenation		Dehydration	
	log k	Σw	log k	Σw
PbS	2.56	3		
ScN	1.31	4	1.43	4
Sc$_2$O$_3$	0.87	3	0.43	3
SiO$_2$	−0.50	1	−2.40	5
Sm$_2$O$_3$	1.59	1	0.66	3
SnO	1.67	8	0.20	7
SnO$_2$	1.01	7	0.44	6
SrO	0.97	7	−0.86	7
Ta$_2$O$_3$	−0.48	1	1.25	2
ThO$_2$	2.18	1	−0.88	1
TiC	0.31	4	1.20	4
TiO$_2$ Anatase	0.55	8	0.80	4
TiO$_2$ Rutile	1.01	15	0.63	15
UO$_2$	0.62	4	−0.06	4
V$_2$O$_3$	0.40	5	−0.27	4
V$_2$O$_2$	0.22	2	0.66	3
WO$_3$	0.36	6	1.04	9
Y$_2$O$_3$	0.75	3	−0.32	4
ZnO	2.00	26	0.33	17
ZnS	2.39	15	1.25	7
ZnSe	2.57	12	1.35	7
ZnTe	4.00	4	2.40	4
ZrO$_2$	0.56	3	0.82	10

obtained on a catalytic system for which only one reference existed and for which no data were given on either catalyst surface area or reaction conditions.

In Fig. 37 is shown the dependence of the logarithm of catalytic activity (for the dehydrogenation of isopropyl alcohol over oxides of the metals of the fourth period) on position in the periodic system of elements. On the abscissa, the even intervals correspond to metal oxides in their primary valence state. In some cases, activity is seen for the oxides of metals in their lower oxidation state; these are situated to the right of the oxides of metals in their higher oxidation state, which is in harmony with the degree of electron occupancy of the d-bands. Of all these oxides, NiO exhibits the highest catalytic activity (electronic configuration d^8). There is not, however, a clear minimum corresponding to d^5. Oxides of Fe, Mn, and Cr show an almost identical activity.

This seems to indicate that catalytic activity increases in the sequence from V_2O_5 toward the end of the period, a trend corresponding to the degree of occupancy of the d-orbitals. This observation differs somewhat from the two-spiked type dependence of Dowden and Wells [173] (Section 1.6). Among the oxides involving cations with a d^{10} structure, the activity declines in the sequence from CaO to V_2O_5, i.e., with increasing acidity of the oxide.

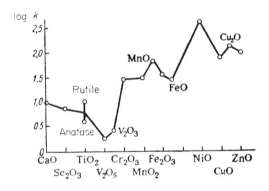

FIG. 37. Variation of the logarithm of activity of the oxides of the metals of the fourth period as catalysts for dehydrogenation of isopropyl alcohol.

Figure 38a presents the dependence of the catalytic activity of oxides for the dehydrogenation of isopropyl alcohol, depending on the position of the metal in the periodic table of elements. In a coordinate system in which the location of a metal in the period is given on the abscissa and their location in the group is shown on the ordinate, the catalytic activity falls on a surface. Asterisks designate those metals whose oxides are characterized in Table III. The equal activity lines encompass such regions of metal oxides for which catalytic activity exceeds a certain value in arbitrary units. For convenience, the unabridged version of the periodic table is used and the oxides of Be and Mg are shown twice— on top of both the main- and the subgroup of the second group. For best visual comprehension, the lines of equal activity are continuous. Figure 38 shows a gradual decrease of catalytic activity from two peak-areas. The first peak-area encompasses oxides of Ni, Hg, and Pb and also oxides of Pd, Ag, and Tl which have not yet been investigated. Assuming that the above-mentioned procedure for data handling is correct, it is anticipated that this group of oxides should exhibit high catalytic activity. The second peak occurs near ThO_2; however, the information on ThO_2 has been assigned low dependability, since only

4.1. DECOMPOSITION OF ALCOHOLS

one study [355] was devoted to it. There is, therefore, a likelihood that the activity maximum will fall on one of the metal oxides in the immediate neighborhood of ThO_2, rather than on it *per se*. The first peak-area contains oxides of the metals from the end of the long periods (i.e., oxides of transition metals with at least half filled *d*-shells) and of the metals immediately following the long periods. In the case of the second peak, the metals have no *d*-electrons in the outer shell and exhibit alkaline properties. Within each of these two peak-areas, catalytic activity increases with increasing atomic weight of the metal.

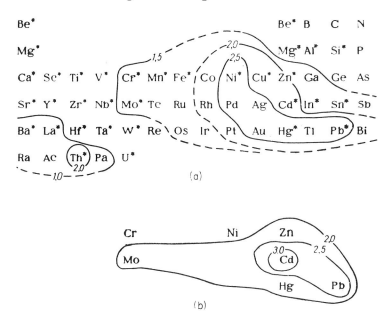

FIG. 38. Lines of equal activity of (a) metal oxides and (b) metal sulfides for the dehydrogenation of isopropyl alcohol.

There are less data available for sulfides, but, as shown in Fig. 38b, their activity chart resembles that for oxides. The catalytic activity of sulfides (except for NiS and CrS) is greater than that of oxides.

The results of such a data treatment procedure show that definite predictions regarding catalytic activity can be made if the body of information is large enough to enable one to fill the activity chart. In such circumstances, the discrepancy in the data on catalytic activity for a given substance become eradicated, the effect of unknown admixtures becomes less important, and the properties of the main component of the material assume the dominant role from a catalytic standpoint.

Data on dehydration of isopropyl alcohol, Fig. 39, were assembled in a similar fashion. In this case, also, there emerged a regular chart, although with some exceptions (MnO, Y_2O_3, HfO_2). Two reasons may account for these irregularities; first, the scarcity of data on dehydration of this alcohol; and, second, in many studies, dehydration represented only a minor reaction and the extent of dehydration was ascertained from the difference in total decomposition. To a certain degree, the chart for dehydration is a reverse of that for dehydrogenation. The oxides of Nb, Ta, and W situated in the saddle point on Fig. 38 represent a maximum so far as dehydration activity is concerned. A second maximum occurs near Al_2O_3; and the minimum falls in the region encompassing the oxides of the transition metals in the second half of the long periods (VII and VIII Groups). In all cases except WO_3-WS_2, catalytic activity for dehydration is greater for sulfides than for oxides.

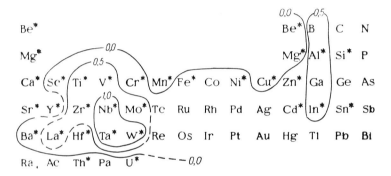

FIG. 39. Lines of equal activity of metal oxides for the dehydration of isopropyl alcohol.

Next, the dependence of the activity of various catalysts for the decomposition of isopropyl alcohol will be reviewed as a function of the other properties of the solid state which were discussed in Part I. The abundance of information permits the methods of correlative analysis [346] to be employed here. In Table IV are presented the results of this analysis, based on data from Table III.

In Table IV, η represents the correlation ratio defined by:

$$\eta = \left[\frac{n \sum_x [(\sum_y n_{xy} y)^2/n_x] - (\sum_x \sum_y n_{xy} y)^2}{n \sum_y n_y y^2 - (\sum_y n_y y^2)} \right]^{1/2} \tag{55}$$

where y is the logarithm of catalytic activity, x is a property of the solid, and η is the number of empirical data points.

4.1. DECOMPOSITION OF ALCOHOLS

TABLE IV

CORRELATION COEFFICIENTS LINKING CATALYTIC ACTIVITY OF BINARY COMPOUNDS FOR THE DEHYDROGENATION OF ISOPROPYL ALCOHOL WITH THE VARIOUS PROPERTIES OF SOLIDS

Property of a solid	Correlation coefficient				Presence or absence of correlation
	η	τ	b	σ	
Lattice parameter d	0.55	0.40	0.27	0.92	+
Width of forbidden zone U	0.58	−0.48	−0.24	0.91	+
Electronegativity difference Δx	0.64	−0.51	−0.70	0.86	+
Work function φ	0.55	0.40	0.20	0.68	+
Value of $1/\epsilon^2$	0.44	−0.08	−0.17	0.86	−

The magnitude of η indicates the degree of dependence between y and x.

$$r = \frac{n \sum_x \sum_y n_{xy} xy - (\sum_x n_x x)(\sum_y n_y y)}{\{[n \sum_x n_x x^2 - (\sum_x n_x x)^2][n \sum_y n_y y^2 - (\sum_y n_y y)^2]\}^{1/2}} \quad (56)$$

The magnitudes of r and η vary from 0 to 1 and $\eta \geqslant r$. When r and η are close to unity, there is a definite relationship between x and y; and when r and η are close to zero, there is no relationship.

In Table IV, b is the coefficient of linear regression or the slope of the straight line $y = f(x)$ constructed using the method of the least squares; σ is the average squared deviation from the regression line. In the preceding figure, there was shown either the presence or the absence of a correlation between y and x. The criterion for such a correlation is expressed by the formula:

$$|r| > t\sigma_r \quad (57)$$

where $|r|$ is the absolute value of the correlation coefficient, t is the deviation of a normally distributed normalized value at the 5% level of significance derived from the tables in ref. [346], and $\sigma_r \approx (1 - r^2)/\sqrt{n}$ is the dispersion of the empirical coefficient of correlation. The following is a detailed discussion of the results graphically presented in Fig. 40 concomitant with their associated regression lines.

As shown in Fig. 40a and as reflected in the values of coefficients r and η, there is a correlation between catalytic activity ($\log k$) and the crystal lattice parameter (i.e., the spacing d separating electropositive and electronegative atoms Me−X is a binary compound). Catalytic activity increases with increasing d. It was shown in Section 1.3 that the width of the forbidden zone U diminishes, all other conditions being constant,

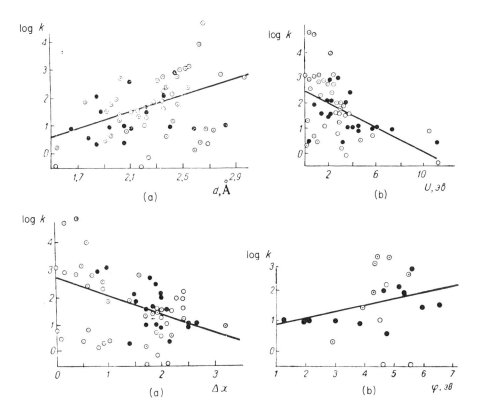

FIG. 40. Correlation between the logarithm of catalytic activity for the dehydrogenation of isopropyl alcohol and (a) the spacing between the metal- and nonmetal atoms in binary compounds; (b) width of forbidden band; (c) difference in electronegativity; (d) work function of the catalyst.
Remark: The identical legend applies to this as well as to the subsequent figures, these are: ●—most reliable data; ⊙—medium reliable data; ○—least reliable data.

with increasing atomic number of the elements composing the semiconductor, hence, with increasing crystalline lattice parameter. In order to ascertain that the dependence between $\log k$ and d does not reflect merely the dependence between $\log k$ and U, the coefficient of correlation $r_{d/U}$ for d and U has been computed for the semiconductors listed in Table IV. This computation yielded a value of $r_{d/U}$ equal to 0.18. In addition, the partial coefficient of correlation between $\log k$ and d has also been calculated, disregarding the relationship between these two parameters and U;

$$r_{\log k d/U} = \frac{r_{\log k/d} - r_{\log k/U} r_{d/U}}{[(1 - r^2_{\log k/U})(1 - r^2_{d/U})]^{1/2}} \tag{58}$$

4.1. DECOMPOSITION OF ALCOHOLS

The value of this partial coefficient was found to be equal to 0.36. This demonstrates that the dependence between d and U exerts little effect on the coefficient of correlation (log k versus d); in other words, this means that between log k and d there exists a true relationship which is not caused by the known dependence of catalytic activity upon the width of the forbidden zone. It is credible that the relationship between log k and d is a result of the active participation of two catalyst surface atoms in the formation of the activated complex. As shown in Fig. 40a, semiconductors of the type Ge, In, As, and others, all with small forbidden-zone widths, and certain oxides of transition metals (NiO) deviate sharply (over and above the mean square deviation) from the regression line toward the more electropositive direction; while dielectrics such as SiO_2, BaO, SrO, and some others whose catalytic activity have been scarcely studied, deviate from the regression line toward the more electronegative direction.

The dependence of log k upon the width of forbidden zone U is shown in Fig. 40b. Catalytic activity declines with increasing values of U. The correlation coefficient between log k and U (equal to 0.48) is greater than the correlation coefficient between U and d; however, the rate of decrease of log k with increasing U is not as large as would be predicted from the theoretical considerations covered in Section 1.3. The data points for semiconductors with small forbidden zone widths fall well above the regression line. This stems from the fact that, despite high values of U (4.4–7.5 eV), solid bases such as CaO, SrO, and BaO exhibit very high catalytic activity. When the overall dependence is extended to include also such solid bases, a smaller slope of the linear regression line results. Conversely, the coefficient of linear correlation increases when solid bases are exempted from the overall dependence. Certain semiconductors exhibit negative deviation with small forbidden-zone widths, e.g., TiC and V_2O_3, for which the extent of dehydration (major reaction) was determined while the extent of dehydrogenation was obtained indirectly from the difference between total alcohol decomposition and the degree of dehydration. It is likely that such a procedure might have resulted, in certain cases, in large inaccuracies in the actual rate of dehydrogenation, and, in particular, values of log k which are too low.

Elimination of several factors which affect the catalytic activity of solids allows for a ready explanation the dependence of log k upon U. As shown in Section 1.3, very little scatter of the data is evidenced (see Fig. 11) when catalytic activity for dehydrogenation of isopropyl alcohol was plotted as a function of the width of forbidden zone for semiconductors belonging to the structural group of sphalerite. If one also neglects the effect of lattice parameter, the functional dependence of $\log k$ upon U within a single isoelectronic series Ge → CuBr becomes simple and well defined, as shown in Fig. 10.

The dependence of $\log k$ upon the difference in electronegativity Δx is shown in Fig. 40c. It is seen that this dependence is similar to that between $\log k$ and U. Also, as shown in Table IV, the values of the coefficients η and r are alike in both cases. The similarity to this dependence between $\log k$ and U arises because, for the oxides of the rare earth elements with high Δx and considerable catalytic activity, the slope of the line indicating the dependence of $\log k$ on Δx is quite steep. It may therefore be concluded that the dependence of $\log k$ upon Δx is determined by the dependence between Δx and U or by the dependence of both of these parameters upon some third parameter such as, e.g., effective change of a cation. This presumption is confirmed by the fact that computation of partial coefficients of correlation $r_{\log k U/\Delta x}$ and $r_{\log k \Delta x/U}$ from equations similar to (58) yield values of $r_{\log k/U}$ and $r_{\log k/\Delta x}$ which are smaller by a factor of 1.5 to 2, but which are still greater than zero. Regrettably, the available data on effective charges of cations are, as yet, too scarce to enable any other comparisons.

Figure 40d and the value of η and r seem to indicate a true dependence of catalytic activity upon the work function φ; there seems to be a tendency for φ to increase with increasing $\log k$. The higher activity of p-type semiconductors, as compared with n-type semiconductors, results from this dependence. Indeed, according to Table III, the average value of $\log k$ for p-type semiconductors is 2.11 and for n-type semiconductors is 1.81. This difference in $\log k$ is, however, too small to claim that, as a rule, p-type semiconductors are more active. This average value of $\log k$ for p-type semiconductors is strongly affected by the oxides of transition metals from the end of the long periods, as shown in Figs. 37 and 38. As discussed in Section 1.6, p-type conductivity of these oxides is often caused by greater mobility, as compared with $3d$-electrons, of the holes in the $2p$-band of the oxygen, rather than by a greater concentration of free holes.

The values of the correlation coefficients (Table IV) indicate that there is no correlation between $\log k$ and $1/\epsilon^2$ (where ϵ is the dielectric constant), although it would be expected from Eq. (13) in Section 1.3, which implies linear dependence between U and $1/\epsilon^2$. Two reasons which may account for this are: (1) low credibility (or accuracy) of the ϵ values, and (2) the possibility that the dependence between $\log k$ and U does actually reflect a hidden dependence of U upon a third parameter which is not related to ϵ in a manner similar to Eq. (13).

Computation of r, η, b, and σ and construction of analogous diagrams

4.1. DECOMPOSITION OF ALCOHOLS

for dehydration of isopropyl alcohol revealed the absence of a correlation between log k on one side and U, d, φ, and $1/\epsilon^2$ on the other (see, e.g., Fig. 41). An actual correlation has been established only between log k and the electronegativity difference (Fig. 42). For this case, the values of the correlation coefficients are: $\eta = 0.58$, $r = -0.49$, $b = -0.73$, $\sigma = 0.83$. The decrease of log k with increasing Δx is readily accountable in terms of the acidic mechanism of alcohol dehydration. It was shown in Section 2.1, that acidic oxides have small and basic oxides have large values of electronegativity difference Δx.

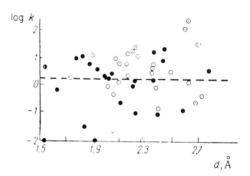

FIG. 41. Absence of the dependence between the logarithm of catalytic activity for dehydration of isopropyl alcohol and the spacing separating the metal and nonmetal atoms.

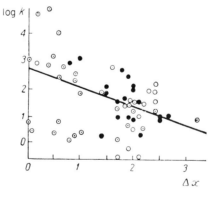

FIG. 42. Correlation between the logarithm of catalytic activity for dehydration of isopropyl alcohol and the difference in electronegativity.

Assignment of a statistical weight to one or another study was the only subjective factor in carrying out the correlative analysis of the available information in the literature. Therefore, in order to minimize this factor, a computation was made using a modified scale of statistical weights. If instead of 4, as was done above, a statistical weight equal to 10 or 2 is assigned to the most reliable data, then the calculated values of the correlation coefficients do change somewhat, but the conclusion as to the presence or absence of correlation remains intact.

In view of this, the above-presented correlative dependencies appear to be true, despite considerable dispersion of the data points. Thus, catalytic activity of binary compounds for dehydrogenation of isopropyl alcohol is a function of not one but several properties of a solid; these are: the crystalline lattice parameter, type of conductivity, work function, and width of forbidden zone. The dispersion of data points on Fig. 40 is due largely to the complex character of the dependencies in question and also, to a certain degree, to the scattering of the experimental catalytic activity data; application of the above method of statistical data averaging leads to considerable convergence of this dispersion.

Inaccuracies in determining the values of φ, U, Δx, and ϵ also plays a role here. The dependence of activity upon any one of these parameters alone is not functional but merely stochastic, i.e., the influence of one or another of these parameters on the scattering of the activity values is speculative.

It should be mentioned that the correlative analysis confirms the majority of conclusions derived from various theories (listed on p. 115) regarding the principles of selecting catalysts for dehydrogenation of alcohols. Only the increase of catalytic activity with increasing lattice parameter cannot be concluded from the multiplet theory of catalysis (Item 2 on p. 115); nevertheless, the general assumption as to the necessity, in this case, of a two-point duplet-type adsorption is probably valid.

In contrast to that, the only firm conclusion, supported by correlative analysis of the data for the dehydration of isopropyl alcohol, is that there exists a dependence of catalytic activity upon Δx for acidic oxides. A lack of correlation of catalytic activity with crystalline lattice parameter may be explained in terms of one-point adsorption of the alcohol, and the absence of a correlation of U with φ in terms of the heterolytic character of the dehydration reaction.

Study of the distribution of isopropyl alcohol in the adsorbed layer on various oxides [237] revealed that active centers for dehydrogenation are different than those for dehydration. Consequently, the selectivity of a given catalyst in the decomposition of alcohols is a function of its relative activity for catalyzing two independent reactions, namely, dehydrogenation and dehydration. The scale of catalytic activity (applied in Table III) was chosen with the view of enabling one to assess roughly which of these two reactions is dominant for a given catalyst.

All the above correlations were established using data on reaction rates; data on activation energies were regarded to be too unreliable. For semiconductors with sphalerite-type structure [78] the values of E

4.1. DECOMPOSITION OF ALCOHOLS

and log k_0 are presented in Fig. 12 (Section 1.3). As mentioned previously, there exists no compensation effect, or in other words there is no linear relationship between E and log k_0. Also, the studies by this author and co-workers [250, 348] of the dehydrogenation of isopropyl alcohol over an isoelectronic series of a system with structures the same as NaCl: KF, CaO, ScN, and TiC, did not detect any compensation effect. The values of E for dehydrogenation were found to be 22.0, 16.0, 19.0, and 14.5 kcal/mole, respectively, and the values of log k_0 varied from 2 to 4 (using rate constants expressed in $min^{-1} \cdot m^{-2}$). Variation in the rates of dehydrogenation and dehydration of isopropyl alcohol is presented in Fig. 43. As shown, the rate constant for dehydrogenation decreases

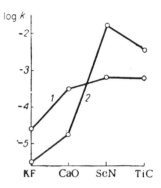

FIG. 43. Changes in logarithm of rate constant for dehydrogenation (1) and dehydration (2) of isopropyl alcohol in the series KF → TiC.

with increasing width of forbidden zone U, but this decrease is quite small considering the 10 eV increase of U. Possibly, the dependence of catalytic activity upon U is reduced due to the compensating effect of another factor, namely, a decrease of crystalline lattice parameter from 5.35 Å for KF to 4.31 Å for TiC (opposite in sign to the coefficient r in Table IV).

In almost all cases, when dehydrogenation and dehydration of isopropyl alcohol were studied concomitantly on a given oxide, E for dehydrogenation was found to be smaller than E for dehydration [78, 347–349], and k_0 for dehydrogenation was found to be 1–2 orders of magnitude smaller than k_0 for dehydration. The values of k_0 for the dehydration of alcohols (10 dehydration of alcohols (10^4–10^6 $min^{-1} \cdot m^{-2}$) agree with those calculated from the theory of absolute reaction rate, provided the activated complex proceeds by translational motion and therefore loses some of its rotational degrees of freedom. Apparently, an adsorbed molecule of alcohol rotates around the axis of its linkage to the catalyst. For such a case, the reaction rate is independent of lattice parameter. Conversely, the smaller values of k_0 for the dehydrogenation of alcohols originate from the additional loss of rotational degrees of freedom in the activated complex, this latter being a result of two- or multipoint bonding to the catalyst. This fact is in harmony with the principles of catalyst selection (considered above), and implies a correlation between catalytic activity and crystalline lattice parameter for the dehydrogenation of alcohols. The data of Rienäcker [372], which show a strong effect of crystal orientation on the catalytic activity of germanium for the dehydration of alcohols, may be explained in terms of two-point adsorption of the alcohol. According to these data, the reaction rate k on various crystal faces follows the order $k_{100} > k_{110} > k_{111}$; and the order for activation energy is $E_{111} > E_{100} > E_{110}$.

Some insight into the mechanism of alcohol decomposition can be gained from considering the principles of catalyst selection. For example, it is safe to say that in adsorption leading to the dehydrogenation of alcohols, a central role is played by the electronegative surface atom. This hypothesis finds support in: (1) high catalytic activity of solid bases, (2) increased heats of adsorption for the alcohol and acetone upon preadsorption of oxygen on the oxide [373], and (3) pertinent spectroscopic [374] and kinetic data.

On ZnO [375], adsorption of an alcohol on an oxygen atom takes place as follows:

$$\text{iso-}C_3H_7OH + \text{---Zn---O---Zn---} \rightarrow H_3C-\underset{\text{---Zn---O---Zn---}}{\overset{CH_3}{\underset{|}{C}}}-O-H + H \qquad (59a)$$

In order to differentiate from two-electron bonding in molecules, the bonding between O and Zn atoms in the surface layer is shown, arbitrarily, as dashed lines. The reaction step is followed by desorption of the reaction products:

$$H_3C-\underset{\text{---Zn---O---Zn---}}{\overset{CH_3}{\underset{|}{C}}}-O-H \xrightarrow{-H} H_3C-\underset{\text{---Zn---O---Zn---}}{\overset{CH_3}{\underset{|}{C}}}-O \rightarrow CH_3COCH_3 + ZnO \qquad (59b)$$

or

$$\underset{H}{\overset{H_2}{\diagdown}}C-\underset{\text{---Zn---O---Zn---}}{\overset{CH_3}{\underset{|}{C}}}-O-H \xrightarrow{-H} H_2C=\underset{\text{---Zn---O---Zn---}}{\overset{CH_3}{\underset{|}{C}}}-O-H \rightarrow CH_3COCH_3 + ZnO \qquad (59c)$$

The energy of activation is greater for the desorption of acetone than for the desorption of hydrogen.

4.1. DECOMPOSITION OF ALCOHOLS

Data on the effect of doping on catalytic activity and reaction kinetics indicate that the first stage of the dehydrogenation of isopropyl alcohol is accelerated by conduction electrons (reaction rate increases with increased Fermi level) while the last stage, desorption of acetone, is enhanced by the holes (rate increases with decreasing Fermi level). Increased catalytic activity with increasing φ (see Table IV) and very high catalytic activity of p-type semiconductors (see Fig. 2) prove that, in most cases, desorption of acetone represents the rate limiting step. However, during adsorption of alcohol on ZnO, the work function φ increases (see Fig. 5); while, under the influence of alcohol, the electrical conductivity increases in the case of electronic semiconductors and and diminishes in the case of hole-type semiconductors. These data show that adsorbed alcohols act as electron donors. This contradiction between data regarding the principles of catalyst selection and those on changes in φ and σ during alcohol adsorption may be explained by assuming that catalysis occurs on surface defects [16, 375]. For example, a surface oxygen atom of ZnO may capture an electron from the bulk and form a $O^{\delta-}$ center. Upon adsorption of an alcohol on it, according to the scheme shown in (59a), the charge δ^- transfers onto the adsorbed molecule. If the local level of the adsorbed molecule $[CH_3COHCH_3]^{\delta-}$ resulting from this process is lower than the local level of the surface nonadsorbing level $O^{\delta-}$, then, during chemisorption, the alcohol will act as if it were an electron donor (even though it is, in essence, an electron acceptor).

Surface levels on germanium-type semiconductors originate from upright projected s-orbitals, which are energetically favored over p-orbitals and are capable of capturing an electron [376]. Possibly, the failure of Frolov and Radshabli [377] to observe any difference in catalytic activity of n- and p-germanium (crushed under vacuum) for the dehydrogenation of isopropyl alcohol, may be attributed to the creation of too large a number of surface defects during crushing.

The significance of surplus charges on surface atoms of the basic solids in respect to alcohol dehydrogenation was discussed in Section 2.1.

In accordance with the information on the principles of catalyst selection, the dehydration of alcohols occurs probably on Lewis acid centers via carbonium–oxonium mechanism; for example, on Al_2O_3 [378–379],

$$
\begin{array}{c}
CH_3CHOHCH_3 \\
+ \\
-O-Al-O-Al-
\end{array}
\rightarrow
\begin{array}{c}
\overset{\delta+}{H_3C-CH-CH_3} \\
| \\
\delta^-\text{OH} \\
\ddots \\
-O-Al-Al-
\end{array}
\rightarrow
\begin{array}{c}
H_2C=CH-CH_3 \\
\downarrow \\
H^{\delta+} \quad H \\
\diagdown_{\delta-}\diagup \\
O \\
\ddots \\
-O-Al-O-
\end{array}
\qquad (60)
$$

and this is followed first by desorption of propylene and then of water which is more strongly bonded to the catalyst. Cyclic complexes, such as those discussed in Section 3.1, do not form on the majority of alcohol dehydration catalysts, since this would result in the dependence of catalytic activity on the crystal lattice parameter of the oxide catalysts.

There is no general agreement among those who have studied the catalytic dehydration of alcohols in heterogeneous systems about which atom of alcohol splits off hydrogen. Possibly, in the majority of cases, there occurs a trans-elimination in accordance with the scheme (60) and the Ingold rule [12] for direction of electronic displacement in a molecule, resulting from elimination of HX from RH. However, it has been found [380] that the dehydration of secondary butyl and other secondary alcohols over ThO_2 yields only α-olefins while dehydration of such alcohols over oxides of W, Mg, Al, Ti, and Zr yields a mixture of α- and β-olefins. It is plausible that, in this case, the formation of a cyclic complex is likely, as proposed in ref. [380]

$$\begin{array}{c} -\text{Th}-\text{O}- \\ \text{H}-\text{O} \quad \quad \text{H} \\ \text{H}-\text{C}-\text{C}-\text{H} \\ \text{H}_2\text{C} \quad \quad \text{H} \\ \text{R} \end{array} \qquad (61)$$

The reason for the difference between catalysis over ThO_2 and other oxides is not clear.

An interaction of the alochol molecule with the excess charge on Al (surface local level, "positive free valence") leads to the formation of a strong bond which impedes catalysis:

$$\begin{array}{ccc} CH_3CHOHCH_3 & & CH_3-CH-CH_3 \\ + & & | \\ | \, \delta+ & \rightarrow & O \quad H^{\delta+} \\ -O-Al-O-Al- & & O-O-Al-O-Al- \\ | & & | \end{array} \qquad (62)$$

Conversely, cancellation of this charge on a metal atom by the action of, e.g., conduction electron, may result in increased concentration of active centers like those shown in scheme (60). This mechanistic picture of the effect of electrical conductivity on catalytic activity in heterolytic reactions is discussed in Section 2.2.

Bronsted acid sites also play a role in the dehydration of alcohols, but the preponderant part of the catalytic activity of oxides for this reaction is to be ascribed to aprotic centers.

4.1. DECOMPOSITION OF ALCOHOLS

The selection of the isopropyl alcohol decomposition reaction was dictated by the availability of suitable data from the point of view of principles of catalyst selection. The literature on decomposition of other alcohols is considerably more scarce. Despite a large number of studies on the dehydrogenation of ethyl alcohol, few studies are reliable and few give rate constants based on unit surface area of catalyst. In general, the principles of catalyst selection for decomposition of isopropyl alcohol apply also to ethyl alcohol. As an example one might cite Fig. 44,

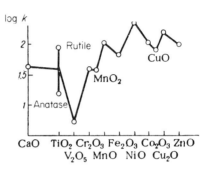

FIG. 44. Variation of the logarithm of catalytic activity of fourth period metal oxides for ethyl alcohol dehydrogenation.

which shows the variation in catalytic activity of oxides of elements of the fourth period for the dehydrogenation of ethyl alcohol, obtained from studies [322, 347, 354, 364–367] using the criteria cited above, for isopropyl alcohol dehydrogenation. As in the case of isopropyl alcohol (Fig. 37), a deep minimum of activity occurs at V_2O_5, and a maximum at NiO. Also, other regularities are preserved here, e.g., for ethyl as well as for isopropyl alcohols, the catalytic activity increases in the series MgO, ZnO, CdO with increasing atomic weight of the metal. In both cases, oxides of alkali earth metals exhibit considerable activity. Because of the scarcity of data, no detailed correlative analysis could be made for ethyl alcohol.

Also in respect to dehydration reactions, the situation for both of these alcohols is identical. Acidic oxides display considerable catalytic activity for dehydration of ethyl alcohol; and this declines with increasing electronegativity Δx.

The regularities of the change of catalytic activity of the oxides of rare earth elements, e.g., oxides of lanthanides, present an interesting question. It was shown in Table III that oxides such as La_2O_3 or CeO_2 are good catalysts for dehydrogenation of alcohols. With increasing atomic number of the element (or with increasing number of 4f-shell electrons) forming an oxide, certain properties of rare earth compounds change in a continuous manner (e.g., lattice parameter increases from

La$_2$O$_3$ to Lu$_2$O$_3$); and other properties change periodically, due to the peculiar features of the electronic configurations f^0, f^7, and f^{14}. In the the first seven elements from Ce to Gd, the spins of f-electrons are parallel; and in the remaining elements from Tb to Lu, the $4f$-shell becomes gradually filled up with electrons with antiparallel spins. This former group of elements represents the so-called cerium subgroup; and the latter represents the yttrium subgroup (this name was assigned to those elements because some of their properties, such as, e.g., ionic radius, resemble those of yttrium).

Several studies were devoted to decomposition of ethyl alcohol over oxides of rare earth elements. Because of contradictory conclusions, no chart of variation of catalytic activity could be drawn up for these oxides. According to some works [347, 366], catalytic activity of the oxide increases with increasing atomic weight of the rare earth element, some other studies [322, 368] claim the opposite. Nevertheless, in all these studies, there prevails an observation that most the catalytically active are oxides of those metals which readily assume higher oxidation states (Ce^{4+}, Pr^{4+}).

Data are very scarce for composition of other alcohols, but the principles of catalyst selection are, in general, similar to those derived for isopropyl and ethyl alcohols. Dehydrogenation of butyl alcohol studied over various oxides supported on MgO [381], occurs at a high rate over ZnO, Cr$_2$O$_3$, and Mn$_3$O$_4$, dielectric oxides being least active. The study on dehydration of butyl alcohol [382] over zeolitic catalyst showed the following activity sequence: Al$_2$O$_3$ > LiNaX > NaX > KNaX > RbNaX. These data allow one to establish a correlation between catalytic activity and e/r^2 of a catalyst. In dehydrogenation of isoamyl alcohol [383], ZnO proved to be the most active of all oxides examined in this study. Investigations of dehydration of propyl alcohol [299] and of isoamyl alcohol [299, 384] indicated oxides with acidic character to be most active. The acidity was found [384] to affect not only the product yield but also the reaction selectivity, e.g., the yields of the individual pentene isomers obtain by dehydration of 3-pentanol. In dehydration of ethylenecyanhydrin to acrylonitrile [385], the acidic and salt-type catalysts were found to be preferable to the semiconductor catalyst—zinc oxide.

4.2. Decomposition of Acids

Formic acid decomposition has frequently been chosen as a model reaction for studying the principles of catalyst selection. This reaction may proceed into two directions:

4.2. DECOMPOSITION OF ACIDS

$$\text{HCOOH} \begin{cases} \nearrow H_2 + CO_2 & \text{dehydrogenation} \\ \searrow H_2O + CO & \text{dehydration} \end{cases}$$

In general, mechanistic considerations, discussed above, for alcohol decomposition apply also to the decomposition of formic acid. There is a widespread view that to dehydration reaction acidic mechanisms apply and to dehydrogenation, electronic oxidation–reduction mechanisms apply.

Most studies on the dehydrogenation of formic acid employed metallic catalysts. Literature on dehydrogenation of HCOOH over nonmetallic catalyst is rather limited [83, 84, 386–391]. Most of these studies dealt with catalysis over oxides of the Group IV metals. The conclusions reached in these studies regarding principles of catalyst selection are not consistent, because of differences in the experimental methods used. The following series of activity of oxides can be compiled from the most credible information sources [388–391]: $NiO > ZnO > Fe_2O > Co_2O_3 > Mn_2O_3 > CuO > MgO > Cr_2O_3 > \alpha\text{-}Al_2O_3 > ThO_2 > \gamma\text{-}Al_2O_3 > TiO_2 > SiO_2$. For the temperature range of 300–400°C., the activity of these catalysts based on 1 m² of catalyst surface varies within 6 to 7 orders of magnitude. As in the case of alcohol decomposition, in general, catalytic activity for dehydrogenation of formic acid increases as one approaches the end of the fourth period, or NiO and ZnO. Relative to other oxides, catalytic activity of Fe_3O_4 (and Fe_2O_3) is greater here than in the case of alcohol dehydrogenation; MgO is more active than such semiconductor catalysts as Cr_2O_3 or ThO_2. In general, basic catalysts are active for dehydrogenation of HCOOH while acidic ones are either inactive or only slightly active.

Studies by Schwab et al. [83] and Penzkofer [84] on catalysis over compounds of the type A^{IV} and $A^{III}B^{V}$ were discussed previously (Section 1.3). According to those authors, p-type semiconductors are more active than n-type semiconductors. Actually, the following activity series evolved from these studies: $n\text{-AlSb} > p\text{-AlSb} > p\text{-InSb} > p\text{-Ge} \sim p\text{-InAs} > n\text{-InAs} > p\text{-Si} \sim i\text{-Ge} > n\text{-InSb} > n\text{-Ge} > n\text{-Si}$. Except for AlSb, there actually exists, in this case, a difference in activity between n- and p-type semiconductors. This led to the postulation of the "donor"-type mechanism for dehydrogenation of formic acid. Some investigators of formic acid decomposition [83, 84] assumed it to be zero-order reaction, and, on this basis, determined the "true" activation energy. The difference in the energies of activation on n- and p-type samples, $\Delta E_{n,p}$, was found, for Ge and AlSb, to be equal to one-half of the width of forbidden zone $U/2$. As shown in Section 1.2, this can be explained by assuming that in p-type semiconductors the Fermi level is very close to the valence zone, while in the case of n-type semiconductors, it is situated in the middle of the forbidden zone; in this latter case, natural conductivity occurs. For InAs, $\Delta E_{n,p} \approx U$. In the case of InSb, $\Delta E_{n,p} > U$, which is difficult to rationalize in view of the understanding of the role of Fermi level in catalysis. It should be mentioned, relative to difference in the values of E, that the difference in catalytic activity of n- and p-type semiconductors was found to be rather small, not greater than 1–2 orders of magnitude.

Catalytic activity of oxides for dehydration of HCOOH is reported [389, 390] to follow the order: $Al_2O_3 > Fe_2O_3 > TiO_2 > Mn_3O_4 > Cr_2O_3 > SiO_2 > MgO > Co_2O_3$. This order may be explained in terms of a correlation between dehydration activity and acidity of an oxide. It is interesting that, despite low yields of dehydration products at low temperatures, the activity of Fe_2O_3 for both dehydration and dehydrogenation of HCOOH was found to be greatest among the catalysts tested. Apparently, order of selectivity of oxides does not coincide with the order of their activity.

From the similarity of principles of catalyst selection for alcohols and formic acid decomposition, a deduction can be made that one mechanism applies to both of these cases.

It was repeatedly postulated [387, 389–391] that dehydrogenation of formic acid proceeds via formation and decay of surface formates or oxalates:

$$\underset{Me}{\overset{HH}{\underset{OO}{\overset{O\diagdown\diagup O}{CC}}}} \xrightarrow{-H_2} \underset{Me}{\overset{OO}{\underset{OO}{\overset{\|\|}{C-C}}}} \xrightarrow{HCOOH} \underset{Me}{\overset{HH}{\underset{OO}{\overset{O\diagdown\diagup O}{CC}}}} + CO_2 \qquad (63)$$

Temperatures of formic acid decomposition on oxide catalysts correlates with temperature of decomposition of the corresponding formates. In the opinion of this author [see schematic (59)], formate- and acetate-like structures participate in alcohol decomposition in the form of complex intermediates. Deeper oxidized structures, e.g., carbonates, should be involved in the form of intermediates in the dehydrogenation of formic acid:

$$\underset{\cdots MeO^{\delta-}-Me\cdots}{HCOOH+} \rightarrow \underset{\cdots MeMe\cdots}{\overset{OO}{\underset{HOH}{\overset{\diagdown\delta-\diagup}{C}}}} \xrightarrow{-H_2} \underset{\cdots MeMe\cdots}{\overset{\delta-}{CO_3}} \rightarrow \underset{MeO}{CO_2+} \qquad (64)$$

4.2. DECOMPOSITION OF ACIDS

Formation of surface carbonates was observed by IR technique during adsorption of formic acid on MgO. It was debated in Section 4.1, whether to this case there is also an application of previous considerations regarding the effect on catalyst activity of such factors as: electronic structure of a catalyst, the presence of local surface states of various types, and the surface basicity. As far as stability is concerned, carbonates of various metals are similar to the formates.

The mechanism of formic acid decomposition resembles that for alcohol dehydration:

$$\begin{array}{c} \text{HCOOH} \\ + \\ \text{—O—Al—O—} \\ | \end{array} \rightarrow \begin{array}{c} \overset{\delta+}{\text{H—C}}\!\!\diagup\!\!\overset{O}{\diagdown}\!\!\text{H} \\ \overset{\delta-}{\diagdown}\!\!\text{O}\!\!\diagup \\ \text{—O—Al—O—} \\ | \end{array} \rightarrow \begin{array}{c} \text{H} \quad\;\; \text{H} \\ \diagdown \quad \diagup \\ \text{O} \quad\;\; \text{O} \\ \text{—O—Al—O—} \\ | \end{array} + \text{CO} \qquad (65)$$

In the case of acetic acid, the ketone-yielding reaction $2CH_3COOH \rightarrow CH_3COCH_3 + CO_2 + H_2O$ corresponds to the decomposition of surface acetate complexes, such as is shown on scheme (63):

$$\text{Me}\!\!\diagup\!\!\overset{\diagup\text{O}\diagdown\text{CO—CH}_3}{\diagdown\text{O—CO}\diagdown\text{CH}_3} \qquad (66)$$

Decomposition of complexes analogous to scheme (64) corresponds to the following methane-yielding decomposition:

$$CH_3COOH \rightarrow CH_4 + CO_2$$

The rules controlling changes in catalytic activities of oxides and carbonates in both of these acetic acid decomposition reactions were studied by Rubinshtein and Yakerson [392, 393]; Yakerson and Fedorovskaya [394, 395]; and Yakerson et al. [396].

The methane-yielding decomposition of acetic acid [392] occurs at 400°C. The rate of this reaction increases in the series $Na_2CO_3 < K_2CO_3 < Rb_2CO_3$, it means it increases with alkalinity of the catalyst.

In the case of ketone-yielding decomposition of CH_3COOH [393], an increase in covalent character of the Me—O bond in the complex [scheme (66)] results in decreased activation energy due to strong electron displacement from methyl to carbonyl group. In Fig. 45, it is shown that only in the individual series such as MgO, CdO, ZnO, $BaCO_3$, $SrCO_3$, and $CaCO_3$ may one speak about the relation between activation energy and the degree of bond covalency (DBC). (The manner in which the crystalline lattice of a carbonate or oxide catalyst may influence the structure of the activated complex is not clear from its schematic representation alone.) Large activation energies for the ketone-yielding decomposition of CH_3COOH on oxides and carbonates of Li, Na, K, Rb, Ca, Sr, Ba, Mg, Zn, and Cd have been explained in terms of the intermediate formation of a finite phase—namely, acetates of the respective metals. On CeO_2, ZrO_2, TiO_2, and SnO_2,

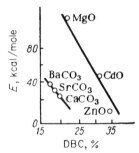

FIG. 45. Dependence of activation energy of ketone-yielding activation of CH_3COOH upon the degree of bond covalency (DBC) [393].

the acetic acid decomposition is in the order of decreasing activation energy, while, on BeO, no connection was found between the energy of activation of the catalytic reaction and the energy of activation for decomposition of the adsorbed acetates. For this latter case cyclic activated complexes of the second type were suggested. For example, according to [392], on oxides of MeO_2 type, the complex involving two molecules of acetic acid becomes activated as a result of hydrogen bonding between surface OH-groups and the carbonyl oxygen of the acetic acid molecule. This is, however, rather doubtful since the high reaction temperatures used (300–500°C) suggest the absence of activation via hydrogen-type bonding.

It was shown in Sections 1.1 and 1.2, that surface OH-groups exhibit only weak acidic or basic properties and generally do not represent catalytically active sites. At 300–500°C they may also not serve as adsorption centers.

The mechanism of ketone-yielding CH_3COOH decomposition reaction is very different on different catalysts. For example, on MeO_2-type catalysts a linear relationship was observed between reaction rate and the value of specific surface area of the catalyst, while on MeO-type catalysts, no such relationship was observed at all. Nevertheless, certain general guiding rules in respect to this reaction may be formulated.

As shown in Fig. 46, there exists for ZnO a tendency for the activation energy (E) for this reaction to decline with increasing Δx and d. Simultaneous occurrence of these two regularities leads to a conclusion that

4.2. DECOMPOSITION OF ACIDS

in the activated complex, metal ions participate along with basic surface centers, e.g., O^{2-}. So far as the degree of bond covalency is concerned, the general opinion is that there is a tendency for E to increase with increasing DBC (or decreasing Δx); and this is the reverse of that deduced in ref. [393], which considered only a limited group of catalyst systems. The actual differences in catalytic activity of the oxides was found to be small, since the change in E was compensated for by an even greater reverse-acting change in log k_0. For example, despite very high activation energy (114 kcal/mole), MgO was found to be one of the active catalysts.

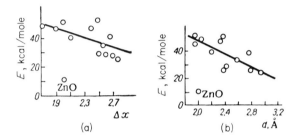

FIG. 46. Dependence of energy of activation of ketone-yielding decomposition of CH_3COOH upon (a) electronegativity differential, and (b) distance between metal and oxygen atoms in the oxide.

Decompositions of formaldoxime, according to Schwab and Leute [397], may follow two directions:

$$H_2C=NOH \begin{matrix} \nearrow HCN + H_2O \\ \searrow NH_3 + CO \end{matrix}$$

In general, dehydration occurs over basic catalysts (MgO, CaO) and decarbonylation on acidic catalysts (Al_2O_3, Fe_2O_3). Apparently, the mechanism of formaldoxime dehydration is different from the mechanism of alcohols and acids dehydration, which proceeds readily over solid acids.

5 • Dehydrogenation and Hydrogenation Reactions

5.1. Dehydrogenation of Hydrocarbons

Many investigators of the dehydrogenation of organic compounds have noted that different rules apply to catalysts for dehydrogenating hydrocarbons than for alcohols. According to the multiplet theory of Balandin, both cases requires a duplet complex (if one does not consider dehydrogenation of six-membered cyclic compounds), but the "reaction index" (defined as the adsorbing and reacting fragment) is

$$\begin{array}{cc} C & O \\ | & | \\ H & H \end{array}$$

for alcohol dehydrogenation while for hydrocarbon dehydrogenation it is

$$\begin{array}{cc} C & C \\ | & | \\ H & H \end{array}$$

Dehydrogenation of butane to butylenes constitutes the major industrial route to monomers for rubber synthesis. Most common catalysts for this reaction are oxides of the sixth-group metals [398]. Among the most active are Cr_2O_3 and V_2O_5, which in industrial practice, are supported on Al_2O_3. Apparently, the d^2–d^3 cation structures are more active for this reaction than any other electronic configurations. Treatment of Cr_2O_3 with hydrogen [399] transformed its conductivity from p- to n-type and led to increased catalytic activity, but this increase was not commensurate with change in conductivity.

5.1. DEHYDROGENATION OF HYDROCARBONS

Cr_2O_3 is also a good catalyst for *dehydrogenation of pentane, isobutane, propane, and hexane*. With increased atomic weight in Group VI, i.e., on going from Cr_2O_3 to the oxides of Mo and W, catalytic activity declines.

Data from the literature do not indicate whether the d^3 structure is a unique configuration for paraffins dehydrogenation. It is possible that a second maximum of dehyrogenation activity exists among the oxides of transition metals at the d^7–d^8 configuration, as was found for H_2–D_2 exchange. This second maximum was explained in terms of crystal field theory (Section 1.6). At butane dehydrogenation temperature range (550–650°C), NiO (d^8) is reduced by the hydrocarbon media. Recently, however, nickel phosphate has been employed as a catalyst for dehydrogenation of paraffins.

Dehydrogenation of cyclohexane frequently serves as a model reaction, in particular when interest centers on the effect of the crystalline lattice structure on catalytic activity. However, in the studies of the rules for catalyst selection for cyclohexane dehydrogenation, only occasionally were the effects of changes in catalyst surface area or diffusional limitation considered. For the purpose of analyzing the literature data from the viewpoint of rules for catalyst selection, it was useful to use the statistical method presented in Section 4.1.

The results of the calculation are shown in Table V. The data on catalysis over oxides and sulfides were drawn from [175, 303, 347, 348, 400–405]. Activity of CrO_3 at 500°C served as a standard, since many data included this case.

Data from Table V were employed by this author in an attempt to relate catalytic activity to various properties of the solids. The results of this attempt are presented in Figs. 47 and 48. Solid points on the graphs represent data whose reliability is 3 or more.

An absence of dependence between catalytic activity and metal–metal spacing is shown in Fig. 47. An analogous conclusion follows from analysis for a correlation between catalytic activity for cyclohexane

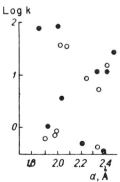

FIG. 47. Absence of correlation between catalytic activity for cyclohexane dehydrogenation and the spacing between metal and nonmetal atoms.

TABLE V

Activity of Catalyst for Dehydrogenation of Cyclohexane

Catalyst	log k	$\sum w$	E, kcal/mole
α-Al_2O_3	−0.21	2	43.0
CaO	−0.46	3	30.5
CdO	<0	1	35.6
CeO_2	1.11	7	35.0
Co_3O_2	1.96	3	25.0
Cr_2O_3	2.00	16	25.4
La_2O_3, Pr_2O_3, Nd_2O_3, Sm_2O_3, Cd_2O_3, Ho_2O_3, Er_2O_3, Tu_2O_3, Yb_2O_3	1.05–1.22	7	35.7–62.2
Mn_2O_3	−0.27	11	32.2
MoO_3	<0	1	46.5
MoS_2	0.78	1	17.7
$NiWO_2$	1.65	3	26
Pt[a]	3.05	3	19.3
ThO_2	1.24	2	38
TiO_2	0.04	5	29.1
V_2O_3	0.60	5	31.3
WS_2	1.50	4	20.0
Y_2O_3	0.98	2	57
ZnO	<0	1	48.6
ZrO_2	1.68	1	

[a] Data for Pt are given for comparison with the oxides.

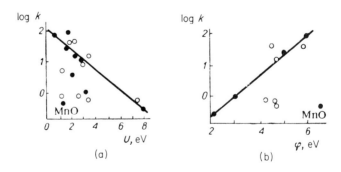

FIG. 48. Correlation between catalytic activity for cyclohexane dehydrogenation and (a) width of the forbidden zone, and (b) work function of the catalyst.

5.1. DEHYDROGENATION OF HYDROCARBONS

dehydrogenation and the spacing between two metal atoms on an oxide support, as should be expected from the duplet concept (Section 3.1). Obviously, no correlation between catalytic activity and lattice structure can be expected if one assumes a sextet-type mechanism of dehydrogenation (see Fig. 32). Cubic and hexagonal crystal, which can conveniently accommodate six-membered cyclic species on their octahedral faces, can be regarded as poor catalysts. Their catalytic activity is inferior to that of oxides with low structural symmetry. A comparison of E (from Table V) with lattice parameter revealed that a decrease in lattice parameter is frequently followed by a reduction in E.

Similarly, no relationship was detected between catalytic activity and either electronegativity difference Δx or dielectric permeability E. A comparison of $\log k$ with the width of forbidden zone U and work function φ, see Fig. 48, is more encouraging: $\log k$ increases with diminishing U and increasing φ. In respect to the latter dependence, p-type semiconductors are more active than n-type semiconductors. None of the oxides with high U and small φ exhibited high catalytic activity. Overall, these correlations are hardly quantitative; and they should only be regarded as rough guidelines.

In considering a relationship between catalytic activity of oxides and the location of the metal in the periodic system, one observes a distinctly high catalytic activity of Cr_2O_3, and this is superior to the activity of the neighboring oxides of the elements of the fourth period (TiO_2, V_2O_3, MnO, ZnO) as well as to oxides and sulfides of the elements of Group VI (Mo, W). The Cr^{+3} cation in Cr_2O_3 has a d^3 electronic configuration. The presence of Cr^{+3} ions (d^3) on the surface of chromia–alumina dehydrogenation catalysts was shown by Shvets and Kazanskii [406] by ESR and differential reflection spectroscopy of powdered samples in the visible range. Van Reijen et al. [407] proposed Cr^2 cations (d^4) on the CrO_3 surface as active dehydrogenation locations and cite ESR spectra of chromia–alumina catalysts and magnetic susceptibility measurements as a basis for this proposal. These same authors name cations with d^1 configuration (V^4 and Mo^5) as active centers in V_2O_5 and MoO_3, both being mediocre catalysts.

Oxides of Co and Ni, with d^6–d^8 cation structures undergo reduction at the experimental conditions used for cyclohexane dehydrogenation. Dixon et al. [175] studied dehydrogenation and disproportionation of cyclohexane over oxides of transition metals and simultaneously measured magnetic properties of these oxides. At low temperature (160°C) (too low to reduce Co_3O_4 to metal), Co_3O_4 exhibited catalytic activity equal to that of Cr_2O_3. The variation in catalytic activity by oxides of metals of the fourth period for the cyclohexane dehydrogenation is shown in Fig. 49. Data for Co_3O_4 and $NiWO_3$ were taken from the study [175] of dehydrogenation and disproportionation of cyclohexane. As shown, there are two activity peaks for cyclohexane dehydrogenation over this series of oxides. Also, changes in activaton energy E are presented in Fig. 49. Catalytic activity of various oxides basically follows the changes in activation energy E, while log k_0 remains almost constant.

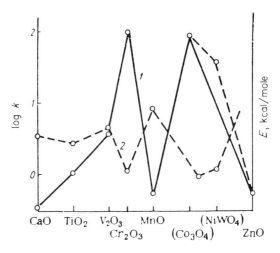

FIG. 49. Change in (1) catalytic activity and (2) activation energy for cyclohexane dehydrogenation over a series for oxides of metals of the fourth period.

The chromatographic pulse technique has been employed to study cyclohexane dehydrogenation over compounds of metals with B, C, and N [348]. Most compounds of this type exhibit metallic conductivity. The order of activity of these compounds in cyclohexane dehydrogenation is: $MoS_2 > WC > Mo_2C > V_3Si > ZrC > Mo_3Si > TiC > CrB_2 \sim LaB_6 > TiB_2 > GdB_6 > W_2C > W_2B_5 > ZrB_2$.

Further studies [408] have shown that, among the refractory metal compounds that have been investigated, the highest catalytic activity for dehydrogenation is exhibited by compounds of chromium: carbides such as $Cr_{23}C_6$, Cr_7C_3, and Cr_3C_2; sulfides such as Cr_2S_3; silicides such as Cr_3Si, Cr_5Si_3, and $CrSi_2$; nitrides such as CrN and Cr_2N; and phosphide, CrP. Cyclohexane dehydrogenation over these catalysts commences at 300–350°C. Among chromium compounds, only borides—CrB, CrB_2, and CrB_6 proved catalytically inert, while Cr_4B showed some activity due to chromium lattice distortions caused by B atoms. Also, TiC, WC, $MoSi_2$, Mo_3Si, V_3Si, and ReSi showed catalytic activity commencing at 300–400°C. About 400–500°C was required for cyclo-

5.1. DEHYDROGENATION OF HYDROCARBONS

hexane dehydrogenation over Mo_2C and ZrN. Almost all the investigated borides as well as $TiSi_2$, TiN, VSi_2, TaN, HfC, NbC, NbN, and W_2C showed very little or zero catalytic activity. As expected, carbides proved to be either inert or only slightly active.

Investigation of high-temperature conductors showed a relationship between catalytic activity and d-electron levels. Nitrides, and silicides of transition metals were found to be catalytically slightly active or inactive. Lack of reliable information on the occupancy of d-shells with electrons in these compounds prevents drawing any conclusions about the actual nature of the relation between catalytic activity and the number of d-electrons. In [348, 408], the increase in catalytic activity in the carbide series TiC, ZrC, and Mo_2C, was explained in terms of decreasing statistical filling of d-levels of transition metal atoms with electrons of the carbon. The more free d-levels (not involved in chemical bonding) there are in a compound the more the catalytic activity for dehydrogenation. According to Samsonov [409] the unfilled d-levels are characterized by the ratio $1/Nn$ (where N is the principal quantum number of the d-level and n is the number of electrons at this level). For TiC, ZrC, and Mo_2C, the values of $1/Nn$ are 0.167, 0.125, and 0.050, respectively. Further increase in activity of $MoSi_2$ may be due to the presence of a significant contribution of the covalent Si—Si bonds and to the corresponding decline in the probability of filling molybdenum d-levels with electrons. In a transition from $MoSi_2$ to Mo_3Si, the degree of cyclohexane conversion declines rapidly, due to diminishing contributions of Si and Si—Si bonds and to the resultant strong effect of the valence electrons on the d-levels of molybdenum. The relatively high activity of V_3Si results from the high degree of vacancy in the d-orbitals of the V atom.

According to recent data [410, 411], however, d-electrons are not prone to participate in chemical bonds in carbides and nitrides (partially also in silicides); and, therefore, no "metallization" of bonds by filling the d-level of a metal with electrons of the nonmetal occurs. Analysis of the fine structure of the X ray K-spectra revealed that in these compounds, the s- and p-electrons of the metal as well as of the nonmetals participate in the valence interaction. Therefore, catalytic activity of carbides, nitrides, and silicides of various metals can be related to the number of d-electrons in the metal ion (which is deprived of its two s-electrons). Except for chromium borides, all the data indicate very high catalytic activity of the compounds of the metals of Group VI (Cr, Mo, and W); this applies particularly to chromium compounds. For example, according to statistical data [410, 411], the vacant d-levels of borides are filled with electrons of the nonmetal (in this case boron, due to its low ionization potential). Filling of d-levels with electrons represents, therefore, a drawback to the occurrence of catalytic activity.

These same studies [410, 411] support the notion that electrons in hydrides transfer completely from hydrogen to the transition metal atom. One may presume that, during C_6H_{12} and H_2 adsorption, the electrons are also transferred to the d-level of the transition metal, and that adsorbate molecules assume a partial positive charge. A displacement of the electron toward the catalyst is also confirmed by the data on the change in the work function [53, 60] observed during adsorption of hydrocarbons on metals and semiconductors.

The data obtained by this author and co-workers [348] on changes in catalytic activity and energy of activation for cyclohexane dehydrogenation of a series of isoelectronic compounds of NaCl-type crystal lattice (KF, CaO, ScN, and TiC) are presented pictorially in Fig. 50.

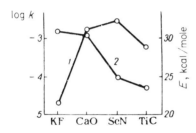

FIG. 50. Dependence of (1) logarithm of catalytic activity and (2) energy of activation, for cyclohexane dehydrogenation at 500°C over KF CaO, ScN, and TiC.

Similarly, in the case of isopropyl alcohol decomposition (Fig. 43), a maximum activity at ScN occurs. A sharp increase in catalytic activity occurs in a series from KF to CaO and it does not take place in the series from CaO to ScN, as would be expected if there existed a dependence between energy of activation E and the width of the forbidden zone. A special study proved that the reduction of catalytic activity in a series from ScN to TiC is not due to formation of a TiO_2 layer. The TiO_2 is 1.5 orders of magnitude less active than TiC.

Excluding TiC, the magnitude of log k_0 for this entire series is similar (4.5, if k is expressed in $min^{-1} \cdot m^{-2}$). This value of log k_0 indicates that the activated complex lost only a fraction of the rotational degrees of freedom. In this connection, Rooney [412] concluded that dehydrogenation of cyclohexane over Pd proceeds via the formation of π-complex (Me denotes a metal atom):

5.1. DEHYDROGENATION OF HYDROCARBONS

$$C_6H_6 + Me \rightleftharpoons \cdots \rightleftharpoons \cdots \qquad C_6H_8 + Me \rightleftharpoons \cdots \qquad (67)$$

$$C_6H_{12} + Me \rightleftharpoons \cdots \qquad C_6H_{10} + Me \rightleftharpoons \cdots$$

It may be assumed that complex similar to the schematic of (67) may form also on nonmetals; and that they preserve some possibility of rotation. This assumption is supported by the absence of a correlation between catalytic activity and lattice parameter, as shown in Fig. 47, and by the coincidence of catalytic properties of metal compounds with different lattice structure but the same cation (e.g., various compounds of Cr).

As shown in Table V, rare earth elements show considerable catalytic activity for dehydrogenation of cyclohexane and other hydrocarbons. Oxides of these elements have similar properties. A variation in catalytic activity of rare earth elements with respect to cyclohexane dehydrogenation [404], tetralin dehydrogenation [363], and pentane dehydrogenation [413] is shown in Fig. 51. The ordinate for curve (2) represents energy of activation in reverse order, because [363] does not contain information on rate constants. Apparently, actual changes in k are substantially smaller than would be expected from the observed large changes in E. Despite some indication of periodicity in the sequence from cerium to yttrium (Section 4.1), the graph (Fig. 51) shows a generally monotonic increase of activation energy with increasing atomic number. It fails to show any influence of the electronic structure of the 4f-orbitals. According to Tolstopyatova and Balandin [363], this variation in activation energy is a result of increasing lattice parameter in the sequence $La_2O_3 \rightarrow Yb_2O_3$.

Regularities in the changes of catalytic activity for *tetralin dehydrogenation* [347, 414] are analogous to those for cyclohexane dehydrogenation. The order of catalytic activity for the oxides of transition metals is: $Cr_2O_3 > ZrO_2 > HfO_2 > Y_2O_3 > Sc_2O_3$.

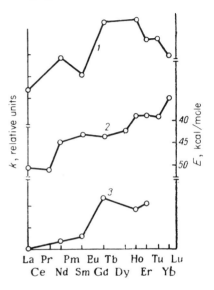

FIG. 51. Variation of (1) catalytic activity (based on 1 m² of surface area) for cyclohexane dehydrogenation [404], (2) activation energy for dehydrogenation of tetralin [363], and (3) catalytic activity for pentane dehydrogenation, for a series of lanthanide oxides.

Data on *pentane dehydrogenation* [402, 415–417] also indicate high catalytic activity of Cr_2O_3, greater than the activity of any of the investigated metal oxides or other compounds of chromium, e.g., $Cr(OH)_3$ and CrO_3. This attests to the peculiar properties of the d^3 electronic structure. Study [416] gives the following activity series for dehydrogenation: $Cr_2O_3 > MOS_2 > V_2O_5 > MoO_2 > ThO_2 > CeO_2 > VO_2 > ZrO_2$. Catalytic activity of oxides involving cation configuration of d^5 and greater has not been studied for the latter two dehydrogenation reactions.

5.2. Hydrogenation

Among hydrogenation reactions, the hydrogenation of ethylene distinguishes itself by being very popular as a model reaction for studying catalysis mechanism over metals and for investigation rules for catalyst selection. Nonmetallic catalysts have hardly ever been employed in such studies because of their inferior activity compared to metals. Therefore, for the purpose of examining the rules for selecting nonmetallic catalysts for hydrogenation reactions, one must resort to the findings of large-scale

5.2. HYDROGENATION

industrial operations, usually performed at evlevated pressures and complicated by diffusional limitations.

Oxides of transition metals, Cr_2O_3, and the oxides of Mo, W, Fe, and some others can be used for ethylene hydrogenation. All such oxides are inferior hydrogenation catalysts to the transition metals, such as Pt, Pd, Rh, Ni, etc.

Schwab et al. [83] found in both hydrogenation of ethylene and dehydrogenation of alcohols (Sections 1.3 and 4.1) that p-type semiconductors are more active than n-type semiconductors. Apparently, such a conclusion can be reached by ordering semiconductors in the sequence of either increasing values of temperature required for reaction initiation or increasing energy of activation. The activity sequence based on temperature of reaction initiation is: p-Ge $>$ p-InSb $>$ n- and p-InAs $>$ p-Si $>$ n- and p-Ge $>$ n-InSb $>$ n-Si $>$ n-Ge $>$ n-InAs. A different arrangement may evolve from analysis of the values of the rate constants. Using this approach, a study [41] conducted at Schwab's laboratory, showed that the activation energy of ethylene hydrogenation is greater on n-type semiconductors than on p-type semiconductors. Nevertheless, the actual values of rate constants in the temperature range used are very close to each other; and the reaction on n-type semiconductors is frequently found to be faster than on p-type semiconductors. The activity sequence of semiconductors, based on rate constants is: n-InAs $>$ p-InAs $>$ p-InSb $>$ n-InSb $>$ Sb $>$ AlSb.

Analogous results were also obtained [83] for *propylene hydrogenation and for hydrogenation of* CO *to* CH_4 (Table IV). These data do not provide any evidence for a relationship between catalytic activity and the width of the forbidden zone. This lack of evidence for such correlation in both studies [83, 85] may possibly be accounted for by diffusional problems and by the necessity of determining catalyst surface area.

Hydrides of alkaline earth metals proved to be highly active catalysts. Catalytic activity for ethylene hydrogenation was greater with BaH_2 than with CaH_2. Wright and Weller [403] claim that high catalytic activity exhibited by BaH_2 and CaH_2 preclude a two-point (or duplet) mode of adsorption, since the spacing separating metal atoms in these systems is 3.93 Å for CaH_2 and 4.34 Å for BaH_2; both being too large for strainless adsorption of the following type:

$$\begin{array}{c} H_2C-CH_2 \\ |\quad\quad| \\ -Me-Me-Me-Me- \end{array}$$

Also hydrides of transition metals have been investigated [418]. The following activity sequence was established for hydrogenation of styrene:

$FeH_n > NiH_n > CoH_n$ (where $n = 1$–3). In this case, however, the hydrides should be regarded not as nonmetallic binary compounds but as metals containing dissolved hydrogen.

Borides also exhibit metallic conductivity. According to Polkovnikov *et al.* [419], borides of palladium, platinum, and rhodium are more active in *hydrogenation of cyclohexene, cyclopentadiene, and crotonic and cinnamic aldehydes* than the respective metals alone. As shown previously, many borides proved to be dehydrogenation catalysts considerably superior to the pure metals.

Studies of the *hydrogenation of butadiene* to butane and the *reduction of nitrobenzene* to aniline over oxides, sulfides, and selenides of Ni, Zn, and Cr showed [420] a significant increase in activity per 1 m² of surface area in the sequence from ZnO to ZnS and further to ZnSe; and also in the sequence from Cr_2O_3 to CrS and further to CrSe (actually, the reported activity data referred to 1 gm of catalyst). In both series, the catalytic activity increased with increased covalent character of the bond and with decreased width of the forbidden zone of the catalyst. Among nickel compounds, NiS exhibited outstanding catalytic activity. Except for NiSe, compounds of nickel were found to be catalytically superior to the corresponding compounds of Zn and Cr. Thus, all this information did not provide a basis for establishing a clear dependence of catalytic activity upon either the width of the forbidden zone or the lattice parameter.

Relatively few catalysts were examined for *hydrogenation of aromatics* [421–426]. Oxides and sulfides of the sixth group metals (Cr, Mo, and W) were found to be the most catalytically active for this reaction. However oxides of these metals showed higher activity: MoO_2 superior to MoO_3. The high catalytic activity of MoO_2, WO_3, Cr_2O_3, V_2O_5, and V_2S_3 suggests that compounds exhibiting the most dehydrogenation activity involve cations with d^2–d^3 electronic structure. As shown in ref. [423], the d^7 electronic configuration is also advantageous: CoS showed higher activity than V_2S_3, NiS, and FeS. Thus, the order of catalytic activity of oxides and sulfides of transition elements for this application is identical to that for the dehydrogenation reactions (Section 5.1). In the series of oxides Cr, Mo, W, and V, the maximum activity coincides with MoO_2—MoO_3. The same conclusion applies to sulfides; in most studies MoS_2 and MoS_3 were found to be superior catalysts to WS_2 and WS_3. All oxides proved to be less active than their respective metals.

It is obvious that the mechanism of hydrogenation is analogous to the mechanism of dehydrogenation. For example, hydrogenation of benzene may follow schematic (67) but the reverse reaction (from right to left) proceeds via π-complexes. It was repeatedly concluded in

5.2. HYDROGENATION

the literature that there is a correlation between catalytic activity of oxides and sulfides and the structure and lattice parameter of their crystals. As shown in Section 3.1, a duplet mechanism applies to hydrogenation over MoO_2 while a sextet mechanism applies to hydrogenation over MoS_2. However, to verify this conclusion, a study encompassing a wider range of oxides and sulfides is required. It should be mentioned that the data on catalytic activity taken from the above-cited references cannot be considered too reliable since they were not referred to unit catalyst surface area.

Recently, studies were made on catalytic activity of salts of transition metals in solution (i.e., homogeneous catalysts) for hydrogenation of olefins and aromatics. For example, very high catalytic activity was exhibited by acetylacetonates of cobalt and chromium [427] for hydrogenation of cyclohexene in solution. Acetylacetonates of manganese, molybdenum, vanadium, ruthenium, and titanium showed lower catalytic activity. Thus, even in a purely homogeneous system, the two-spiked diagram for catalytic activity of the compounds of the fourth group of metals applies. Obviously, in this case, formation of intermediate π-complexes is well established, and neither duplet nor multipoint active centers play any role.

Hydrogenation of carbon disulfide (more precisely, hydrogenolysis) was studied in detail by Ivanovskii *et al.* [428], using a circulation flow-type technique and calculating reaction rate constants on the basis of 1 m² of surface area. In this case also, CoS showed catalytic activity superior to that of FeS and NiS, which indicates the same catalytic activity pattern for transition metal compounds as that observed in hydrogenation of aromatics. The least active catalyst (FeS) also had the lowest energy of activation (17.4 kcal/mole, while CoS and NiS had 21.6 and 21.3 kcal/mole, respectively) and lowest value of k_0. Dowden and Wells [173] found that the most active of the series of sulfides of transition metals of the fourth period for hydrogenation of CS_2 are sulfides of Ni and Cu.

Industrial production of organic chemicals from CO and H_2 usually employs metallic catalysts, although certain oxides are also capable of yielding high conversions of CO into hydrocarbons and oxygenated compounds. From the rules of catalyst selection for this application [429], it can be deduced that the mechanism for this case is similar to that applying to other hydrogenation reactions. Most active among typical oxide catalysts are MoO_3, WO_3, V_2O_5, and MnO. Less active are SiO_2, MgO, ThO_2, Al_2O_3, and Cr_2O_3. This reaction is also catalyzed by nitrides, carbides, and borides of transition metals. Among borides [430], the order of decreasing activity is: Ni_2B, Co_2B, and Fe_2B.

Cobalt carbide is less active than cobalt metal. In contrast to that, carbides, nitrides, and carbonitrides of iron are more catalytically active than the metallic iron alone [431]. As shown in Section 5.1, for carbides and nitrides, the d-levels of the metal may be partially filled with electrons of the nonmetallic moiety. Obviously, the most advantagous electronic configuration of the cation is d^7, which is analogous to that of Co. In carbides and nitrides of iron the number of filled up d-levels is greater than in the metal iron (d^6); and, therefore, these compounds are more active than the metal itself.

5.3. Decomposition of Inorganic Hydrides

Decomposition of gaseous inorganic hydrides represents a rather narrow class of reactions. Because of their simplicity, they are occasionally used as model reactions for studying the rules of catalyst selection.

Decomposition of ammonia is often used in studies directed to elucidation of the mechanism of ammonia synthesis, an important reaction from the standpoint of chemical technology. Iron catalyst (electronic structure $3d^64s^2$) is an established catalyst in the ammonia synthesis process. In this connection, the question arises as to whether the characteristic features of the d^6 electronic configuration are preserved in binary compounds. Phosphides and nitrides of transition metals with known metallic conductivity were employed in studies on NH_3 decomposition [432, 433]. In some of these compounds, e.g., in nitrides of Ti, Cr, and Mn, the bonds exhibit a strong ionic character [408–411]. The lattice parameter of all these compounds (see the Appendix) is very similar. Therefore, a significant change in activity in a series of these compounds must be related to changes in the electronic properties and not to the surface geometry.

Figure 52 is a graph of changes of the logarithm of catalytic activity (in arbitrary units) of nitrides and phosphides of the metals of the fourth group (at 470°C). Catalytic activity of nitrides is considerably greater than that of the corresponding phosphides. In general, the graphs of activity changes in nitrides and phosphides coincide—a minimum at Ti, Cr, and Co compounds, a maximum at compounds of V, and an even greater maximum at Fe_2N for nitrides and at MnP for phosphides. Electronic configuration of d-levels of the metal in Fe_2N is, apparently, close to d^7 and in MnP to d^6. Thus, the previously mentioned high catalytic activity of the d^6 structure in metals manifests itself also in nitrides and phosphides. An explanation of the first maximum (on

5.3. DECOMPOSITION OF INORGANIC HYDRIDES

compounds of V) shown in Fig. 52 is difficult, without knowing precisely the effective charges in these compounds. Application of the formal ionic scheme of crystal field theory (Section 1.6) is futile in this case because of the preponderance of metallic and covalent bonds.

Changes in activation energy E coincide precisely (although with opposite sign) with changes in the logarithm of the rate constant (at 470°C). Changes in log k_0 of phosphides, are symbatic to changes in E, and this indicates a compensation effect between E for nitrides, log k_0 changes very little, i.e., increase of catalytic activity follows a decrease in E and is not affected by increased k_0.

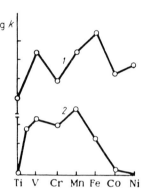

FIG. 52. Change of the logarithm of catalytic activity (at 470°C) of (1) nitrides, and (2) phosphides of transition metals of the fourth period for ammonia decomposition.

There are few data on the catalytic activity of oxides for the decomposition of ammonia. They are less active than metals. Vrieland and Selwood [434] studied NH_3 decomposition over oxides at very high temperature (700–900°C). Considerable activity was exhibited by MnO. Catalytic activity of oxides of rare earth elements (Yb_2O_3, Er_2O_3, Gd_2O_3, Lu_2O_3, Eu_2O_3) was 2 orders of magnitude smaller than that of MnO.

Decomposition of hydrazine proceeds readily over many solids; and it takes two courses [435]:

$$N_2H_4 \rightarrow N_2 + 2H_2 \qquad (68a)$$

$$3N_2H_4 \rightarrow 4NH_3 + N_2 \qquad (68b)$$

In refs. [77, 78, 435] it was shown that the first path is characteristic for solid (and liquid) bases, while the second path is followed on acidic, semiconductor, and certain metallic catalysts. Apparently, in contrast to the decomposition of alcohols (Section 4.1), the acidic catalysts belong to the same group as semiconductors.

The general sequence of activity of hydrazine decomposition catalysts is: metals > semiconductors > solid bases > solid acids > solid salts. Among semiconductors of the isoelectronic series of Ge, there is a regular decrease in catalytic activity with increasing width of the forbidden zone. In Fig. 53, the temperature for reaction initiation serves as a yardstick for catalytic activity of semiconductors: Ge, Ga, As, Ga_2, Se_3, and CuBr. Such a presentation is valid because the surface areas of the samples used were similar. Obviously, this dependence should not be viewed as truly quantitative. Nevertheless, the observed qualitative increase in activity (in two cases [77, 80]) with increasing U indicates that this dependence is not trivial.

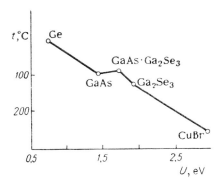

FIG. 53. Dependence of the temperature required to initiate hydrazine decomposition upon the width of the forbidden zone in semiconductors belonging to the isoelectronic series of Ge and its neighboring elements.

So far as catalytic activity for hydrazine decomposition is concerned, oxides of transition metals and NiS are similar to metals and Ge. Also, organic semiconductors involving an atom of a transition metal coordinates with organic ligands—so-called polychelates—exhibit considerable catalytic activity. Among the most active are polychelates of Cu, Fe, Mn, Ni, Co, and Pd; polychelates based on nontransition metals such as Zn and Cd were found to be catalytically inert.

Doping of Ge with a donor (As) and an acceptor (In) did not lead to any change in the rate of hydrazine decomposition: n- and p-type Ge catalysts proved to be equally good catalysts [438]. The same conclusion was reached when NiS was doped with donors and acceptors [436]. Introduction of an acceptor dopant Li_2O into NiO resulted in some increase in the rate of N_2H_4 decomposition. On the other hand, doping with a donor InO_3-poisoned NiO catalyst for N_2H_4 decomposition [436].

Hydrazine decomposition by the second reaction path (68b) is characterized by high stoichiometric coefficients. To account for the fact that this process occurs on the walls of a beaker, Szwarc [439] postulated the formation of a surface trimolecular complex:

5.3. DECOMPOSITION OF INORGANIC HYDRIDES

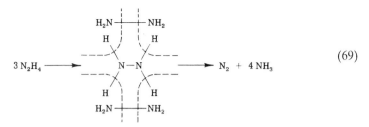

(69)

The necessity of orienting three molecules in a complex such as shown in diagram (69) results in very small values of the pre-exponential factor. Apparently, such a situation occurs at the silica–alumina and silica. On the other catalyst systems (metals, semiconductors), k_0 assumes "normal" values (e.g., on Ge, log k_0 = 24.5–26.5, if k_0 is expressed in molecules/min · cm^2). Also, the activation energy for hydrazine decomposition is lowest on the most active catalysts (a few kilocalories per mole). On these catalysts, decomposition takes place even near ambient temperature.

The concept of surface chain reaction [435] has been employed to account for large stoichiometric numbers, high k_0 values, low E values, and the empirical evidence that the reaction is homogeneous—an extension of the N_2H_4 decomposition from the surface into the solution [440]. Reaction initiation may result from cleavage of N_2H_4 into $NH_2 \cdot$ radicals. Then, the chain reaction takes over:

$$N_2H_4 + NH_2 \cdot \rightarrow NH_3 + N_2H_3 \cdot$$
$$N_2H_4 + N_2H_3 \cdot \rightarrow 2NH_3 + N_2 + H \cdot \quad (70)$$
$$N_2H_4 + H \cdot \rightarrow NH_3 + NH_2 \cdot$$

The chain termination is a result of interaction between two adsorbed radicals.

Studies of electrical conductivity σ and electron work function φ during adsorption and decomposition of hydrazine over semiconductors [435] showed that, during chemisorption, the N_2H_4 molecule is a donor. It assumes a positive charge ($N_2H_4^+$). During the chain reaction (70), it assumes a negative charge; and the reaction is accelerated by electrons. It may be shown, using specific assumptions concerning the chain termination, that the overall reaction rate constant will involve partial constants, one of which will be proportional to the concentration of electrons and the other to the concentration of holes. Thus, as shown in Section 1.3, the reaction rate will not depend on concentration of the dopants, but it will decline with increasing width of the forbidden zone, even in the range of conduction induced by the impurities.

Another possible explanation of the independence of the rate of N_2H_4 decomposition (on Ge and NiS) on both the number and the charge of the current carrier is based on the presence of a large number of surface states during hydrazine adsorption.

It could also be claimed that current carriers are not involved in the elementary steps of N_2H_4 decomposition. However, the aggregate information—high rate of N_2H_4 decomposition on metals and semiconductors, its correlation with U, change of σ and φ during the reaction, as well as reaction extension into solution, conflicts with such a claim and supports the electronic (homolytic) reaction mechanism.

In the case of densely populated surface states on the catalyst during adsorption of N_2H_4 molecules, both radicals resulting from cleavage may be strongly bonded to the surface. It may well happen that the chain propagation steps (70) will have greater energy of activation than initiation or termination steps. In such a case, the surface chain reaction will not take place; and the process will follow course (68a) via further interaction of surface $NH_2 \cdot$ radicals. A process like this apparently takes place on solid bases (e.g., on CaO) which contain a large number of quasi-isolated surface levels (Section 2.1).

The following sequence of diminishing catalytic activity was obtained [441] for *decomposition of* HBr into bromine and hydrogen: $Sb_2O_3 > Al_2O_3 > Fe_2O_3 > CuBr_2 > BaCl >$ glass. The acidic mechanism is evidently operative in this case.

According to Taylor [442], *decompositions of* GeH_4, AsH_3, *and* SbH_3 represent true topochemical reactions which proceed on surfaces of of elemental Ge, As, and Sb and result from the decomposition of GeH_4, AsH_3, and SbH_3, respectively. In the series proceeding from the semiconductor Ge to the metal Sb, the reaction rate increases very sharply.

6 • Hydrogen–Deuterium Exchange and Other Simple Reactions

Reactions considered in Chapters 5 and 6, are characterized by either the accepting or donating of hydrogen. General rules of catalyst selection for these reactions reveal that the main function of a catalyst is its ability to activate the hydrogen molecule. In this context, it is interesting to examine some simple catalytic reactions involving only molecular or atomic hydrogen. Such reactions are: H_2-D_2 exchange, *ortho–para*-hydrogen conversion, and hydrogen atom recombination.

From numerous studies of *hydrogen–deuterium exchange* [9, 174, 180–182, 339, 403, 443–450], $H_2 + D_2 = 2HD$, only those involving oxides of transition elements of the fourth period [174, 443] are suitable for analysis. For these, catalytic activities based on 1 m² of catalyst surface area were reported, and experiments were conducted outside the diffusion limited range. As pointed out in Section 1.6, this data served Dowden and Wells as a basis for formulating their concept of the role of the crystal field in catalysis. Occurrence of maxima of catalytic activity on NiO and Cr_2O_3 and of a minimum on iron oxides are confirmed by the data of work [9]. In the oxides of nontransition metals and such oxides as TiO_2, V_2O_5, ZnO, and Ga_2O_3, which ions have d^0 and d^{10} electronic configuration, i.e., configuration of nontransition metals, the catalytic activity declines. In Section 1.6, the most recent data [180–182] on H_2-D_2 exchange over oxides of the fourth period metals were also discussed. These data led to an improvement of the graph shown in Fig. 24.

Among sulfides of fourth period metals, the most active is CoS with a d^7 metal electronic structure [450]; less active are FeS, CuS, Cu_2S, and NiS. Most active among the halides of fourth period metals are $CoCl_2$ and $NiCl_2$ [339], but as shown in Section 3.2 this is possibly due to their reduction to the metallic state.

By examining the catalytic activity of binary compounds of metals of a single group, one may notice that catalytic activity increases with increasing atomic weight of the metal. This dependence is confirmed in the series from MgO to ZnO [446, 448], from TiO_2 to ZrO_2, from CeO_2 to ThO_2 [446, 448], from MoS_2 to WS_2 [447], from CaH_2 to BaH_2 [403, 448], and from VSi_2 to $MoSi_2$ [445]. A decline in catalytic activity is observed only in the series MoO_3, WO_3, and VO_3 [9].

The study by Boreskov and Kuchaev [444] established a lower catalytic activity for a semiconducting Ge-film in H_2–D_2 exchange as compared to transition metals Ni, Au, Co, and Fe.

The following activity sequence was observed [449] among oxides of rare earth elements: $Dy_2O_3 > Er_2O_3 > Gd_2O_3 > Nd_2O_3$.

Figures 54a and 54b summarize the results of several studies [9, 174,

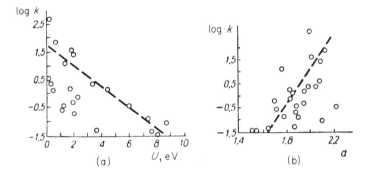

FIG. 54. Correlation between logarithm of catalytic activity for H_2–D_2 exchange and (a) width of forbidden zone, and (b) distance separating metal and oxide atoms in oxides.

180–182, 443, 446, 448] on catalytic activity for a series of oxides in H_2–D_2 exchanges as a function of the width of forbidden zone U and the cation–anion spacing in the oxides. Activity was expressed as the logarithm of the average rate constant based on 1 m² of catalyst surface area. Because of the paucity of the primary data, comparisons of this type should be regarded only as first approximations. It can be seen from Fig. 54a that there exists a relationship between catalytic activity and the width of the forbidden zone; it is impossible to obtain high catalytic activity for H_2–D_2 exchange from compounds with high values of U. The reverse conclusion, however, is not true: Among compounds

6. H_2–D_2 EXCHANGE AND OTHER SIMPLE REACTIONS

with small U there are some with high and some with low catalytic activity.

Overall, the correlation of catalytic activity with the lattice parameter is not very good. However, it can be seen in Fig. 54b that activity increases with increasing d.

Clark [9] offered a conclusion that n-type semiconductors are more active for H_2–D_2 exchange than p-type semiconductors. This conclusion was supported by his data showing reduced oxides to be more active than the oxidized oxides. If one, however, takes a look at a variety of oxides, then, on the average, p-type oxides are one order of magnitude more active than n-type oxides.

Various investigators of the mechanism of H_2–D_2 exchange on semiconductors considered it in terms of either associative ($H_2 + D_2 \to H_2D_2 \to 2HD$) or dissociative ($H_2 + D_2 \to 2H + 2D \to 2HD$), and either homolytic or heterolytic mechanisms. Many such mechanisms were discussed in Part I of this monograph. Particular attention was focused on the ideas of Hauffe [23] regarding the relation between catalytic activity of oxides for exchange reaction and the type of conductivity (Section 1.1), and on Balandin's concept [297] concerning the role of the crystalline lattice parameter [see schematic (47) in Section 1.1].

The above discussion of the rules for catalysts selection for H_2–D_2 exchange does not lead to a clear-cut conclusion regarding a connection between catalytic and electronic properties. In particular, no conclusion emerges for the dependence of catalytic properties upon the location of Fermi level in the crystal. The dependence of catalytic activity upon width of the forbidden zone cannot be accounted for by assuming that catalysis is due to intrinsic conductivity (Section 10.3), because H_2–D_2 exchange over oxides was studied in a temperature range (from -80 to $+200°C$) favoring a mixed type of conductivity. This dependence is apparently related to the connection between U and the effective charge of the surface cation, the latter being responsible for cleaving H_2 molecules (Section 1.5).

Apparently, also anions (electronegatively charged surface atoms) play a role in the H_2–D_2 exchange. This may account for the fact that the catalytic activity of the ions of transition metals (in H_2–D_2 exchange conducted in a homogeneous system and in other reactions conducted in solution and involving molecular hydrogen) does not concur with the order of activity observed over corresponding oxides. As shown [450], in homogeneous reactions, the most active catalysts for H_2–D_2 exchange are ions with electronic structures d^9 and d^{10}; these are Cu^{2+}, Cu^+, Pd^{2+}, Ag^+, and Hg^{2+}. The rules of selection of heterogeneous non-metallic catalysts for H_2–D_2 exchange can be viewed in terms of heterolytic adsorption of the H_2 molecules on surface metal and oxygen atoms followed by desorption of couples originating from two neighboring complexes [173]:

$$\begin{array}{ccccc} H^- & \ldots (H^+ & \ldots D^-) & \ldots D^+ \\ O^{2-} & Cr^{2+} & O^{2-} & Cr^{2+} & O^{2-} \end{array} \qquad (71)$$

As shown in Section 1.6, electronic transition in a complex according to the scheme of (71) may cause changes in configuration of the metal ion (for example, $Cr^{3+} \rightarrow Cr^{2+}$, i.e., $d^3 \rightarrow d^2$) which results in energetic stabilization of the crystal field so that it is favorable for a catalytic reaction. A change in electronic configuration of the cation in the course of reaction is, however, not a necessary condition for H_2–D_2 exchange, as this reaction occurs on such oxides as MgO and Al_2O_3. The cations of these do not change their valence states. Certain correlations between catalytic activity and Me–O distance in binary compounds (Fig. 54b) may also be explained by the mechanism of (71), i.e., by two-point adsorption. Nevertheless, the question of the effect of lattice parameter on reactions involving hydrogen is not completely resolved as yet.

Hydrogen transfer reactions between organic compounds may follow either a homolytic or heterolytic mechanism. Consequently, factors of importance for selection of a suitable catalyst are electronic and acid–base properties of the solids. For hydrogen transfer between C_2H_4 and D_2O, the following sequence of activity was established in study [451]: ZnO > Cr_2O_3 > γ-Al_2O_3 > MgO > CaF_2 > $CaSO_4$ > SiO_2, AlSb, $AlPO_4$, $Al_2(SO_4)_3$.

According to Schwab et al. [451], the most active catalysts for this reaction are Lewis acids. On this basis, it is difficult to account for the high activity of ZnO and the low activity of the group of aluminum compounds. It is more fruitful to talk about the role of the semiconductor properties in this reaction. This may provide a basis for explaining high activities of ZnO and Cr_2O_3.

In hydrogen transfer between alcohols and isovaleric aldehyde [452], metals are generally more active than oxides; among the superior oxides are the solid bases, in particular MgO.

Ortho–para-hydrogen conversion, p-$H_2 \rightleftharpoons o$-H_2, follows the same mechanism at high temperatures (400°C) as the H_2–D_2 exchange. It was, shown [449] that not only the order of catalysts according to activity but also the absolute rate of *ortho–para*-hydrogen conversion over Nd_2O_3, Sm_2O_3, Gd_2O_3, Dy_2O_3, and Er_2O_3 are identical as those for H_2–D_2 exchange.

A different characteristic mechanism applies to *ortho–para*-hydrogen conversion at low temperatures. This mechanism is based on interaction between nuclear spin of an H_2 molecule and the inhomogeneous magnetic field of the paramagnetic catalyst particles. This interaction results in in internal reorientation of the nuclear spin of a molecule without splitting it into atoms. According to a theory of Wigner [453], the effectiveness of the collision of an H_2 molecule with a paramagnetic ion (i.e., the rate constant related to one active center) is equal to:

$$\Phi = \frac{8\mu_a^2 \mu_p^2 I \pi^2}{q n^6 h^2 k T} \tag{72}$$

where μ_a is magnetic moment of the adsorption center; μ_p is magnetic moment of a proton; I is moment of inertia of an H_2 molecule; r is distance separating the H_2 molecule from the active center during the interaction; h is Planck's constant; k is Boltzmann's constant; T is reaction temperature in degrees Kelvin.

The approximate formula for effective magnetic moment of a paramagnetic ion is:

$$\mu_a = [n(n+2)]^{1/2} \tag{73}$$

where n is the number of unpaired electrons.

In reality, because of effects exerted by the crystal field (Section 1.6) and various chemical effects [454], the empirical values of μ_a differ from the theoretical ones.

It could be deduced from Wigner's theory that, in the *ortho–para* conversion of hydrogen ions at low temperature, a particularly high catalytic activity should be exhibited by systems with unfilled d- and f-shells. The highest values of μ_a occur for certain rare earth ions (for Dy^{3+} it is equal to 10.6). Figure 55 shows the first-order dependence of the rate constants at 77°K upon the square of the magnetic moment (μ_a^2) of the rare earth ion in various oxides [449]. As shown, a linear relationship exists between these parameters. At $\mu_a = 0$ the k should

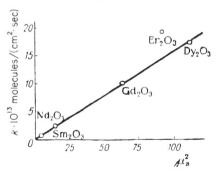

FIG. 55. Dependence of the rate constant of *ortho–para*-hydrogen conversion on rare earth elements as a function of the square of the magnetic moment of the cation [449].

be equal to zero. However, $k = 6.45 \times 10^{13}$ molecules/cm^2 sec was found for Al_2O_3 at 90°K. Ashmead et al. [449] explain this result by the formation of paramagnetic centers on the Al_2O_3 surface—"free valencies" resulting from dehydration of the Al_2O_3 surface Section 2.2). This explanation might account for deviation of Er_2O_3 from the straight line shown in Fig. 55.

As shown in Fig. 56, prepared from the data of Buyanov [455], the

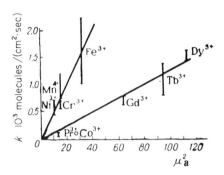

FIG. 56. Dependence of rate constant of *ortho–para* hydrogen conversion on hydroxides of both metals of the fourth period and rare earth elements upon the square of magnetic moment of the cation [455].

rate constant for *ortho–para* conversion of hydrogen on paramagnetic hydroxides is also linearly dependent upon μ_a^2. However, this dependence holds only within only one period. For example, hydroxides of rare earth elements are considerably less active than hydroxides of the elements of the fourth period. This author explains this fact in terms of the greater magnitude of the radii of ions of rare earth elements and also in terms of the decrease of the Φ constant for collisions [see Eq. (72)], assuming that r in this formula is a function of the ionic radius. According to Eq. (72) reaction rate declines with increasing temperature. It was observed in study [455] that, both on paramagnetic oxides and on hydroxides at 65–78°K, the rate increases somewhat increasing temperature; the apparent activation energy changes from 45 to 250 cal/mole.

Absolute rates of *ortho–para* hydrogen conversion, according to the above theory, did not concur with those obtained experimentally. Better agreement with experiment results, according to [456, 457], if one assumes that reaction occurs not as a result of collisions in the gas phase, but in an adsorbed layer in which H_2 molecules vibrate perpendicularly to the plane of the catalyst surface. The number of active centers on rare earths (10^{13} to 10^{14} per cm^2) is equal to the number of ionic vacancies which provide access to the paramagnetic ions of the catalyst. Each vibration of a H_2 molecules is equivalent to a collision with effectiveness equal to Φ.

In the series of hydroxides of metals of the fourth period, the activity increases and reaches a maximum at Mn and Fe and then gradually falls to the lowest level at Co. The same kind of change in activity applies also to the corresponding oxides [455, 458] (Fig. 57). However,

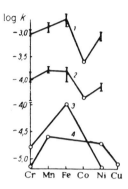

FIG. 57. Dependence of catalytic activity (1,2) of hydroxides and (3,4) of oxides of metals of the fourth period for *ortho–para*-hydrogen conversion: 1—at 78°K [455]; 2—at 22°K [455]; 3—at 78°K [455]; 4—at 78°K [458].

the range of absolute magnitude of activities, in this case, is significantly lower than in the former case. According to Buyanov [455], this reduction of catalytic activity is due to the fact that, during dehydration of hydroxides into oxides, there occurs an aggregation of the paramagnetic ions, which prior to that were diluted. As a result of this aggregation, the paramagnetism of hydroxides transforms into the antiferromagnetism of the corresponding oxides.

Recombination of hydrogen, oxygen, and nitrogen atoms on various surfaces is of interest [459–468] because of its theoretical utility as a simple catalytic reaction and also because of its practical importance in space research [463]. In this case, the coefficient of recombination γ (which is equal to the ratio of the number of atoms recombining into molecules per unit surface area per unit time to the total number of atoms impinging onto this surface in the same time interval) serves as a measure of catalytic activity.

Voevodskii and Lavrovskaya [459] found the following sequence of activity for hydrogen atom recombination at 20°C for various surfaces: Pt > Zn·Cr$_2$O$_3$ > MoO$_3$ > PbO > Zn > Cr$_2$O$_3$ > SiO$_2$ > KCl, K$_2$B$_4$O$_7$. In this series, the value of γ for Pt is equal to 1; and, for the terminal members of the series (solid salts), it is smaller than 10^{-4}. A similar situation was found to apply the recombination of oxygen and nitrogen atoms. In general, metals are most catalytically active at room temperature ($\gamma = 10^{-2}$ to 1), next come semiconductors and alkaline oxides ($\gamma = 10^{-3}$ to 10^{-1}), and least active are salts and acidic oxides ($\gamma = 10^{-5}$ to 10^{-3}). Figure 58a shows a graph drawn from [463] and supplemented with more recent data of studies [461, 468]. It shows values of γ for recombination of oxygen and nitrogen atoms. On the basis of his graph, one may calculate the rate constant k (cm sec^{-1}) which, for first-order reactions, is functionally related to γ: $k = \gamma(RT/2\pi A)^{1/2}$; where A is atomic weight and $R = 8.31 \times 10^7$ erg · deg^{-1} · mole^{-1}.

Coefficients of recombination of various atoms on a single surface frequently follow the order: $\gamma_{H+H} > \gamma_{O+O} > \gamma_{N+N}$, but the actual differences in their values are rather small [465]. With increasing temperature, the coefficient of recombination asymptotically approaches a value of 1; in the 300–700°K range, the energy of activation increases increases from 1–2 kcal/mole for Pt to 10–12 kcal/mole for such surfaces as SiO_2 or LiCl.

The coefficient of recombination of radicals, OH + OH, on surfaces of Pt, $ZnO \cdot Cr_2O_3$, Al_2O_3, KOH, K_2CO_3, NaP_3O_4, and K_2SiO_3 is close to the corresponding values of γ_{H+H} [462]. However, KCl catalyzes recombination of OH radicals, but is inactive for recombination of H atoms. The coefficient of recombination γ_{H+OH} is smaller than either γ_{H+H} or γ_{OH+OH}; and this is reflected in the suppressed rate of H atom recombination in aqueous media.

Linnett and Marsden [461] noted that the catalytic activity of oxides for O atom recombination increases with increasing atomic weight of the metal. This, however, was not always verified experimentally. For example, WO_3 is more active than MoO_3; and Bi_2O_3 is more active than Sb_2O_3, which is in agreement with study [461]. On the other hand, Al_2O_3 is more active than Ga_2O_3, SnO_2 is more active than PbO_2, and MgO is more active than both CaO and SnO, which is contrary to the proposal of the above-mentioned authors. Catalytic activity also declines in the seies LiCl, NaCl, KCl, RbCl, and CsCl. In Fig. 58b is shown a change in catalytic activity (log γ) in the series of oxides of elements of the fourth long period for recombination of oxygen atoms. It is noteworthy that from right to left in this group, the activity of oxides increases in the series from MnO to Cn_2O, while the values log γ simultaneously decline sharply. Except for Fe_2O_3, all these oxides are p-type semiconductors. Close to them in catalytic activity are the alkaline oxides, MgO and CaO, which do not have semiconducting properties.

Data on log γ for O atom recombination at room temperature [461], which are also shown in Fig. 58, were confirmed subsequently in the study by Dickens and Sutcliff [468]. Only NiO was found to have a very low value of γ (1.5×10^{-3}). It was shown that oxygen atoms oxidize Cr_2O_3 to CrO_3. Thus, the low activity of Cr_2O_3 was caused by its transformation into an n-type semiconductor; and this involved a

6. H_2–D_2 EXCHANGE AND OTHER SIMPLE REACTIONS

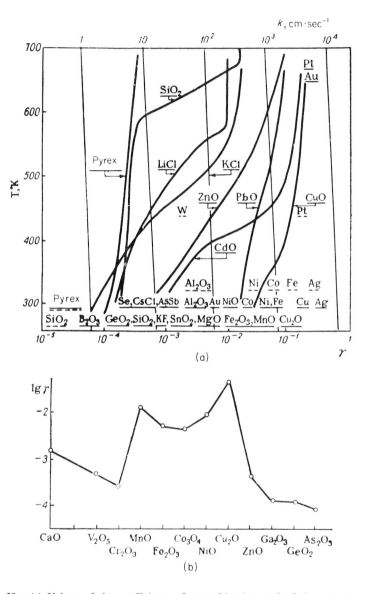

FIG. 58. (a) Values of the coefficients of recombination and of the corresponding rate constants for O atom recombination (underlined—solid line) and for N atom recombination (underlined—dashed line) on various surfaces (from [461, 463, 468] and other sources). (b) Changes in log γ for O atom recombination on oxides of metals of the fourth period at room temperature.

change in the cation structure from d^3 into d^{10}, i.e., conversion of the cation electronic structure into a structure of a nontransition metal. Particularly high activity was exhibited by Cu_2O, CuO, and MgO. Changes in γ of these oxides with increasing temperature were considerably smaller than those implied in [468]. Some of these data are included in Fig. 58. Increasing the temperature resulted in the occurrence of inflections on the Arrhenius curves. For example, for Mn_2O_3 in the 296–460 and 460–620°K ranges, E equals 6.7 and 2.6 kcal/mole, respectively; while for CdO in the 296–350, 350–460, and 460–620°K ranges, E is 1, 11, and 2.6 kcal/mole, respectively. Above 460°K, on all the oxides that were investigated, the activation of energy ranges from 1.3 to 3.0 kcal/mole; and the Arrhenius curves at $T \to \infty$ converge to $\gamma_0 = 1$, which indicates absence of the compensation effect.

According to Roginskii *et al.* [466, 467], recombination on metals occurs when an atom from the gas phase impinges at the chemisorbed atomic layer. A similar mechanism has also been proposed for oxides. For example, in study [464], the following mechanism for O atom recombination on NiO was postulated:

$$NiO + O \longrightarrow NiO \cdot O_{ads} \xrightarrow{O_{gas}} NiO + O_2{}^*$$

The liberated O_2 molecule is in an excited state.

For oxides, a strong chemical interaction is very likely. For example, atomic hydrogen may react with surface metal and oxygen atoms to form hydride and hydroxyl groups, respectively. Atoms of oxygen may reduce the surface oxide layer by removing oxygen from it:

$$MeO + O \to Me + O_2$$

Thereafter, a new oxygen atom adsorbs on the surface defect resulting from the above reduction reaction, and the process is carried on.

On the basis of such mechanistic ideas, one may develop rules for catalyst selection for such reactions. In this case, catalytic activity may be related to the location of the Fermi level in a semiconductor (this can be derived from the magnitude and type of conductivity); crystal field stabilization on the surface around ions of transition metals plays a minor role. Such semiconducting oxides of transition metals as CdO, PbO, and Bi_2O_3 are catalytically superior to the oxides of transition metals. Also, the nontransition metals are just as active as the transition metals. High catalytic activity of the solid bases may be explained in terms of greater number of "free valencies" (excess of isolated charges) on their surfaces (Section 5.1).

In the case of atomic recombination reactions, it would be pertinent

6. H_2–D_2 EXCHANGE AND OTHER SIMPLE REACTIONS

to examine the relation between catalytic activity and the energy of the bonds of the atoms with the surface. No progress resulted from attempts [468] to relate values of E for recombination with such parameters as the ionization potential of the ion, the electron affinity of the anion, and the energy of polarization involved in changes of the charge of surface atoms. The experimental errors often exceed the range of changes of the activation energy of recombination E. This range is only a few kilocalories per mole.

Certain simple reactions involving molecular oxygen will be discussed in the next chapter, which deals with oxidation reactions.

7 • Reactions of Oxidation and Decomposition of Oxygen-Containing Compounds

7.1. Simple Oxidation Reactions

Until now the discussion has revolved, in general, around the rules for selection of catalysts for reactions which require activation of molecular hydrogen. Hereafter, the reactions discussed will involve molecular oxygen.

The *oxidation of hydrogen* on metals has been studied in detail by several investigators, but such studies on oxides are less numerous. In Fig. 59, data of Boreskov and Popovskii [65, 469] on the oxidation of hydrogen on oxides of metals of the fourth period are presented.

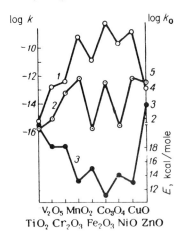

FIG. 59. Changes (1) in logarithm of catalytic activity of oxides of the fourth period, (2) in logarithm of the pre-exponential factor, and (3) in activation energy, for oxidation of hydrogen at 150°C [65, 469].

7.1. SIMPLE OXIDATION REACTIONS

These authors concluded both that the catalytic activity of oxides is not unambiguously determined by the electronic structure of the cation and that the hole-type semiconductors are more active than the electronic semiconductors. However, in Fig. 59, one may notice p-type semiconductors (Co_3O_4, NiO) as well as n-type semiconductors (MnO_2) among the most active catalysts. So far as the effect of electronic structure of the cation is concerned, Dowden's two-spiked diagram of catalytic activity of the fourth period metal oxides (Section 1.6) has been verified. Deep minima of activity occur at oxides whose cations have stable d-shells: d^0 (TiO_2), d^5 (Fe_2O_3) and d^{10} (ZnO). The large difference in the catalytic activities of Cr_2O_3 and MnO_2 is puzzling, since their cations both possess an identical electronic structure (d^3).

It is interesting that, in the case of oxides, the dependence of catalytic activity is more pronounced upon changes in E than upon changes in k_0 (curves 2 and 3 on Fig. 59). Values of k_0 calculated by the present author only vary over three orders of magnitude. At the same time, for the most active catalysts a simultaneous increase in k_0 and decline in E occurs. A very wide range of catalytic activity (seven orders of magnitude) speaks for itself in favor of the reliability of the determined values of reaction rates. It is known, for the case of diffusional limitation, that the differences in activity over various catalysts are frequently rather small. An investigation of hydrogen oxidation over oxides of lanthanides [470] revealed only a very small variation in their specific catalytic activity at 450°C. Only PrO_2 and Nd_2O_3 showed activity well above that of the other oxides. Their activity was greater in a hydrogen atmosphere than in an oxygen atmosphere. In hydrogen, the activation energy of hydrogen oxidation varied within a 10–18-kcal/mole interval, while, in an oxygen atmosphere, the spread in activation energy for this oxidation was even greater. In comparison with the highly active fourth period metal oxides ($k = 10^{-10}$–10^{-5} moles · min^{-1} · m^{-2} at 300°C), the same rare earth oxides exhibit only mediocre catalytic activity ($k = 10^{-7}$–10^{-6} moles · min^{-1} · m^{-2} at 400°C).

The oxidation of CO and the decomposition of N_2O are the two most popular reactions employed both in studies on electronic theory of catalysis and in investigation of catalyst selection rules for oxidation–reduction reactions. Contradictory data [32–37] regarding the mechanism of CO oxidation and the rules of catalyst selection for this reaction, were discussed in Section 1.1. Most authors [11, 25, 32] consider it a donor reaction, for which p-type semiconductors should be superior to n-type semiconductors. The validity of this conclusion, as well as the occurrence of certain relationships, will now be examined by the method of correlative analysis. This method is applicable in view of the availability of sufficient fundamental data for averaging [48, 329, 471–482]. Because many data are available for the NiO system, data on CO oxidation on NiO were taken as a reference for the purpose of comparison. Whenever it was possible, a comparison was made using the values of activity at 150°C (or values extrapolated to this temperature). This procedure was used to obtain the data on average catalytic activity (in logarithmic scale) presented in Table VI.

Using these data, Fig. 60 shows that there is a dependence of the catalytic activity of oxides of the fourth period metals upon the location

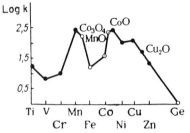

FIG. 60. Changes in the logarithm of catalytic activity of metal oxides of the fourth period in CO oxidation.

of these metals in the periodic system. The diagram of changes in catalytic activity for oxidation of CO (Fig. 60) closely resembles that for H_2–D_2 exchange (see Fig. 24); maxima occur at Mn, Co, and Ni, and the minimum at Fe_2O_3. One difference in this case is the relatively low activity shown by Cr_2O_3 (analogous to the case of H_2 oxidation).

TABLE VI

RELATIVE CATALYTIC ACTIVITY OF OXIDES FOR OXIDATION OF CO

Catalyst	Log k	Σw	Catalyst	Log k	Σw
Ag_2O	2.09	4	HgO	0.75	3
Al_2O_3	−0.04	7	MgO	0.11	1
BeO	0.37	2	MnO	2.21	1
CdO	2.10	2	MnO_2	2.43	9
CeO_2	0.85	8	NiO	2.00	14
CoO	2.42	6	SiO_2	−0.50	1
Co_2O_3	1.58	2	SnO_2	1.89	6
Co_3O_3	2.32	8	ThO_2	0.43	7
Cr_2O_3	1.02	12	TiO_2	1.26	6
CuO	2.08	12	V_2O_2	0.84	6
Cu_2O	1.71	8	WO_3	0.44	3
Fe_2O_3	1.22	13	ZnO	1.33	6
GeO_2	0.00	1	ZrO_2	1.04	6

7.1. SIMPLE OXIDATION REACTIONS

Figure 61 shows the dependence of catalytic activity of oxides superimposed on the entire periodic system of elements. The notations are identical to those used in Fig. 38 (Section 4.1). In Fig. 61, two regions of maximum activity are shown: one around the oxides of Co, Ni, Ag, and Cd; and the other around the oxides of Mn. These maxima are separated by a minimum at Fe_2O_3. Numerous studies have confirmed the presence of such maxima and minima of activity; and they are therefore assigned a high value of statistical weight (Table VI). In that it constitutes a deviation from the general relationship, the minimum observed in HgO is less credible, and its occurrence is rather accidental.

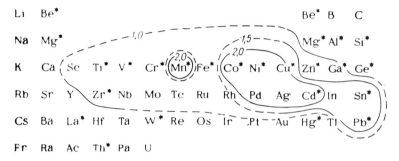

FIG. 61. Lines of isoactivity of metallic oxides for CO oxidation.

Calculation on the basis of Table VI, of the correlation coefficients (r) relating the logarithm of catalytic activity (y) to various properties of solids (x) using formula (56) leads to the following condition: There is no correlation between $\log k$ and the spacing (d) separating the cation from the anion ($r = -0.092$). Application of criterion (57) indicates that this value of r may well result from a random selection of the data for the calculations. Figure 62a shows the dependence of catalytic activity of oxides upon the width of the forbidden zone U. The regression line here, as well as in Fig. 62b, c, and d, corresponds to the coefficient of linear regression calculated by the least squares technique. The value of $r = -0.74$ indicates that a very strong dependence of $\log k$ upon U exists in the oxidation of CO over oxides, a much stronger dependence than that which was found in the dehydrogenation of alcohols. This dependence, shown not to be caused by the relation between $\log k$ and d (surface geometry factor), is apparently related either to the fact that catalysis occurs within the range of natural conductivity or to the effective charge of the cation, as discussed in Sections 1.1 and 1.5. The latter alternative is more likely, since it is in line with a strong dependence of $\log k$ upon the location in the periodic system of the metal forming the oxide (Figs. 60 and 61). However, the dependence of log k upon electronegativity difference, Δx, was found to be smaller than that of log k upon U. As a result, the correlation coefficient, $r_{\log k/\Delta x} = -0.43$, will be smaller than $r_{\log k/U}$. Figure 62b shows the dependence of log k upon Δx. For all investigated catalysts, the values of Δx fall, as shown, within a relatively narrow range between 1.5 and 2.5. Calculation of the partial correlation coefficient $r_{\log kU/\Delta x}$ according to formula (58) shows

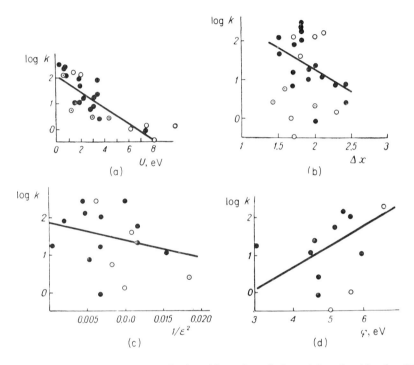

FIG. 62. Correlation between the logarithm of catalytic activity of oxides for CO oxidation and (a) the width of the forbidden zone, (b) the electronegativity difference, (c) the value of $1/\epsilon^2$, and (d) the work function.

that, leaving aside the dependence upon Δx, there is a true dependence of log k upon U. Neglecting the correlation between Δx and U, the correlation coefficient for log k with U, $r_{\log k\Delta x/U} = 0.20$, is small. Apparently, this is caused by the uncertainty of the Δx values for transition metal oxides. A better correlation was obtained in alcohol dehydrogenation (Section 4.1) because, in this case, the range of Δx for the catalysts studied was considerably wider (0. to 3.2).

Figure 62c shows the dependence of log k upon $1/\epsilon^2$ for CO oxidation

7.1. SIMPLE OXIDATION REACTIONS

(where ϵ is dielectric permeability). The correlation coefficient $r = -0.29$. However, when the partial correlation coefficients $r_{\log k1/\epsilon^2/U}$ is calculated according to formula (58); its value is found to be zero. Therefore, the correlation between $\log k$ and $1/\epsilon^2$ is fully dependent upon the relationship between $\log k$ and U, according to formulas (8) and (13) in Section 1.3.

In Fig. 62d, the dependence of catalytic activity, $\log k$, upon the work function φ is shown. The correlation coefficient for this dependence is small, $r = 0.26$. According to Eq. (57), this value of r is only slightly greater (due to the scarcity of available data) than that characteristic of the random data; nevertheless, this increased value of r is real. This dependence points out that p-type semiconductors should exhibit superior activity. In reality, the statistically averaged activity of p-type oxides in CO oxidation (Table VI) was found to be 1.18. This difference is, however, too small to justify condemnation of studies [33, 34] (Section 1.1) that report contradictory data.

Apparently, oxidation of CO follows different mechanisms on different catalysts. Near room temperature, the reaction proceeds over such catalysts as NiO, CuO, and MnO_2 with a low energy of activation (3–6 kcal/mole). At high temperatures (200–400°C), the activation energy on these and other catalysts is as high as 15–25 kcal/mole. Certain proposed mechanisms of CO oxidation have been rejected in view of the results obtained by correlation analysis. For example, the concept of two-point adsorption of CO [329] (Section 3.1) was not supported because no correlation was found between the catalytic activity of oxides and their surface geometries. Of more credibility [48, 483] are the conclusions that catalysis occurs through the formation of surface carbonyl complexes (74a) and surface bicarbonate complexes (74b), as established by infrared spectroscopy in the case of NiO:

$$\begin{array}{cc} \text{O} & \text{O}^{\ominus} \\ \| & \diagdown\ \diagup \\ \text{C} & \text{C} \\ | & | \\ \text{O} & \text{O} \\ | & | \\ \text{Ni} & \text{Ni} \\ \text{(a)} & \text{(b)} \end{array} \qquad (74)$$

At low temperatures, such complexes may form directly from adsorbed CO molecules and molecular or atomic oxygen. An anionic vacancy is required for the reaction to make the transition metal ion accessible to absorbing ligands. At high temperatures, surface oxygen atoms may participate in the reaction. Decomposition of complexes with liberation of CO_2—such as (74b)-type, is accelerated by positive holes present in most catalysts; this liberation (or desorption) of CO_2 represents a limiting step of the reaction, as revealed by the above discussion of the rules of catalyst selection.

Investigation of CO oxidation over rare earth elements [482] revealed a very strong effect on catalytic activity by catalyst pretreatment with H_2 or CO. The activation energy was found to decrease through the series from La_2O_3 (12 kcal/mole) to Nd_2O_3 (4 kcal/mole) and further on to Dy_2O_3 (2.2 kcal/mole). The activity trends for this reaction are generally similar to those observed for dehydrogenation of oxides and cyclohexane (Sections 4.1 and 5.1.).

For the oxidation of SO_2 *to* SO_3, an important industrial process, the rules of catalyst selection have not been studied recently. Among oxide catalysts for this reaction, V_2O_5 occupies a prominent position. This oxide has no d-electrons. In Section 3.2, the activity of V_2O_5 for SO_2 oxidation was explained in terms of a characteristic crystal structure. Possibly, V^{4+} ions represent active centers (cation structure d^1); these ions may result from partial reduction of the V_2O_5 catalyst in the SO_2 atmosphere. Similarly, TiO_2 (d^0) increases its catalytic activity for SO_2 oxidation upon its reduction to Ti_2O_3 (d^1). The catalytic activity order of the most studied oxides of the fourth period elements is [484–487]: $V_2O_5 > Cr_2O_3 > Fe_2O_3 > CuO$; and these are followed by oxides of Ce, Th, Ti, W, As, Mo, Sn, and U.

Oxidation of ammonia to NO. Figure 63 shows averaged data [179, 488–493] on the catalytic activity of fourth period metal oxides for NH_3

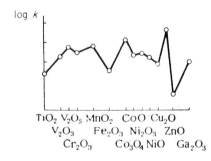

FIG. 63. Change in catalytic activity (in logarithmic scale) in a series of oxides of fourth period metals for ammonia oxidation.

oxidation. In general, activity differences were not found to be particularly large, although, in this case also, the minimum of activity falls around Fe_2O_3 and ZnO; and the maximum falls around MnO, Co_3O_4, and CuO. Among the oxides of fifth period elements, the most active are MoO_2 and Ag_2O; and among oxides of sixth period elements, the most active is Bi_2O_5. There is a tendency, within a series, for the catalytic

7.1. SIMPLE OXIDATION REACTIONS

activity to decline as the atomic weight of the metal increases. Very high catalytic activity of oxides with p-type conductivity is indicated in the study [493]; however, the actual difference between those oxides and the oxides with n-type conductivity is small. Reliability of most of the data on NH_3 oxidation over oxides is not good: neither changes in activation energies nor in k_0 were determined; and, in some cases, even the specific surface area of the oxide was not determined. Most reliable is the fact that Co_3O_4 is a very active catalyst, because in study [179] it was proposed as a substitute for platinum catalyst for NH_3 oxidation. Study [494] of *nitric acid synthesis* from NO, O_2, and H_2O revealed the highest catalytic activity for oxides of Cr and Co and an inferior activity for Fe_2O_3. Al_2O_3 exhibited low activity.

According to many studies [314, 473, 495, 496], the regularity in changes of catalytic activity of oxides in oxidation reaction may be explained in terms of changes in the strength of the metal–oxygen bond in oxides. In Fig. 64, the Rienäcker [473] diagram illustrating this

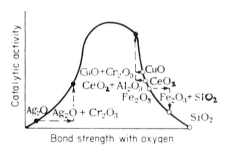

FIG. 64. Diagram showing the dependence of catalytic activity in oxidation reactions upon the strength of bonding of oxygen in oxides [473].

Rienäcker [473] diagram illustrating this hypothesis is shown. Most active catalysts are characterized by having a certain optimum strength of such bonding. As compared with the optimum strength of bonding of oxygen to metal in oxides, an increase (CuO and Fe_2O_3) as well as a decrease (Ag_2O) results in a decline in catalytic activity. The effect of dopants is explained in a similar fashion: Dopants that increase the the strength of Me—O bonds in oxides to the left of the maximum increase the activity of the oxide; while those that increase the bond strength to the right of the maximum decrease this activity. Other authors found a linear relationship. For example, according to Komuro and Yamamsto [495], the higher the heat of formation of the oxide, the smaller is its catalytic activity for hydrogen oxidation.

The rate of isotopic exchange of oxygen from the gas phase with oxide oxygen: $MeO^{16} + O_2^{18} \rightleftarrows MeO^{18} \rightarrow O^{16}O^{18}$, may serve as a yardstick of reactivity of oxygen in the oxide surface layer.

The *homomolecular exchange of oxygen*, $O_2^{16} + O_2^{18} = 2O^{16}O^{18}$, does not require participation of the lattice oxygen. Both of these reactions on oxides of fourth period elements have been investigated [495–503]. As shown in Fig. 65, the patterns of catalytic activity variation for both

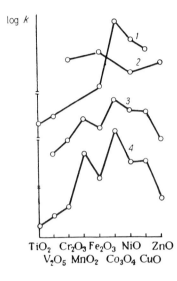

FIG. 65. Changes in logarithm of catalytic activity in the series of fourth period metal oxides for isotopic oxygen exchange with (1–3) surface oxygen of the oxides and (4) in homomolecular exchange: 1—data from [496], 2—data from [497], 3—data from [498], 4—data from [498].

reactions are similar. Catalytic activity increases from TiO_2 to a maximum level at Co_3O_4 (electronic structure d^6–d^7) and then declines toward ZnO.[1] A somewhat smaller maximum occurs at MnO_2; and a minimum is found at Fe_2O_3. In contrast to works [498, 501], it was found in study [497] that Fe_2O_3 is considerably more active than NiO. The dependence of rate constants of homomolecular exchange [498] upon width of the forbidden zone (Fig. 66a), and upon work function (Fig. 66b) is presented. In both cases there is a correlation: U increases as k decreases; and φ increases as k increases. In connection with the dependence shown in Fig. 66b, p-type semiconductors are more active than n-type semiconductors. In Part I (see Figs. 8 and 15), it was shown that values of φ and U change regularly in the series of oxides of fourth period metals: There occur maxima and minima at d^0, d^5, and d^{10}. Therefore, in order to verify such correlations of activity with φ or U, it would be pertinent to investigate the catalytic activity in oxygen exchange of metal oxides of the other periods. Information is given in

[1] On this, as well as on other diagrams, data of various authors taken under various conditions and at various temperatures are shown. Such diagrams permit judgment of either similarity or deviation of the patterns of changes in catalytic activity but do not compare the absolute levels of such activities.

7.1. SIMPLE OXIDATION REACTIONS

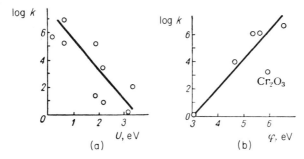

FIG. 66. Correlation between catalytic activity of oxides in homomolecular oxygen exchange and (a) width of forbidden zone, (b) electron work function.

study [497] showing surprisingly high activity of MgO despite its dielectric properties, high U (8.7 eV) and low φ (3.1–4.4 eV).

In Fig. 67, E and $\log k_0$ are presented for homomolecular oxygen exchange, according to work [498]. The dashed line corresponds to the value of k_0 calculated on the assumption of nonlocalized adsorption of O_2. As shown in Fig. 67, the most active oxides have low values of of both energy of activation and pre-exponential factor. On the basis of these data, one can speak about certain counteracting tendencies in the changes of these parameters, but not about complete compensation. Energies of activation of this reaction vary over a wide range (15–20 kcal/mole); k_0 changes over only a small interval (within four orders of magnitude), while the rate constant in the 200–400°C range changes by eight orders of magnitude.

FIG. 67. Values of activation energy and of logarithm of pre-exponential factor for homomolecular oxygen exchange on oxide catalysts [502].

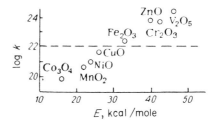

With increasing degree of exchange between gaseous oxygen and surface oxygen of the oxides, the reaction rate declines and the energy of activation increases. As shown by Boreskov and Popovskii [500], the rates of this reaction coincide precisely with the rates of homomolecular oxygen exchange on these same oxides. The energies of activation and the reaction order in oxygen also coincide for both of these reactions. Possibly, the homomolecular oxygen exchange, as well as oxygen exchange with the oxide, proceeds through an intermediate monoatomic ion O^- and then O^{2-}, which is identical to the ion in the lattice.

Isotopic oxygen exchange requires considerably higher temperatures than does oxidation of H_2 and CO. Energies of activation of oxygen exchange are higher than those of oxidation. The rules of catalyst selection (relative order of activity of oxides) remain the same, as indicated by the comparison of Figs. 59–62 (for oxidation reactions) and Figs. 65 and 66 (for oxygen exchange reactions). On this basis, it was concluded [496, 498] that catalytic activity in both exchange and oxidation reactions depends upon the same parameter—the energy of the Me—O bond in the surface layer of the oxide. However, as shown in Chapters 4 and 5, for most reactions not involving oxygen, the rules of catalyst selection are almost identical, i.e., minimum activity corresponds to d^0, d^5, and d^{10}, while maximum activity corresponds to d^3 and d^7.

Oxides of rare earth elements exhibited almost identical specific catalytic activities in isotopic exchange reactions. The most active proved to be Pr_6O_{11} and Nd_2O_3. Thus, the orders of activity in this case and in hydrogen oxidation are identical.

According to Winter [497], the sequence of oxide activity, based on rates of *oxygen chemisorption* is as follows: Fe, Cr, Mg, Zn, and Ni; this sequence is similar to that for oxygen exchange. Oxides with n-type conductivity were found to be somewhat more active than oxides with p-type conductivity. Studies of chemisorption of oxygen on evaporated films of metals and elemental semiconductors [504] led to the following activity sequence: Bi > Sb > As > Se, Te. These data indicate that metallic bismuth is more active than the semiconductors. Among semiconductors, the following rule applies: The lower the activity the larger will be the width of the forbidden zone. The authors of [504] believe that there is a correlation between the catalytic activity of such films and the work function.

7.2. Oxidation of Organic Compounds

In contrast to the simple oxidation reactions discussed in Section 7.1, most organic compounds either oxidize completely to give CO_2 and H_2O or, under milder conditions, give valuable oxygen-containing products: alcohols, acids, aldehydes, aliphatic oxides, and others. The literature is filled with studies of catalyst selection rules for oxidation of organic compounds. The majority of such studies was concerned with choosing selective catalysts for mild oxidation. Obviously, changes in activity and selectivity do not follow the same direction. Therefore, no

7.2. OXIDATION OF ORGANIC COMPOUNDS

general rules concerning selection of catalysts for oxidation of organic compounds can be derived from data on selection of catalysts for preparing a specific oxygen-containing compound.

For the oxidation of methane [502, 530, 531] to CO_2 and H_2O, as well as for many other oxidation reactions, Co_3O_4 proved to be superior among oxide catalysts. Other active catalysts are: Cr_2O_3, MnO_2, CuO, and NiO. Low activity is exhibited by ZnO, V_2O_5, MoO_3, and TiO_2. In the series of oxides of the fourth period elements, the minimum again occurs at Fe_2O_3. The most active oxide catalysts for oxidation are only slightly inferior to the platinum group metals and are more active than the nonplatinum group metals. Changes in energy of activation were, to a large degree (but not completely), compensated for by corresponding changes in log k_0. As shown in Fig. 68, the linear correlation between E and log k_0 is better satisfied if one considers only data for oxides and rejects all data for metals and spinels.

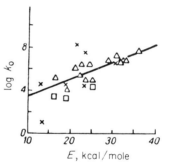

FIG. 68. The relation between the logarithm of the pre-exponential factor and the energy of activation for methane oxidation over various catalysts: ×—metals; □—spinels; △—oxides.

On most oxides of transition metals, ethylene oxidation proceeds to CO_2 and H_2O [457, 502, 532]. In complete oxidation of ethylene [475], studied at 450–600°C, Cr_2O_3 and CuO proved to be the most active of the oxides of the fourth period metals.

Studies of *oxidation of ethylene* at lower temperatures (250–450°C) led to a somewhat different activity diagram, in that Cr_2O_3 was among the most active catalysts. For example, at 262°C, the following order of activity was obtained [532]: $Co_3O_4 > Cr_2O_3 > Ag_2O > Mn_2O_3 > CuO > NiO > V_2O_5 > CdO > Fe_2O_3 > MoO_3 > WO_3 > TiO_2 > ZnO$. Thus, here also, one obtains a two-spiked diagram of changes in catalytic activity for the series of fourth period metal oxides. A minimum value of E (9.0 kcal/mole) for CuO and a maximum value (17.1 kcal/mole) for TiO_2 were obtained. The value of k_0 is smaller on CuO, and on Mn_2O_3 it reaches a maximum corresponding to an activation energy for C_2H_4 oxidation equal to 13.5 kcal/mole.

Out of all these oxides, only on Ag_2O (which reduces to Ag) does there occur a mild oxidation yielding ethylene oxide.

For the complete oxidation of propane [533], Cr_2O_3 was found to be superior to NiO, MgO, and V_2O_5.

The complete oxidation of propylene to CO_2 and H_2O was investigated in detail over various oxides [534]. Using these data, the changes in catalytic activity of the fourth period metal oxides were diagramed in Fig. 69. The ordinate represents values of $1/T$ (where T is the temperature at which an arbitrarily selected reaction rate is achieved). In general, the order of activity resembles that obtained for other oxidation reactions (Sections 7.1 and 7.2). Low activity of NiO was not confirmed in studies [502, 535] which report it to be close to that of MnO_2.

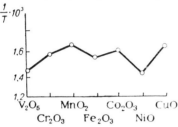

FIG. 69. Changes in the catalytic activity of oxides of the fourth period metals in propylene oxidation [534].

However, in a more recent work [536], NiO was again found to have low activity; and the change in specific catalytic activity over the series of fourth period metal oxides was the same as that shown in Fig. 69. Among oxides of fourth period metals, U_3O and CdO exhibited similar activities, which were higher than those of MoO_3, ZrO_2, and CeO_2. Of relatively low activity are ZnO, BaO, PbO, Bi_2O_3, WO_3, and Al_2O_3. Highly active catalysts such as MnO_2, NiO, and Cr_2O_3 led to complete oxidation without yielding any products characteristic of mild oxidation. For mild oxidation of propylene to acrolein, Cu_2O [537, 538] and such compounds of copper which, in the course of oxidation, could convert into Cu_2O, proved to be the most selective. Such oxides as V_2O_5, MoO_3, and WO_3 directed the reaction toward the formation of other products characteristic of partial oxidation.

Activity and selectivity of the fourth period metal oxides in the oxidation of propylene to acrolein were covered in study [537]. If the amount of acrolein produced in a unit of time over 1 m² of catalyst surface is used as a yardstick of activity, one obtains the series: $Cu_2O > Co_3O_4 > Fe_2O_3 > NiO > V_2O_5$. Cobalt oxide is almost as active a catalyst as the copper oxides; however, its selectivity is somewhat inferior, as it tends to lead to complete oxidation. In contrast, V_2O_5 exhibits low activity, but is highly selective to acrolein because of its inability to promote complete oxidation. These examples indicate that the order of

7.2. OXIDATION OF ORGANIC COMPOUNDS

catalysts based on activity is different from their order based on selectivity; an identical conclusion applied also to the decomposition of alcohols (Section 4.1).

Complete oxidation of pentane, 2-pentane, 1-pentene, 2-methylbutane, 2,3-dimethylbutane, hexane, cyclohexane, and benzene was investigated by Stein et al. [539] by a chromatographic flow-type technique. The results comparing the catalytic activity of oxides of the fourth period metals in complete oxidation of these hydrocarbons are presented in Fig. 70.

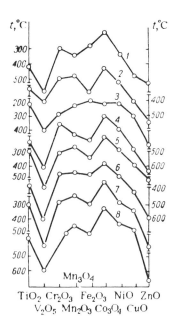

FIG. 70. Comparison of the catalytic activity (in terms of the temperature required for 80% conversion) of oxides of fourth period metals [539] for complete oxidation: 1—for pentane; 2—for 2-pentene; 3—for 1-pentene; 4—for 2-methylbutane; 5—for hexane; 6—for 2,3-dimethylbutane; 7—for cyclohexane; 8—for benzene [539].

Since surface areas of the catalysts were similar, the activity was expressed as temperature required for 80% conversion. These different hydrocarbons all oxidized between 200 and 600°C. As shown in Fig. 70, all catalysts have identical patterns of activity change. With only one exception, an activity maximum occurred on Co_3O_4. A second maximum occurred near the oxides of Cr and Mn. Between these maxima, there is a minimum at Fe_2O_3 (with two exceptions). For all the reactions studied, a strong diminution in activity coincided with ZnO_2 and V_2O_5. In this case, correlation of catalytic activity with other lattice parameter or electronegativity difference was not possible. The connection between catalytic activity and the width of forbidden zone is shown in Fig. 71. The ordinate is the temperature required for 80% conversion; this is an arithmetic average taken over the oxidation temperatures of all the various hydrocarbons. It is found that a linear dependence is shown by most of the oxides. The exceptional activity of γ-Al_2O_3 results from the high magnitude of its specific surface area. In view of their high value of U, both MgO and SiO_2 exhibited poor activity, despite their large surface areas. The deviation from linear dependence becomes somewhat smaller if one takes into account the fact that the CuO, WO_3, V_2O_5, and ZnO employed in study [539] had very small surface areas; this consideration, however, does not entirely explain such poor performance by these oxides. Almost all these oxides are n-type semiconductors. In general, the superiority of n-type semiconductors, in terms of activity, is not very significant. A clear-cut correlation of catalytic activity with the work function φ was not observed.

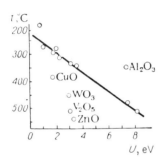

FIG. 71. Dependence of the average temperature required to initiate complete oxidation of various hydrocarbons [539] upon width of forbidden zone.

One could argue against the general validity of the conclusions reached in [539] regarding the rules of catalyst selection. The same catalyst samples were used in the oxidation of all the hydrocarbons in question; and, possibly, the activities of these particular samples were above normal. However, the above-mentioned rules of catalyst selection were subsequently verified by other workers. For example, it was found [480] that for *complete oxidation of pentane and benzene* the maxima observed at oxides of manganese and cobalt are separated by a minimum at Fe_2O_3. Also, V_2O_5 showed poor activity. Unexpectedly low activity was exhibited by NiO. According to study [532], the following oxides are highly active for the oxidation of benzene: NiO, Mn_2O_3, and Cr_2O_3.

The general rules for selection of oxidation catalysts were also confirmed by the results of *complete oxidation of iso-octane* [189, 540]. It was indicated in Section 1.7 that intensely colored compounds proved to be good catalysts for this oxidation reaction. Among the most active catalysts are oxides of Mn, Cr, Co, and Ni. Also chromites of these metals and of Cu and Mg showed high catalytic activity. In the case of complete oxidation of *iso*-octane on chromite catalysts, Margolis and Todes [540]

7.2. OXIDATION OF ORGANIC COMPOUNDS

found a strong compensation effect—a linear dependence of E upon $\log k_0$.

In both mild *oxidation of benzene to maleic anhydride* [541, 542] and oxidation of crotonaldehyde to crotonic acid [542] the activity of oxides follows the order: $V_2O_5 > Cr_2O_3 > MoO_3 > Co_2O_3$. A similar order applied also to the selectivity of the oxides. For all mixed oxides based on V_2O_5, the selectivity was considerably greater than that for the individual oxides alone.

In *complete oxidation of toluene and napthalene* [543, 544], as well as in oxidation of some other hydrocarbons, the oxides of Ni and Co exhibit the highest activity, these are followed by oxides of Cr, Cu, Ce, Th, and V. White oxides with a large width of forbidden zone (MgO, BeO, B_2O_3, and SiO_2) are catalytically inert. Within the oxides of Group V, V_2O_5 was found to be more active than Nb_2O_5 and Ta_2O_5, while among oxides of Group VI, WO_3 proved to be more active than MoO_3.

Different rules of catalyst selection apply to the partial *oxidation of toluene to benzaldehyde*. Catalysts which are suitable for complete combustion, Co_3O_4 and MnO_2, are of mediocre suitability for partial oxidation. Catalysts with little activity for complete oxidation (V_2O_5, WO_3, and MoO_3) showed [543] high selectivity and produced high yields of benzaldehyde from toluene.

For oxidation of such hydrocarbons as butylenes, butadiene, naphtalene, phenantrene and other organic compounds, superior catalysts, characterized by selective formation of mild oxidation products (e.g., (e.g., aldehydes, ketones, and acids), are V_2O_5, Cr_2O_3, MoO_3, and WO_3.

To sum up the discussion of reactions involving oxygen, it should be mentioned that the rules for catalyst selection are similar for the oxidation of simple molecules, for isotopic oxygen exchange, and for the complete oxidation of hydrocarbons. This indicates that a similarity exists between the mechanisms operating in all these reactions. In addition, this observation reflects also on the situation regarding the rules of catalyst selection for reactions involving hydrogen (Section 4.1, and Chapter 5): for most cases, a two-spiked diagram of change of catalytic activity of fourth period metal oxides applies; the activity declines with increasing width of forbidden zone; the activity increases with increasing atomic weight of the metal involved in the formation of the oxide, etc. Apparently, the closeness of the catalytic activity in above mentioned cases relates to the same factors, such as similar electronic transitions between the zones and similar changes in crystal field stabilization energy.

Some deviations observed can be explained easily. For example, in the majority of reactions involving oxygen, the relative activity of Cr_2O_3 is considerably lower than that observed in reactions involving hydrogen; but its level is lower than that of MnO_2 with an identical electronic structure of the cation, d^3. Apparently, with excess oxygen, the Cr^{3+} oxidizes to higher oxidation states Cr^{5+} (d^1) and Cr^6 (d^0), which are not at all, or only slightly stablized by the crystal field (Section 1.6). Possibly, the deviation of Cr_2O_3 from the general dependence in homomolecular oxygen exchange (see Fig. 66b) is due to these causes. Thus, when Cr_2O_3 is used under conditions precluding its oxidation, such as in work [532], its catalytic activity for oxidation is high. In general, assuming the same type of coordinate system, a transformation of the cation into a higher oxidation state in reactions involving oxygen should result in a shift to the right on the diagram of changes of catalytic activity of fourth period metal oxides. This shift will be in the opposite direction for reactions involving hydrogen.

For reactions involving oxygen, more frequently than for those involving hydrogen, the catalytic activity is related to the type of conductivity of the catalyst: p-type semiconductors are frequently more active than n-type semiconductors. This is apparently due to the fact that formation of oxygen ions O^-, O^{2-}, or O_2^- (which involves an electron), is a facile process not requiring high activation energy. Another alternative explanation is that these ions do not participate in reactions of complete oxidation. The limiting step may represent the interaction of a hydrocarbon or some other oxidizable molecule on the surface with adsorbed oxygen or oxygen from the gas phase, a process which involves free holes of the catalyst. The connection between catalytic activity and the type of conductivity is not identical for oxidation reactions. Sometimes (Fig. 69 and also Figs. 73, 75, and 78) p-type semiconductors are not superior to n-type semiconductors. In such cases, a different limiting step must apply.

In partial oxidation reactions, oxides with low symmetry, such as Cu_2O, Ag_2O, V_2O_5, and MoO_3, exhibit high catalytic activity. Of possible application to such cases are the considerations presented in Section 3.2, regarding the relation between activity (and selectivity) and the type of crystal lattice. For such reactions, it is possible that hydrocarbons react with oxygen preadsorbed in molecular form on the catalyst surface. Adsorption of the hydrocarbon on top of the oxygen layer may possibly represent the rate limiting step in the case of partial oxidation.

In the *oxidation of hexafluoroethylene*, C_3F_6, other types of catalysts, namely fluorides of alkali metals, exhibit highest activity.

The very high energies of activation (above 100 kcal/mole) indicate that a basically different reaction mechanism is involved in these cases. According to ref. [545], such mechanisms involve interaction of the fluorine atoms of the reactant with the catalyst surface. It is not accidental that the order of activity for this reaction does not coincide with the order of activity for the fluorine exchange reaction between metal fluoride and $C_3F·$. The order of activity is: $CsF > RbF > LiF > KF > NaF$, while, according to the energy of activation, the order should be: $LiF > NaF > KF > Rb > CsF$. Around the Tamman temperature (corresponding to loosening of the crystal lattice), a discontinuity occurs on the Arrhenius curves; and, above this temperature, the energy of activation falls to a very small value (less than 10 kcal/mole).

It could be postulated that the halogenation of paraffins, by analogy to the oxidation of paraffins, will be catalyzed by metal halides. Experiments show that, in actuality, halides of transition metals are active catalysts for halogenation reactions. In *hexane chlorination* [546] pure $CoCl_2$, as well as $CoCl_2$ and mixed with chlorides of other metals, is an active catalyst. Catalysts tested in this study were supported on clay, pumice, asbestos, and activated carbon. Chlorides of alkaline earth metals proved to be best the promoters for $CoCl_2$. The activity of $CoCl_2$ is followed by that of $FeCl_3$, $MgCl_2$, $CuCl_2$, $NiCl_2$, $BaCl_2$, and $AlCl_3$.

Investigation of low temperature halogenation of organic compounds in solution led to somewhat different rules of catalyst selection. In the *chlorination of benzene* [547], the following activity order was obtained: $FeCl_3 > AlCl_3 > TiCl_4 > SnCl_4$. Also, the data on *bromination of benzene and toluene* [548] do not show any superiority for halides of metals with variable valence. $FeBr_3$ and $BiBr_3$ proved to be active along with $AlBr_3$; and $ZnBr_2$ showed little activity. This indicates a relationship between catalytic activity and Lewis-type acidity: most active are the metal bromides, whose cations (Al^{3+} and Fe^{3+}) exhibit strong polarizing ability. It should be pointed out that, according to data of Tronov and Pershina [548], of all the catalysts investigated, only $ZnBr_2$ and $CuBr_2$ are heterogeneous, while the remainder are used in solution. A heterolytic mechanism applies to these cases. Similar results were obtained in studies of chlorobenzene and nitrobenzene bromination.

7.3. The Decomposition of Oxygen-Containing Substances

The principles of catalyst selection for the decomposition of unstable oxygen-containing compounds point the way to the principles of selection for oxidation reactions. These reactions are often used for modeling oxidation–reduction processes.

The decomposition of hydrogen peroxide occurs readily on numerous solid surfaces in aqueous solution at close to room temperature. Therefore it is very convenient for the purpose of evaluation of predictions for catalyst selection based on impurity conductance and Fermi level. According to the data of Schwab *et al.* [83] and Greger [85], the initial velocity of H_2O_2 decomposition on n-AlSb is higher than on p-AlSb which indicates the acceptor character of the reaction (Section 1.1). In the study of Clopp and Parravano [505] on the contrary, p-type samples of InSb, GaSb and Bi had lower energies of activation than n-type. However, the temperature interval in which the reaction was studied was only 15°C and the range of values of log k was 0.3 (i.e., k differed by a factor of two). Therefore the conclusions about the relationship between reaction velocity and the state of the Fermi level given by the authors are not very convincing.

The calculated logarithm of the average activity of the oxides of the p-type was 0.14 and for oxides of the n-type it was 0.48 according to data of Zhabrova *et al.* [506], i.e., on the basis of a statistical average the n-type semiconductors have some advantage. MnO_2 and V_2O_5 are particularly active. It is interesting to note that in other reactions—dehydrogenation of iso-$C_3H \cdot OH$, oxidation of $CO - V_2O_5$, to the contrary, is the least active of the oxides of the fourth period elements.

Thus conclusions based on the acceptor mechanism of H_2O_2 decomposition have some basis. Somewhat higher activity of n-type semiconductor oxides is given by Hart *et al.* [507] who studied H_2O_2 decomposition in the vapor phase. From the data of ref. [508] p-type semiconductors: NiO, Co_3O_4, and CuO were better than n-type semiconductors: CdO, PbO, Fe_2O_3, and ZnO, and only n-MnO_2 was more active than the other catalysts.

According to Roginskii [509] decomposition of H_2O_2 proceeds simultaneously on two types of regions of the semiconductor: acceptor and donor. In the first the H_2O_2 molecule reacts with the hole and forms oxygen and protons

$$H_2O_2 + p \rightarrow H_2O_2^+ \xrightarrow{p} 2H^+ + O_2$$

and in the second with an electron forming hydroxyl ions

$$H_2O_2 + e \rightarrow H_2O_2^- \xrightarrow{e} 2OH^-$$

In such a scheme of supplementing processes it is desirable to have high electroconductivity; catalyst activity must increase with the

7.3. THE DECOMPOSITION OF OXYGEN-CONTAINING SUBSTANCES

decrease in width of the forbidden zone. Figure 72 indicates a very satisfactory correlation.

Changes in activity of a number of fourth period metal oxides are plotted in Fig. 73, employing data of ref. [506] supplemented by refs. [507 and 508]. Maximum activity occurs on the oxides MnO_2 and Co_3O_4, with minimum on TiO_2 (d^0), MnO, Fe_2O_3 (d^5), and ZnO (d^{10}).

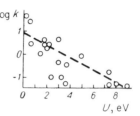

FIG. 72. The correlation between the catalytic activity in the reaction of H_2O_2 decomposition and the width of the forbidden zone.

The oxides of transition metals are considerably more active than the oxides of nontransition metals. This also holds for the carbonates. For example $MnCO_3$, $Fe_2(CO_3)_3$, and $NiCO_3$ are more than an order of magnitude more active than $MgCO_3$, $ZnCO_3$, and $CaCO_3$, and (surprisingly) more active than the corresponding oxides. Among the oxides of nontransition metals PbO has an especially high activity close to that of MnO_2 and Co_3O_4. CaO also exhibits considerable activity as well as other alkali and alkaline earth oxides. The acid oxides TiO_2, Al_2O_3, and SiO_2 show low activity. The high activity of the semiconductors and solid bases should lead to inverse correlation of oxide

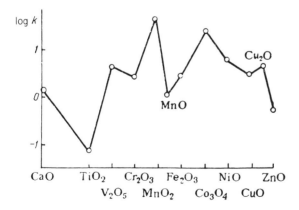

FIG. 73. The change of the logarithm of catalytic activity in a number of oxides of fourth period metals in the reaction of H_2O_2 decomposition according to the data of refs. [506–508].

activity in the reaction of H_2O_2 decomposition with the work function φ. Decrease of activity with increase of φ was rarely observed in the data studied.

Wolski [510] studied H_2O_2 decomposition on the hydroxides of La, Sm, Nd, and Pr. It was not possible to find a correlation for the change of catalytic activity of the rare earth hydroxides.

Hart and Ross [511] studied *the decomposition of tertiary butyl hydroperoxides* in the gas phase at the same conditions at which they decomposed H_2O_2. The following order of activity was found: $Cu_2O > CoO > Ag_2O > Al_2O_3$. The energy of activation did not follow the same sequence and was equal to: 4.6 kcal/mole for Ag_2O; 8.7 kcal/mole for CoO; 13 kcal/mole for Cu_2O with a considerably higher value for Al_2O_3. These values are very close to corresponding values of activation energy of H_2O_2 decomposition which indicates the same mechanisms in both cases.

The decomposition of nitrous oxide to $N_2 + \frac{1}{2}O_2$ was considered as a model reaction in a number of studies concerned with the theory of catalysis. The principles of catalyst selection for this reaction were already discussed in Section 1.1. The majority of the authors [21, 25, 26, 47, 48, 497] consider that N_2O decomposition occurs by a donor reaction, i.e., as the limiting step we have electron transfer from the molecule of N_2O to the catalyst or the reverse transfer of a hole. Let us consider how much experimental data corresponds to these conclusions. The experimental data for N_2O decomposition in the literature are quite numerous [25, 26, 47, 48, 512–515], although they are not very reliable since they are, in general, old references lacking information about the surface, regime, etc.

The statistical analysis of the literature data described previously has confirmed the advantage of the p-semiconductors over n-semiconductors. In the general, the interval of change in the logarithm of the catalytic activity from -2 to $+3.5$ (the mean logarithm of the activity of p-semiconductors) was 1.50 and of n-semiconductors was -0.43. Therefore, on the average, p-semiconductors are more active than n-semiconductors by an order of magnitude. This is shown clearly in Fig. 74 for the case of the oxides of the fourth period where the order of the p-type is several orders of magnitude greater in activity than the oxides of n-type. There are no data given on the activity of MnO_2 in Fig. 74. The study of the activity of this oxide—semiconductor of n-type could furnish the answer to the question of what is more important —the type of conductivity or the electronic structure of the cation.

Cr_2O_3 appears to be a low activity catalyst for N_2O decomposition. However, Winter [497] shows that carefully degassed Cr_2O_3 is very

7.3. THE DECOMPOSITION OF OXYGEN-CONTAINING SUBSTANCES

active but is swiftly poisoned by oxygen. Insulators are even less active than n-semiconductors in spite of the opinion of a number of the above-mentioned authors [21, 25, 26]. The lowest activity occurs with SiO_2, Ga_2O_3, GeO_2, and BeO. The basic oxides were more active than the acid. With the increase in the atomic weight of the metal, which forms an oxide, the activity increases in the same group. For example, consider the transfer from BeO to BaO or from TiO_2 to ThO_2.

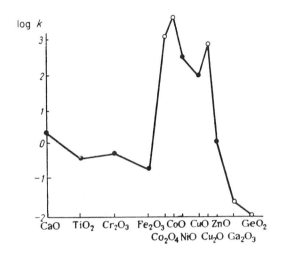

FIG. 74. The change of the logarithm of catalytic activity for a number of oxides of fourth period metals for the reaction of N_2O decomposition.

The effort to obtain a correlation of catalytic activity of oxides in the reaction of N_2O decomposition with lattice parameters has been unsuccessful. There is also no correlation, as noted by Saito et al. [47], between catalytic activity and the difference in the electronegativity, Δx. It is possible that this is caused by poor reliability of the data considered. A correlation between log k and the work function φ, as can be seen from Fig. 75, also does not exist due to the presence of two counteracting tendencies. These are the increase of log k with the increase of φ caused by p-conductivity and the increase of log k with the decrease of φ due to the basic properties of the surface. The unique correlation that exists in this situation results in a correlation of log k with the width of the forbidden zone U (Fig. 76), reflecting the smoothness of this correlation for the high activity of alkali-earth oxides which have highest U.

The values of the activation energy of N_2O decomposition given in the literature are even less reliable than the data of the catalytic activity.

They vary in the wide interval from 16–25 kcal/mole for Co_3O_4, NiO, CuO, and ThO_2, to 50–60 kcal/mole for such oxides as Ga_2O_3, GeO_2, and SiO_2. For the first oxides, reaction occurs at 200–300°C and for the second 700–800°C, i.e., at the temperatures close to the initial homogeneous N_2O decomposition. It is interesting to note that the changes in catalytic activity (which are dependent basically on activation energy as can be seen from the values of k_0) do not differ greatly.

The study of N_2O decomposition on the rare earth elements [449] was given in a chart of the change of catalytic activity which is not reproduced here.

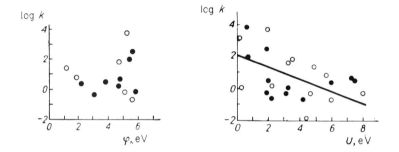

FIG. 75. (left) The absence of a correlation between the catalytic activity and the reaction of N_2O decomposition and the work function.

FIG. 76. (right) The correlation between the catalytic activity in the reaction of N_2O decomposition and the width of the forbidden zone.

The mechanism of N_2O decomposition on oxides was discussed in Section 1.1. On the basis of the principles of catalyst selection, the limiting step of the reaction is the desorption of oxygen with the transfer of a positive hole of catalyst to it. The data showing large poisoning effects on N_2O decomposition speak in favor of this mechanism. At high temperatures, as shown by the data on isotopic exchange [497], this oxygen cannot be distinguished from the oxygen at the surface film of the oxide. According to Stone [515], at high temperatures (equal to or greater than 400–500°C), the reaction can occur without participation of electrons and holes of conductivity by means of an ionic mechanism, for example by means of exchange of oxygen ions between surface and and gas phase. In the case of insulators, stable defects contribute to this mechanism, for example F-centers:

$$N_2O + F\text{-center} \rightarrow N_2 + O^{2-}$$
$$2O^{2-} \rightarrow O_2 + 2F\text{-centers}$$

7.3. THE DECOMPOSITION OF OXYGEN-CONTAINING SUBSTANCES

The decomposition of the nitric oxide to $N_2 + O_2$ is a reaction that takes place at high temperatures. In this reaction [516] at 750–1050°C the oxides of the transition metals Cr_2O_3 and Fe_2O_3 are less active than the oxides of nontransition metals Ga_2O_3, Al_2O_3, and CaO. An increase in the width of the forbidden zone leads, in general, not to a decrease but to an increase in the activity. The reason for such unusual dependency is hard to explain. The change of the activity is basically due to the change of E from 19 kcal/mole for Ga_2O_3 to 31.5 kcal/mole for ZnO and 40–60 kcal/mole for ZrO_2; k_0 changes but little—for the entire interval by only an order of two.

The recently published work by Yureva *et al.* [517] deals with this reaction at considerably lower temperature (250–750°C) on metal oxides of the fourth period. In Fig. 77 is given the graph of the

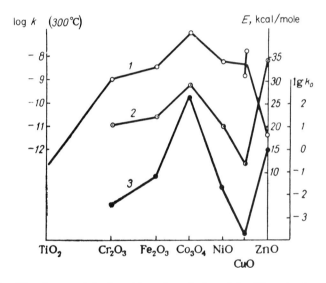

FIG. 77. The change of the logarithm of catalytic activity (1), activation energy (2), and the logarithm of pre-exponential multiplier (3), for a number of oxides of fourth period metals in the reactions of NO decomposition [517].

change of the catalytic activity at 300°C. Most active is cobalt oxide Co_3O_4, less active are ZnO and TiO_2 at the end of the period (in the study of Fraser and Daniels [516] the activity of TiO_2 was also small). The high activity of Co_3O_4 is due to its high activation energy (29 kcal/mole), important at high reaction temperature. It is interesting that the extrapolation of the data given in [517] to temperatures in the range of experiments of previous works (1040°C) shows that at this temperature the difference in the activity of oxides studied: NiO, CuO, Fe_2O_3, Cr_2O_3, and ZnO (except Co_3O_4) should be small, in the range of 1 to 2 orders. This result is in agreement with the data of work [516]. The values of k_0 here are changed in relatively narrow ranges—about three orders, not taking into account Co_3O_4 for which the value k_0 is considerably greater.

Schwab and Hartman studied *ozone decomposition* [518] and concluded that the electron structure of O_3 and N_2O are very similar; and consequently, the principles of catalyst selection should be the same. In fact, the high activity of the solid bases (BaO, KOH) and oxides of the transition metals as compared with nontransition metals was confirmed. In the series of the oxides of fourth period metals, as in the case of N_2O decomposition, the activity increases in going from Fe_2O_3 to oxides of nickel (Fig. 78). Also note that the interval of the

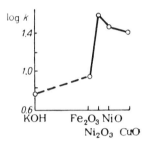

FIG. 78. The change of catalytic activity of the metal oxides of the fourth period and KOH for ozone decomposition [518].

activity change was much smaller—about two orders, although the reaction took place at considerably lower temperatures (20–100°C) than for N_2O decomposition. Very low values of the activation energy (2–5 kcal/mole for the solid bases, oxides of fourth period, and Tl_2O_3 and even for Au_2O_3—13.8 kcal/mole and for PbO_2—20.5 kcal/mole) indicate the possibility of diffusion influence. The change in the activation energy was compensated for, in this case, by a corresponding change in log k_0. Besides, in the case of ozone decomposition in contrast with N_2O decomposition the higher oxides were more active than the lower; and the latter were more active than the metals, for example: $NiO_3 > NiO > Ni$; $Fe_2O_3 > Fe$; $Ago > Ag_2O > Ag$; $CuO > Cu_2O > Cu$; $Au_2O_3 > Au$; $PtO_2 > Pt$; $Tl_2O_3 > Tl_2O$. Therefore, for ozone decomposition, *n*-semiconductors are more active than *p*-semiconductors [519].

The decomposition of solid oxygen-containing salts proceeds with the formation of the new solid phase and the separation of oxygen (topochemical reaction). The kinetics of these reactions differ from the pure gaseous or fluid-phase reactions that were considered above. Nevertheless the principle for catalyst selection for these materials is

7.3. THE DECOMPOSITION OF OXYGEN-CONTAINING SUBSTANCES

often the same as that for the gaseous oxidation–reduction reactions. Indeed, it happened historically that the first reactions, from which the electronic approach to catalysts started to develop, were reactions of the decomposition of oxygen-containing salts [14].

In Section 1.7 it was noted that the most active catalysts for the decomposition of *potassium permanganate* are the oxides of transition metals, in particular colored oxides. From the data of ref. [14], a graph (Fig. 79) of the change in the catalytic activity (at the temperature

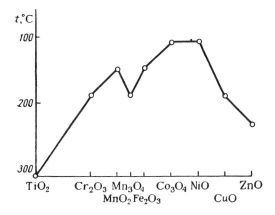

FIG. 79. Change in temperature at the beginning of reaction of $KMnO_2$ decomposition in the series of the oxides of fourth period [14].

corresponding to the beginning of the reaction) was constructed for the oxides of fourth period metals. The maximal activity is exhibited by oxides of p-type which are located at the end of the period (Co_3O_4, NiO, and CuO); but n-oxides (MnO_2 and Fe_2O_3) also have considerable activity, so we cannot conclude emphatically that there is an advantage for p-type semiconductors. The dielectrics MgO and SiO_2 show low activity. One can be convinced easily that here, also, there is a correlation between the activity and the width of the forbidden zone. The data on permanganate decomposition with different cations [191] indicates the dependency of decomposition velocity on polarization action of the cation. The most active are the permanganates of the metals whose cations have higher values e/r: Be^{2+}, Cu^{2+}, Zn^{2+}, and Li^+. According to these data, Fig. 80 gives the change of the catalytic activity for a number of permanganates of the alkali metals. The larger the radius of the ion of alkali metal, the smaller will be the velocity of the decomposition of the corresponding permanganate.

An analogous dependency was obtained for catalysts for the *decomposition of potassium chlorate*. Here, also, the catalytic activity at the middle of the series of oxides of the fourth period increases when going to the second half of the period. The most active [14, 316] was Fe_2O_3—a situation which seldom occurs in heterogeneous catalysis (more often the activity with Fe_2O_3 is at a minimum). According to the data of ref. [520], the most active catalyst is MnO_2, which is the same as for Fe_2O_3 an n-semiconductor. High activity is also exhibited by the oxides of Co, Ni, Cu, Sn, Ag, and Pb.

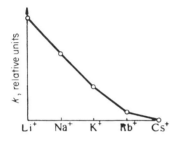

FIG. 80. The decrease of the velocity of decomposition of the permanganate $MeMnO_4$ with the increase of atomic weight of the metal Me [191].

For the *decomposition of potassium perchlorate* in the solid phase, the following order of activity was obtained [521]: $NiO > CuO > MgO > TiO_2 > ZnO$. An especially high activity, more than three orders higher than the rest of the oxides, was obtained with NiO. By the introduction of additives into NiO, simultaneous changes in E and $\log k_0$ are obtained with a compensation effect. If we take different oxide catalysts, always including less active samples of NiO, we can conclude that, in this case, the change in the catalyst activity was due to change of E: from 35 kcal/mole for NiO to 48–53 kcal/mole for Mgo, TiO_2, and ZnO. The value of $\log k_0$ stays relatively constant, exhibiting a maximum change of 1.5 orders (for NiO with additives of 5 orders). Possible reasons for such changes of E and $\log k_0$ were discussed in Section 1.3 (Fig. 13).

In the reaction of *decomposition of calcium hypochlorite* $Ca(OCl)_2$, and also $Ca(ClO_2)_2$, $Sr(ClO_2)_2$, and $Ba(Cl_2)_2$, studied in aqueous solution with hydroxide catalysts [522, 523], the most active was hydroxide of Fe, hydroxides of Cu, Co, and Ni were lower: The order of of activity was: $Fe(OH)_3 > Cu(OH)_2 > Ni(OH)_2 > Co(OH)_2$, obtained analogously [524] by the decomposition of hypobromite NaBrO.

In the decomposition of the perborates $NaBO_3$, $NaBO_4$, and others in aqueous solution the more active catalyst was $Co(OH)_2$, with a lower order for $Cu(OH)_2$ and $Ni(OH)_2$ [525].

For the *decomposition of the oxides of mercury* in the solid state, similar principles of catalyst selection were obtained as for decomposition

7.3. THE DECOMPOSITION OF OXYGEN-CONTAINING SUBSTANCES

of solid $KClO_2$ or $KMnO_4$. The colored oxides exhibited high activity. The oxides of *p*-type of the fourth period should be noted (Fig. 81). In this case, a minimum was noted for Cr_2O_3, which is close to the activity of dielectrics Al_2O_3, SiO_2, and MgO. In the series MgO, ZnO, and CdO, the activity increases with an increase in the atomic weight of the metals.

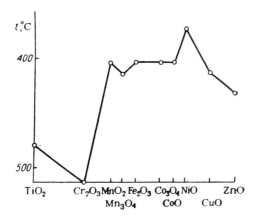

FIG. 81. The temperature change at the beginning of the reaction of decomposition of HgO for the series of oxides of fourth period metals [526].

The nitroparaffins—compounds of a different class, differ in their behavior from the previously considered oxides or salts. Nevertheless, in the *decomposition of nitroparaffins* CH_3NO_2, $C_2H_5NO_2$, and $C_3H_7NO_2$, similar principles of catalyst selection can also be found. The change of initiation temperature for decomposition of nitroparaffins for the oxides of fourth period is given in Fig. 82. The maximum is found for the oxides Cr, Co, and Ni; and the minimum for V_2O_5 and Fe_2O_3. Considerable activity is shown by the basic oxides (CaO and MgO); the acid oxides (Al_2O_3, WO_3, and P_2O_5) show low activity. Hermoini and Salman [527], on the basis of data they obtained, arrive at conclusions regarding the advantage of *n*-semiconductors, as compared to *p*-semiconductors, for this reaction. They suggest a two-step mechanism for the reaction of which the first step is adsorption on the base (obviously on the ion O^{2-}), and the second, limiting step, the transfer of an electron from the catalyst to the reagent. This conclusion, however, does not follow from the data. Actually, *n*-semiconductors can be found among the most active (Co_2O_3, PbO_2, and Ni_2O_3), and also among the least active (Sb_2O_3, V_2O_5, and WO_3) oxide catalysts. The mean statistical temperature at the beginning of decomposition for n-type (214°C) and for p-type semiconductors (210°C), is thus essentially the same. Correlation of the catalytic activity with the width of the forbidden zone is, as can easily be confirmed, satisfactory in these cases.

Patai et al. [528] studied the *catalytic reactions of solid chlorates, bromates, and iodates with polyvinylbenzene*. The reaction was studied using pressed pellets of polydivinylbenzene with the salts ($KClO_3$, $KBrO_3$, KIO_3, and $KClO_4$) and the catalyst. The first step is the decomposition of the solid salt rich in oxygen and the second is the oxidation of the organic substance. In general, the catalysts used were

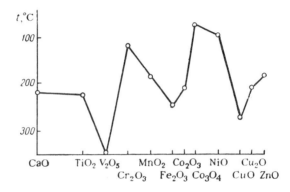

FIG. 82. The change in initiation temperature for decomposition of nitroparaffins for the series oxides of fourth period metals [527].

salts; and only some were oxides of metals or pure metals. The latter did not show any advantages in activity in comparison with the salts and oxides. As there is no information in the literature on such salt properties as the width of the forbidden zone, work function, or even the difference in electronegativity (in this case fewer observations than for binary compounds) a question arises relative to the correlation of the catalytic activity with the properties of the cations. Considering this problem, Patai et al. [528] came to the conclusion that the Lewis acids (i.e., the ions of the metal having a high ratio e/r) are good and the Lewis bases are poor catalysts for these reactions. Actually, the principles are more involved and are complicated by interaction of the cations with anions. For example, one lithium salt (LiCl) is grouped with the better and another (Li_2CO_3) with the worse catalysts. The high activity catalysts were the sulfates of Al and K and also these salts of ammonia: NH_4Cl and NH_4Br, which are probably decomposed at the temperature of the catalysis (250–400°C). Some compounds of the transition metals also

7.3. THE DECOMPOSITION OF OXYGEN-CONTAINING SUBSTANCES

are catalysts. It is curious that among the better catalysts are the salts of Fe^{3+}, and Mn^{2+} and also V_2O_5, i.e., the configurations d^0 and d^5. Less active are the salts of Co^{2+}, Ni^{2+}, Mn^{4+}, and Cu^+. It is possible that, in this case, the stabilization of the crystal field is the factor that is unfavorable for catalysis. One can assume, for example, that oxygen produced by decomposition of the solid salt is firmly connected to the cation of the transition metal and does not participate in further reaction of the polymer oxidation. With weaker bonding of oxygen to metal the reaction may proceed. The least active solids, which do not accelerate but retard this reaction, are the salts having cations with the structure d^{10}: Zn^{2+}, Ag^+, and also some salts of Pb, Ca, and Ba.

The decomposition of *potassium azide* KN_3 probably proceeds by a mechanism similar to that for decomposition of unstable oxygen-containing compounds. In this reaction [529], the following order of the decrease of the catalytic activity of the oxides of fourth period was obtained: $Fe_3O_4 > Co_3O_4 \sim NiO \sim Cu_2O > MnO_2 > V_2O_5 > Cr_2O_3 \sim TiO_2$.

8 • Acid–Base Reactions

8.1. Addition and Removal of Water and Hydrogen Halides

With the exception of dehydration of alcohols and acids, and possibly also halogenation of benzene and toluene, we have thus far considered reactions of the homolytic or oxidation–reduction type. Considerably less reliable data are available in the literature concerning the principles for catalyst selection for heterolytic or acid–base reactions, which we shall consider now. There is almost no single study giving data on the value of the surface of solid catalysts or about the regime in which reaction proceeds.

Heterolytic processes are related to reactions of hydration; and close to them in mechanism are reactions of acceptance of the molecule HX (where X is Cl, Br, I, or Cn) and the corresponding reactions of dehydration and dehydrohalogenation.

The catalysts for *hydration of ethylene and propylene* [549] are the acid oxides: Al_2O_3, W_2O_5, and ThO_2 and the phosphates of Al, Cd, Zn, Ti, Cr, etc. There are no reliable data on the catalytic activity calculated on the basis of surface area. Probably, relative activities of the oxides will correlate with relative activities for the dehydration of isopropyl alcohol (see Table III in Section 4.1). The reaction must proceed through the same intermediate compounds as the reaction of dehydration (diagram 60 in Section 4.1). As was shown, this reaction proceeds on Lewis acid centers. When the surface of Al_2O_3 is free from OH groups, i.e., corresponding to an increase in the number of Lewis centers, the catalytic activity of Al_2O_3 for ethylene hydration [550] is considerably increased.

8.1. ADDITION AND REMOVAL OF WATER

The reactions of addition and removal of the molecule H_2O and hydrogen halide HX are often used as models for checking geometric relations in catalysis. In Section 4.1 it was shown that the catalytic activity for the dehydration of alcohols is not related to the lattice parameter, though on ThO_2 it was possible to form 2 point activated complexes with *cis*-split off of H_2O.

Noller and Ostermeier [326, 327] studied *dehydrohalogenation of alkyl chlorides*: C_2H_5Cl, C_3H_7Cl, *iso*-C_3H_7Cl, and C_4H_9Cl, secondary C_4H_9Cl, tertiary C_4H_9Cl and chlorocyclohexane on a number of catalysts: chlorides of alkali metals, oxides, carbides, and nitrides of Ti, V, Nb, Ta, and Cr. Disregarding the low reliability of these results (absence of surface measurement, data on reaction regime, and small value of the temperature interval 20–40°C), the authors give a number of conclusions on the mechanism of the reaction and the principles for catalyst selection. In their opinion [see diagram (52) in Section 3.1], due to the formation of ring donor–acceptor complexes the minimum activation energy is achieved when the distance between cation and anion is equal to 2.57 Å, i.e., equal to the distance between the atoms of H and Cl belonging to neighboring C atoms in the molecule of the alkyl chloride. A decrease in the energy of activation should contribute to a large difference in the electronegativity Δx of the atoms of metal and the metalloid which form the catalyst. In the case of the chlorides of alkali metals, both these factors, according to the authors, reinforce each other. In the case of carbides and nitrides these factors act in opposite directions: TiC is geometrically more suitable (2.16 Å), than TiN (2.11 Å), but N has larger electronegativity (3.0) than C (2.5); therefore their activity is approximately the same. Ta_2O_5 and Nb_2O_5, and also AgCl, according to Noller and Ostermeier show very high evergy of activation which does not fit into the ion-geometric scheme.

Let us consider how much, in fact, basic principles confirm the model of Noller and Ostermeier.

The change of activation energy for dehydrochlorination for different alkyl chlorides and alkali halides [326] is shown in Fig. 83. On the same graph are shown the data for AgCl, but they are placed between NaCl and KCl (closer to KCl) in agreement with the size of the ionic radius Ag^+ because by such a location a more regular degree of change in E was obtained. On Fig. 84 are given reaction initiation temperatures for the different halides. It can be seen that for the lower alkychlorides and chlorocyclohexane, catalytic activity increases (E and t decrease) and for the higher alkylchlorides it decreases (E and t increase) in passing from RbCl to LiCl. This fact raises the possibility that dehydrochlorination of lower and higher alkylchlorides proceed by different mechanisms.

If we consider all data on all catalysts (chlorides, oxides, carbides, nitrides) and we plot on the abscissa the value of the difference in the electronegativity for each catalyst Δx, and on the ordinate the average (for different alkylchlorides) value of E, we find that there is no

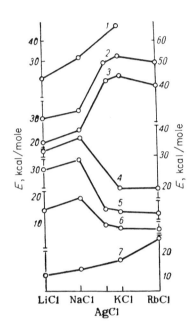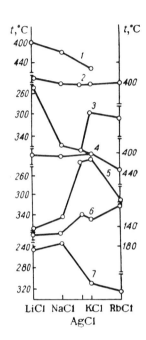

FIG. 83 (left). The change of activation energy for HCl removal in the presence of chlorides of alkali metals and AgCl [326]: 1—C_2H_5Cl; 2—n-C_3H_7Cl; 3—iso-C_3H_7Cl; 4—n-C_4H_9Cl; 5—sec-C_4H_9Cl; 6—$tert$-C_4H_9Cl; 7—chlorocyclohexane.

FIG. 84 (right). The change of initiation temperature for splitting off HCl in the presence of the chlorides of the alkali metals and AgCl [326]: 1—C_2H_5Cl; 2—n-C_3H_7Cl; 3—iso-C_3H_7Cl; 4—n-C_4H_9Cl; 5—sec-C_4H_9Cl; 6—$tert$-C_4H_9Cl; 7—chlorocyclohexane.

regularity that can be seen from Fig. 85. The same conclusion is reached by plotting t instead of E on the ordinate.

In Fig. 86 is shown the relationship between E and the distance d between metals and nonmetals in the lattice grid of the catalyst. It is obvious that, in this case even though the scatter of the points is very great, the same conclusions hold for presence of the minimum of activation energy at the point 2.57 Å, predicted by Noller and Ostermeier [326, 327]. If we consider the relationship between t and the distance d, we obtain a completely different principle (Fig. 87): a monotonic decrease of the catalytic activity with increase in lattice parameter. The second

8.1. ADDITION AND REMOVAL OF WATER

dependency is more realistic, in our opinion, that the first one. Because of the small change in the velocity of the reaction in small temperature intervals (in a number of cases 15–20°C), it is hardly possible to expect great accuracy in the determination of E.

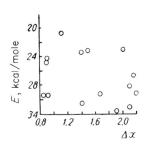 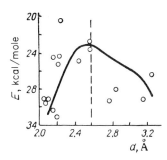

FIG. 85 (left). The absence of correlation between mean [326] activation energy of HCl removal for different alkyl chlorides and the difference in electronegativity between the atoms of the catalyst.

FIG. 86 (right). The connection between the mean activation energy of HCl removal for alkylchlorides and the distance between the atoms of the catalyst.

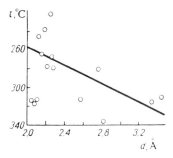

FIG. 87. The correlation between the mean [326] temperature of initiating removal of HCl from alkyl chlorides and the distance between the atoms of the catalyst.

If we compute the arithmetic mean values of E and t (for different alkylchlorides) and consider their dependency on the atom of nonmetal in the compound, we obtain much poorer correlation values. Based on temperatures of the beginning of the reaction (°C) one obtains the following order of the activity: oxides (250) > carbides (269) > nitrides (287) > chlorides (312); for the activation energy (kcal/mole): carbides (25.0) chlorides (27.6) nitrides (28.0) oxides (33.4). It is possible, that the dependency of the catalytic activity (on t) is caused by trivial reasons. Thus it is certain that oxides with high surface area are much easier to obtain than chlorides. If we consider the order of the cations for the carbides, nitrides and oxides as a function of activation energy, we get the order: Ti > Nb > Ta > V > Cr, and in terms of the temperature of the reaction initiation: Ta > Ti > Nb > Cr > V.

The activation energy (for the different catalysts and alkylchlorides) changes from 8 to 53 kcal/mole. Even the same catalyst decomposes different alkylchlorides with completely different energies of activation. For example, on RbCl, E necessary to remove HCl from tert-C_4H_9Cl is equal to 8 kcal/mole, and the E for C_3H_7Cl is 51 kcal/mole. Such a large difference is hard to explain. The range of change of k_0 in this study is ~15 orders. For different reagents on the one catalyst, and for the same alkylchloride on different catalysts, one can find a compensating effect (almost linear) apparently a relationship between E and log k_0. Probably, this could also be due to the low accuracy of measurement.

Noller et al. [551] studied the HCl split-off from isopropylchloride. The velocity of the reaction as calculated on 1 m² of the catalyst surface changes in the order $TiO_2 > Al_2O_3 > SiO_2$.

In ref. [328] isomerization and dehydrochlorination of chlorostilbene (cis and trans isomers) to tolane (diphenylacetylene) was studied:

$$\begin{array}{c} C_6H_5 \\ \diagdown \\ Cl \end{array} C=C \begin{array}{c} C_6H_5 \\ \diagdown \\ H \end{array} \rightleftarrows \begin{array}{c} C_6H_5 \\ \diagdown \\ Cl \end{array} C=C \begin{array}{c} H \\ \diagdown \\ C_6H_5 \end{array}$$

$$\updownarrow \qquad\qquad \updownarrow$$

$$C_6H_5\text{—}C \equiv C_6H_5 + HCl$$

It was shown that cis–trans isomerization on the catalysts studied proceeds very slowly. The order of increase of activation energy for HCl removal from the cis-chlorostilbene is: $CaCl_2 < MnO_2 < CdO, MgO, CaO < SiO_2$, from trans-chlorostilbene: $CaO < CaCl_2, CdO < MgO < Al_2O_3 < BaCl_2 < MnO_2$. This is also the order of k_0 change. The data on the area of the catalysts studied in ref. [328] are not given. In Fig. 88 is given the dependency of the activation energy of dehydrochlorination of the cis and trans isomers on the distance between anion and cation.

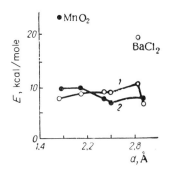

FIG. 88. The dependency of activation energy of dehydrochlorination on the distance between the atoms of metals and nonmetals in the catalyst: 1— cis-chlorostilbene; 2— trans-chlorostilbene.

8.1. ADDITION AND REMOVAL OF WATER

It is obvious that for the majority of the catalysts the value E changes but little; and it is close to the same for both isomers. At higher values of lattice parameters, E for trans removal is somewhat lower than for cis removal. This is contradictory to the previously given rule [316] that, for cis removal of HCl, it is desirable that the distance of metal to nonmetal in the catalyst should equal 2.57 Å. The higher values of E for trans removal for MnO_2 and $BaCl_2$ are inexplicable.

In a study of *dehydrochlorination of 2-chlorobutane* it was found [328] that the energy of activation in the formation of 2-butene was equal to 32 kcal/mole on $CaCl_2$, 26 kcal/mole on CaO, and 10 kcal/mole on CaH_2; for the formation of 1-butene 37 kcal/mole on $CaCl_2$ and 5 kcal/mole on CaO.

Andreu et al. [328] came to the conclusion that in many reactions involving removal and addition of HCl, intermediate compounds can be π-complexes (I), formed by the proton of HCl and double bond of the olefins, and σ-complexes (II):

$$
\begin{array}{cc}
\begin{array}{c} H_3C \\ | \\ HC \\ \parallel \rightarrow H^+\text{---}Cl^- \\ H_2C \end{array} & \begin{array}{c} CH_3 \\ / \\ CH \\ / \quad \oplus \\ H_2C \\ | \\ H\text{---}Cl^- \end{array} \\
\text{(I)} & \text{(II)}
\end{array}
\tag{75}
$$

The role of the catalyst is reduced to the polarization and protonization of HCl molecule. According to data obtained, for the dehydrochlorination of chlorostilbene and chlorobutane, it is difficult to establish any conclusion either on the relationship between the catalytic activity and the dielectric permeability or with the acid–base properties of the catalyst which should be possible if the scheme shown in (75) holds.

The *dehydrochlorination of methylchloride* with the formation of 3-menthene proceeds [552] according to the mechanism of trans removal The most active catalyst is bentonite, followed by NaF $>$ $CaSO_4$ $>$ γ-Al_2O_3 $>$ NaCl $>$ $CaCl_2$ $>$ KF, KI $>$ KBr. The low values of activation energy were found on the catalysts NaCl (6 kcal/mole) and NaF (8 kcal/mole). A high value was found on γ-Al_2O_3 (20 kcal/mole). The data on the specific area of the catalyst surface are not given.

A study of the *dehydrochlorination of ethylchloride* on chlorides of the alkali earth metals [191] showed that the path for change in E was analogous to that for the same reaction on the chlorides of the alkali metals [326]: the values of E increase with an increase in atomic weight of the metal (from 15.6 kcal/mole for $MgCl_2$ to 21.0 kcal/mole for $BaCl_2$). A compensating effect between E and k_0 was also observed in this case.

The *dehydrochlorination of dichloroethane* to vinyl chloride [553] proceeds with the highest velocity on compounds Group IV metals: Ti, Zr, Th, and those close to them and also on the salts of Fe, Al, Cd, Zn, and Co. The proton acids were less active.

The nonprotonic acid mechanism for dehydrochlorination agrees with the test data of ref. [552] and is not in disagreement with the data for the studies of Noller and Ostermeier as well as others [326–328, 551, 552]. An analogous scheme (60) can be written for the catalytic dehydration of alcohols (Section 4.1).

Reactions with acetylene participation: hydration, hydrohalogenation, synthesis of nitriles of acrylic acid $CH \equiv CH + HCN \rightarrow CH_2 = CHCN$, and synthesis of vinylacetate $CH \equiv CH + CH_3COOH \rightarrow CH_2CHOOCCH_3$ proceed on the same type of catalysts. It seems that in all reactions of this type, acetylene enters into the active complex. The activation of the HX molecule is also required though this factor is not as important as activation of the molecule C_2H_2.

Principles for catalyst selection for these reactions were considered in Section 1.6. The most active are Hg^{2+} salts. It is certain that even as early as the ninteenth century the hydration of acetylene to acetaldehyde was typically carried out in the presence of Hg salts (Kuchenov reaction). The salts of Cd^{2+}, Zn^{2+}, Ag^+, and Cu^+ are also active. This was explained (Section 1.6) by the ability of the cations with filled d-orbitals (i.e., structure d^{10}) to participate in donor interaction [184, 185]. As far as anion influence is concerned, the activity of the corresponding salts Hg^{2+}, Cu^+, and Ag^+, in reactions with acetylene participation decreases in the order: $ClO_4^- > NO_3^- > SO_4^{2-} > Cl^- > Br^- > I^- > CN^-$. The stronger, according to Flid [184], the covalent character of the bond between cation and anion the smaller will be the catalytic activity of the cation.

The catalytic activity of cations is decreased by the transfer to the left or right from the first and second side subgroups in the long periods. For example, Zn^{2+} is more active than Ni^{2+}; Ni^{2+} is more active than Co^{2+} and Fe^{2+}. In the vertical direction, the activity decreases from the bottom up (from Hg^{2+} to Zn^{2+} and further to Mg^{2+}). The exception is the first side subgroup where order is: $Au^+ > Cu^+ > Ag^+$. Silver compounds are prone to the formation of stable acetylides and not complex compounds of Ag, and are, therefore, less active.

The order of activity of salts in solution is usually preserved upon applying them to carriers: active carbon, pumice, etc. For example, in the hydrochlorination of acetylene on solid chlorides applied on active carbon, the most active was $HgCl_2$. The reaction was also catalyzed by $CuCl_2$, $FeCl_3$, $PtCl_4$, and $BaCl_2$ [554].

8.1. ADDITION AND REMOVAL OF WATER

Oxides, phosphates, tungstates, sulfates, and other salts are also employed. According to the data of [555–559], for hydration of acetylene the most active of a number of phosphates are: $Cu_3(PO_4)_2$, $Cd_3(PO_4)_2$, and $Zn_3(PO_4)_2$; the phosphates of Hg are somewhat less active. From a number of tungstates and molybdates, the most active are Cd compounds. ZnO was also an active catalyst [559].

From a study of the vapor phase synthesis of acrylonitrile over the oxides of Zn, Cd, Hg, and Bi, the most active was also shown to be ZnO [185]. In the synthesis of vinyl acetate, the salts of Zn were more active than Cd and Mg salts [560].

The study of anion influence on the vapor phase hydration of acetylene over Cu compounds [558] led to the following order of the activity: $CuSO_4 > Cu(CH_3COO)_2 > CuVO_4 > Cu_3(PO_4)_2 > Cu(OH)_2 > CuI_2 > CuCl_2$.

Except for the cations with filled d-orbitals, a number of other cations, such as Bi^{3+}, are also active. Flid [184] explains the high catalytic activity of $SbCl_3$ and especially $BiCl_3$ when applied on active carbon, by placing s-electrons into outer orbitals of cations of the carrier and establishing conditions by which the cation will have filled outer d-orbitals.

The reactions of acetylene can proceed also by the proton-acid mechanism. In particular, H_3PO_4 applied on carbon, catalyzes the hydration and hydrochlorination of acetylene [557].

By catalytic hydration and hydrochlorination of acetylene in the presence of oxides and salts of alkali earth metals, donor bonds of the cation with the molecule C_2H_2 cannot be formed because of the absence of the outer filled d-orbitals. In this case, and also in catalysis by the salts of Group VIII metals [185], catalytic activity is determined by the acceptor properties of the cation, i.e., nonprotonic acid catalysis takes place. Actually the order of decrease of the activity of the oxides, phosphates and other compounds in reaction of acetylene: $Mg^{2+} > Ca^{2+} > Sr^{2+} > Ba^{2+}$ corresponds to the order of decrease of e/r ratio of the cations. Very low activity of BeO in the synthesis of vinyl acetate [561] and $Be_3(PO_4)_2$ in the hydration of acetylene [559] is unexplained. Note that the activity of typical Lewis acids: $AlPO_4$ and Al_2O_3 does not exceed but is often lower than that of compounds of alkali earth metals.

It is possible that one should use catalysts which activate the triple bond for promoting the reaction of dehydrochlorination of chlorostilbene discussed above. In ref. [328] the catalyst with the electron configuration of the cation d^{10} (CdO) also had no advantage compared to the others.

In the reaction of *hydrolysis of chlorobenzene* [562, 563] the most active catalysts are the salts of $CuCl_2$ applied on a carrier (silica gel is effective). The proton acids (H_3PO_4) and the acid oxides of the fourth group metals (SiO_2, TiO_2, ThO_2, and ZrO_2) are also active.

The *hydrolysis of prussic acid* [564] $HCN + H_2O \rightarrow CO + NH_3$ also proceeds by an acid mechanism. The oxides and the sulfides of metal ions with a high value of e/r: Al, Ce, Th, Ti, and Zr are very active catalysts for this reaction. The low catalytic activity of Fe_2O_3, as compared with its neighbors in the fourth period, oxides Mn and Co (Fig. 89), is surprising. Note that according to the crystal field theory and ideas about Lewis acids (Section 2.2) the systems Fe^{3+}, not having a stable crystal field, should be stronger acids and catalysts than Co^{3+} and Mn^{4+}. The catalytic activity, in this case, as in the majority of other investigations of heterolytic reactions, was determined on the basis of total product.

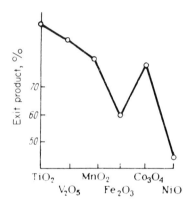

Fig. 89. The change in catalytic activity of the oxides of fourth period metals on the output of the product in the reaction of hydrolysis of the prussic acid [564].

Splitting of glyceryl phosphate and the conversion of sodium meta- and pyrophosphates into ortho-phosphate [565] occur in water solutions of pH = 7–9 by the catalytic action of various hydroxides. The highest catalytic activity was found among the hydroxides of rare earth elements which certainly have basic properties. The strong bases: $Mg(OH)_2$, $Ca(OH)_2$, and TlOH, were catalytically inactive or only slightly active. Hydroxides of the transition metals Ni, Mn, Co, Fe, etc. also have low activity. The active hydroxides are Th, Zr, Y, and Ti. It is obvious that, in this case, catalytic activity is related to Lewis acidity, i.e., with polarization action of the cation on which adsorption of the reacting molecule takes place.

We conclude this survey of the reactions involving removal of H_2O and HX, by noting that the most widely accepted mechanism for these reactions, judged on the basis of the principles of catalyst selection, is

8.2. The Reactions of Condensation and Polycondensation

The reactions described in this section are considerably more complex in mechanism than the ones considered so far. Therefore, the amount of data available on any one of them is limited

A large number of solid catalysts for *aldol reactions*:

$$CH_3NO_2 + HCHO \rightarrow CH_2CHNO_2 + H_2O \qquad (a)$$
$$CH_3CHO + HCHO \rightarrow CH_2CHCHO + H_2O \qquad (b)$$
$$CH_3COCH_3 + HCHO \rightarrow CH_2CHCOCH_3 + H_2O \qquad (c)$$
$$CH_3CN + HCHO \rightarrow CH_2CHCN + H_2O \qquad (d)$$

were studied by Malinowsky and his co-workers [566–568]. As is the case for homogenous fluid phase reactions of aldol condensation they are accelerated by the presence of bases. For example, in reaction (b) acrolein synthesis from acetaldehyde and formaldehyde, the catalysts employed [566, 567] were alkali and alkaline earth hydroxides applied on silica gel. Catalytic activity increases with an increase in the atomic weight of the metal (Fig. 90). The velocity of reactions (b), (c), and (d) on a silica gel catalyst increases linearly with an increase of Na^+ content on the SiO_2.

The authors propose that the active centers are the O atoms in the formation $\leqslant$Si—O—Na. This is in perfect agreement with ideas about the mechanism of catalysis on solid bases discussed in Section 2.1. On such a center, heterolytic removal of the proton from the carbon atom of a neighboring carbonyl group can proceed with the formation of a carbanion or polarized complex. For example:

$$\begin{array}{c} Si \\ \diagdown \\ O^{\delta-} \cdots H^{\delta+} \cdots \overset{\overset{\displaystyle R}{|}}{CH} \\ \diagup \qquad \qquad | \\ Na \qquad \qquad HC=O \end{array} \qquad (76)$$

This step of the reaction is limiting, subsequently the fast combination of the second molecule occurs with the split-off OH-group. It is possible that the latter is split off from the enol form of the carbonyl compound. The detailed scheme is difficult to present at the present time because of the lack of data on the mechanism of the reaction and on the principles of the catalyst selection. It is possible that, for the joining of the second molecule, some geometrical condition must be satisfied between the atoms of the surface layer of the catalyst.

In the reaction of *condensation of propionic aldehyde* with the formation of 2 methyl-2-pentanal the following order of activity was found: $Ca_3(PO_4)_2 > MgSiO_3 > Li_3PO_4 > Ca(OH)_2 > CaCO_3$.

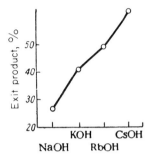

FIG. 90. The change of the catalytic activity (on the yield of the product) for a series of bases applied on silica gel in the reaction of aldol condensation [567].

For the *condensation of cyclohexanone* into a cyclic ketone with two six-membered rings, the most active catalyst was $Ca(OH)_2$. Next follow catalysts with more acid properties: $Ca_3(PO_4)_2$, Al_2O_3, Li_3PO_4, Na_3PO_4, and $Li_2P_2O_7$ [569]. Available data confirm that the aldol condensation fits into the class of reactions which are base catalyzed.

In the reaction of the *condensation of acetaldehyde* with furfural, MgO is shown to be more active than the solid catalysts with more basic character [570]: CaO, K_2CO_3.

By studying the *crotonic condensation of acetaldehyde*, $CH_3CHO + CH_3CHO \rightarrow CH_3CHCHCHO + H_2O$ [571], it was found, on the contrary, that the acid oxides and the hydroxides of Ti and Al are more active than hydroxides of Mg and Ca and also than the oxides of the transition metals: Mn, Fe, and Ni. In the series of phosphates of the metals of Group II [572], a regular lowering of the activity with increase in the atomic weight of the metal occurs, i.e., with the decrease in value of e/r of cation. It is interesting that the phosphates of the metal oxides of the side subgroups Zn and Cd do not differ much in catalytic properties from the phosphates of the metals of main subgroups Ca, Sr, and Ba (Fig. 91).

A lower value of the catalytic activity which does not correspond to general principles was observed with phosphates of Be. Maximum catalytic activity exists in the case of Mg phosphates.

For *acetone synthesis* by vapor phase conversion of acetaldehyde [572], oxides with a basic character were considerably more active than those with acidic properties. The exception was Fe_2O_3, which showed high activity. The oxides of the transition metals MnO_2, NiO, and CuO

8.2. REACTIONS OF CONDENSATION

had no advantage over the oxides of nontransition metals: ZnO, PbO, and CaO.

Condensing decomposition of ethyl alcohol leads to the *synthesis of divinyl (butadiene)*—this is the process which, after discovery of active catalyst by Lebedev was used for a long time as the basic process to obtain this monomer for synthetic rubber production:

$$2C_2H_5OH \rightarrow C_4H_6 + 2H_2O + H_2 \qquad (77)$$

The following order of activity of oxides can be given for production of divinyl [367]: $MgO > ThO_2 > ZrO_2 > Fe_2O_3 > Al_2O_3 > CaO > TiO_2 > Cr_2O_3 > ZnO > SiO_2 > NiO > Co_2O_3 > CuO, SrO$.

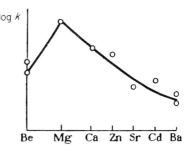

FIG. 91. The change of the catalytic activity for the series of phosphates of the Group II metals for protonic condensation of acetaldehyde [572].

Consideration of the principles of catalyst selection shows that the order of activity is not identical to that obtained for dehydrogenation and dehydration of alcohols (Section 4.1). The active catalysts for dehydrogenation of alcohols, alkaline earth oxides and oxides of the transition metals, exhibit low activity for divinyl synthesis. The most selective catalyst was MgO, whose condensation properties combine dehydrogenation and dehydration activity which is required in Eq. (77). The amphoteric oxides ZnO_2, Fe_2O_3, and Al_2O_3, also are selective. In the other studies (see, e.g., [573]) it was found that fairly active catalysts for divinyl synthesis are the oxides Ta_2O_5 and MoO_3.

In Section 4.1, it was shown that MgO is a relatively poor catalyst for dehydrogenation and an even poorer catalyst for dehydration. Its ability to catalyze three reactions simultaneously leads to its high selectivity. In agreement with the principles given in the first part of the book, the dehydration properties of MgO are due to the Lewis acid—Mg^{2+}, with a relatively high value of e/r; for condensation properties, the proton acceptor center O^{2-}; and for dehydration properties the presence of high charges on the surface ions contribute to the activity of the solid base to catalyze homolytic reactions.

If we assume that the reaction (77) proceeds in several steps, then, for each step, one could select separately a more active catalyst than MgO. Therefore, mixed catalysts should be more active. Actually a highly selective catalyst for butadiene synthesis [574] was found to be a mixture of the oxides Al_2O_3 and ZnO. The first is an active catalyst for dehydration and the second for dehydrogenation; and both can promote also the reaction of condensation. More active catalysts than MgO for the synthesis of divinyl include also mixtures of Al_2O_3 with MgO, Cr_2O_3, or CuO.

It is interesting that Ta_2O_5 and MoO_3 appear to be active catalysts for divinyl synthesis by other methods. For example, from 4-oxybutanone-2 and isopropyl alcohol or the method of one step condensation of acetone, formaldehyde and isopropyl alcohol [575]. MgO also is used in a number of active catalysts for these processes.

In the *Knoevenagel condensation*, the reaction of cyclohexanone with ethyl cyanacetate [576], the catalytic activity of the fluorides progresses in the following way: RbF > CsF > KF > NaF > LiF. Thus the most active were fluorides having a highly basic character.

For the *esterification of rosin* [577] basic catalysts exhibit somewhat lower activity than ZnO.

By the *dehydrogenative dimerization of methallyl alcohol* $CH_2=C(CH_3)-CH_2OH$ catalytic activity increases with increase of alkali properties [578]. KOH was the most active catalyst.

The cyanization of allyl alcohol with ammonia [579]

$$CH_2=CHCH_2OH + NH_3 \rightarrow H_2O + CH_2=CHCH_2NH_2$$
$$CH_2=CHCH_2NH_2 \rightarrow CH_3CH_2CN + H_2$$

proceeds in the presence of metallic and oxide catalysts. The catalytic activity of ZnO was close to that of metals. With an increase in acid properties of the oxide a decrease in activity occurs.

The caustic alkalies KOH, NaOH, and also $Ca(OH)_2$ were active catalysts for the *condensation of chloroform with acetone* to form 1,1,1-trichlor-2-methyl-propanol-2. LiOH, Na_2CO_3, and NaCN [580] were less active.

In some condensation reactions acid catalysts are more active. For example, this is the case for the condensation of aldehydes and ketones with diols to form dioxanes and dioxalanes [581]. In all cases, the most active are Lewis acids: $TiCl_4$, $SnCl_4$, etc.

The condensation (alkylation) of aniline with ethanol to form ethylaniline and diethyl aniline [582] proceeds fastest on aluminum silicates, Al_2O_3 and $AlPO_4$. ZnO is less active. In ref [583], as in the majority of

8.2. REACTIONS OF CONDENSATION

references studied for the reaction of condensation, there are no data for the specific surface of the catalyst involved.

The dehydrocondensation of trialkyl and triaryl silanes with oxiorganic compounds:

$$RR_2'SiH + R''OH \rightarrow RR_2'SiOR'' + H_2$$

was studied [583] in the presence of metal halides as catalysts. In spite of the small ionic radii of the metals which form these halides, they exhibit considerably different catalytic activities. There were no regular differences in the activity for compounds of transition metals from the nontransition. The order of chloride activity was: $NiCl_2 > ZnCl_2 > GeCl_4 > SnCl_2 > SnBr_2 > CoCl_2 > CrCl_3$.

The condensation reaction leading to the formation of high molecular weight products are called polycondensations. For the *condensation of formaldehyde*[1] into polyoxioxo compounds (sugars) a unique active catalyst appears to be PbO. The remaining oxides studied, among which were acidic (TiO_2 and ThO_2), basic (BaO, SrO, and MgO), oxides of the transition metals (Cr_2O_3 and CoO) and others, showed no catalytic activity [584]. Butlerov who discovered this reaction, and other investigators showed that it can be accelerated by $Ca(OH)_2$.

The polycondensation of aminoenanthic acid [585] in the solid phase proceeds also most readily in the presence of MgO. Thus MgO appears to be the best catalyst of the group for condensation. The acid catalysts TiO_2, $SnCl_4$ are less active than the basic though boric acid displayed considerable activity.

The reaction of the polyisocyanates with polyesters to form polyurethanes was carried out in the pressence of solid acetate catalysts [586]. The very alkaline acetates of Na and K were shown to be very active catalysts. A great number of catalysts—the salts of transition and nontransition metals and metallo-organic compounds—were tested in the study [587]. Here, the following order of the catalytic activity of the salts was given: Bi > Fe > Sn > Pb > Ti > Sb > strong bases > Co, Zn salts > amines. The metal ion according to the authors opinion, [587] coordinates with the oxygen atom of both reacting molecules: isocyanate R—N=C=O and hydroxyl containing compound R'OH:

$$\begin{array}{c} R-N=C^{\delta+}=O^{\delta+} \\ \diagdown \\ Me^{\delta+}X_2^{\delta+} \\ \diagup \\ H-O \\ \diagdown \\ R' \end{array}$$

---

[1] In this case, in the condensation $nCH_2O \rightarrow C_nH_{2n}O_n$ the split-off of low molecular products does not occur but, rather a shift in the light (hydrogen) atom for every step of the condensation mechanism.

In the absence of the metal ion the reaction can proceed by an acid–base mechanism.

The *polycondensation of benzylchloride* was carried out in the liquid phase in the presence of soluble chlorides of the metals of Group IV [588]. In the main subgroup the catalytic activity increases with increase of the atomic weight of the metal: $SiCl_4 < GeCl_4 < SnCl_4$ and decreases in the side subgroup: $TiCl_4 > ZrCl_4 > ThCl_4$. One can propose that the catalytic activity of metal chlorides is, in this case, related to their Lewis acidity. Completely analogous results were obtained by the study of *polycondensation of* α-*chloromethyl naphthalene* [589]. The most active catalysts ($TiCl_4$ and $SnCl_4$) led to higher molecular weights of the polycondensation products. Very low activity of $SiCl_4$ is explained by its weak acid properties and by steric hindrance due to formation of intermediate complexes with the benzyl chloride or with chlormethylnaphthalene [589].

The present short survey indicates that the reactions of condensation proceed by different mechanisms. It can also be said that increase in the catalytic activity with the increase in the basicity of solid catalyst surface is the most frequently observed rule for reactions of this type.

8.3. Isomerization Reactions

In spite of the great practical importance of the isomerization reactions, there are few scientific data which have appeared in the literature on the principles of catalyst selection for these reactions. The majority of investigations deal with the study of conversion of individual hydrocarbons.

Skeletal isomerization of paraffins and olefins [590] takes place on two types of catalysts. For example, the isomerization of pentane into isopentane or of butane into isobutane proceeds on acid catalysts: alumosilicate, $AlCl_3$, $AlBr_3$, and H_3PO_4, applied on crushed firebrick or active carbon, etc. The mechanism of the reactions, according to the majority of investigators, is by carbonium ions (Section 2.1). This follows from the fact that $AlCl_3$ or Al_2O_3 in the absence of traces of H_2O or protonic acids are not very catalytically active.

Recently, wide industrial use has been made of paraffin isomerization in the presence of hydrogen (plat-forming, hydroforming, etc.). This process probably proceeds by a second mechanism: split-off of H_2, and isomerization of olefin, followed by hydrogenation. Therefore, polyfunctional catalysts are applied. For example, with $Pt-Al_2O_3$, one component, i.e., the metal or semiconductor, catalyzes split-off and

8.3. ISOMERIZATION REACTIONS

recombination of H_2, and a second, i.e., acid, catalyzes isomerization of the olefins. For the semiconductor component, catalysts such as MoS_2, NiS, WS_2, and others are used. Data on the catalysts for pentane isomerization [9] show that the oxide of Mo exhibits simultaneously both semiconductor and acid properties and is more active than the "pure" semiconductor catalyst Cr_2O_3 or "pure" acid catalyst Al_2O_3.

The isomerization of cycloparaffins into olefins apparently proceeds by the same mechanism as olefinic isomerization—through the formation of carbonium ions. The *isomerization of cyclopropane into propylene* on the protonic acids proceeds at a high velocity [591]. The individual oxides (Al_2O_3 and SiO_2), which have small proton acidity, are catalytically inactive. The regularity of change of the catalytic activity in the reaction of *isomerization of cyclohexane* into methyl cyclopentane (as is the case for cyclohexane cracking) is also explained by Kelechits [422] from the point of view of carbonium-ion theory. Relatively low isomerization activity of MoS_2 and WS_2 can be increased tremendously by applying these components on an acid carrier.

For *isomerization of 1-butene into 2-butene*, strong protonic acids: H_2SO_4, H_3PO_4, alumosilicate, and activated clays are the most active catalysts [324, 590, 592–594]. The reaction occurs on them appreciably at room temperatures. Somewhat lower activity is shown by aprotonic acid catalysts: Al_2O_3, $Al_2(SO_4)_3$, $CuSO_4$, and ThO_2. The metals Ag and Ni, and semiconductors Cr_2O_3, ZnO, and the others have no advantage in activity over the dielectrics, the latter are even better. From among the metal silicate catalysts, the most active is alumosilicate, followed by magnesium silicate, nickel silicate, and zirconium silicate [594]. Among the semiconductors [593], the most active is NiO, followed by (according to a number of studies) Cr_2O_3 and ZnO.

Turkevich [324] proposed that the principles for catalyst selection for this reaction can be related to the presence of an optimal distance between proton-donor and proton-acceptor groups of the catalyst and the formation of cyclic complexes with the reacting olefin (Section 3.1). The chemical nature of the catalyst according to his view is of lesser importance. So nickel hydroxide falls into the same group in activity as H_2SO_4. From the data [592], however, it appears that the former is considerably less active. Nevertheless it is uncertain on an *a priori* basis which groups are to be taken as proton donors and which as proton acceptors. This is especially the case for catalysts of oxides, salts, and metallic types which do not contain hydrogen.

On the alkali catalysts (LiOH, NaOH, KOH, CaO, and Na/Al_2O_3) isomerization of 1-butene is very selective at 300–500°C (from 75 to 96%) in the direction of formation of *cis*-2-butene. In this case, dual site adsorption is more probable than in the cases considered by Turkevich. Here, the increase in the basic properties of the surface, for example, in going from LiOH to NaOH and KOH, results in increased reaction velocity. The overall velocity of isomerization is also considerably lower than on the acid catalysts. On solid acids, for such oxides as Cr_2O_3 and ZnO and also on metals, on the contrary, a considerable amount of *trans*-2-butene is formed.

It appears that solid bases are not at all able to catalyze skeletal isomerization of the olefins and cis–trans isomerization. The isomerization of the olefins, together with a shift of double bond, can proceed simultaneously, as has been shown by Pines and Schaap [595] and Gostunskaya and Kazanskii [596], on solid bases with a high velocity. The most active low-temperature catalyst for such isomerization appears to be $CaNH_2$. According to ref. [597], the reaction proceeds through the intermediate formation of carbanions:

$$R-CH_2-CH=CH_2 + CaNH_2 \rightarrow [R-\overset{-}{C}H-CH=CH_2]Ca\overset{+}{N}H_3$$

$$[R-CH=CH-\overset{-}{C}H_2]Ca\overset{+}{N}H_3 \rightarrow R-CH=CH-CH_3 + CaNH_2$$

At low temperatures, the isomerization of 1-butene into 2-butene is also catalyzed by $Mg(OH)_2$ [598]. BeO has more acidic properties. On it both isomerization with shift of C=C bond and skeletal isomerization proceed at high temperatures.

According to ref. [599], the catalytic activity for isomerization of *cis*-2-butene into *trans*-2-butene (and vice versa) decreases in the following order: protonic acids > solid aprotonic catalysts > semiconductors > solid bases.

The isomerization of 1-pentene on the hydrides of Ca and Ba [403] proceeds with shift of the double bond and formation of 60–70% *trans*-2-pentene and 20–30% *cis*-2-pentene. Skeletal isomerization was not observed. BaH_2 was more active than CaH_2 due to its more basic properties.

In the *isomerization of 2-methyl-1-pentene* the connection of catalytic activity with acid properties of the surface was studied [600]. The order of activity of the catalysts studied calculated on the basis of 1 m² of surface was: cation exchange resin KU−1 > H_3PO_4 applied on Kieselguhr > alumosilicate > Al_2O_3 > SiO_2. Silica gel was, in general, inactive. Of these catalysts, alumosilicate exhibited the highest activity based on a unit acid site.

The semiconductor catalysts (V_2O_3, ThO_2, CrO_2, and MoO_3) at moderately high temperatures (250–450°C) effect isomerization of

8.4. CRACKING AND ALKYLATION OF HYDROCARBONS

C_4–C_8 olefins with shift of the double bond, but do not catalyze skeletal isomerization [598]. In the case of oxides of metals of Groups VII and VIII applied in their pure form (without carrier), isomerization cannot be studied conveniently because of the simultaneous occurrence of cracking and olefin polymerization reactions.

Data on isomerization catalysts for the organic compounds (non-hydrocarbons) are very few. In ref. [601] there were studied heterogenous catalysts for cis–trans *isomerization of 2-benzylidene-3-keto-2, 3-hydronaphthene*. Based on their activity, they can be ordered in the following manner: Al_2O_3, BaO, CaO, and MgO > ZnO, TiO_2, $Ba(OH)_2$, $Ca(OH)_2$, and $Mh(OH_2)$ > Cu_2O > $Al(OH)_3$, > $B(OH)_3$, SnO, Fe_2O_3, and PbO.

The velocity of *isomerization of p-dichlorobenzene* [602] in the presence of $AlCl_3$ can be changed considerably under the influence of solid cocatalysts (or carriers). It appears that the highest activity is exhibited by the Lewis acids ($MgSO_4$, Al_2O_3, and TiO_2) or Bronsted (H_2SO_4 and H_3PO_4). Less activity was shown by semiconductors.

8.4. Cracking and Alkylation of Hydrocarbons

Judged on the basis of volume of industrial production, catalytic cracking is the largest commercial catalytic process. Corresponding to the importance of the process there are a large number of studies devoted to it. The majority of these studies deal with catalysis on the alumosilicates and their compounds: Al_2O_3 and SiO_2 [219, 220], and rarely on $AlCl_3$ and the other chlorides.

The high catalytic activity of the alumosilicates is explained, by the majority of investigators, on the basis of the presence of protonic acids in them. The suggested structure of the active center was given in the schemes (35) and (36) in Section 2.1. The cracking reaction proceeds by a heterolytic mechanism through the carbonium ion. The Lewis acids, for example, Al_2O_3, are catalytically less active. The cracking of hydrocarbons is carried out on them by a completely different, possibly homolytic, radical mechanism. This is confirmed, in particular, by the considerable dehydrogenation activity of Al_2O_3 under conditions of cracking [227].

Cracking can proceed not only on alumosilicate but also on zirconium silicate and magnesium silicate. The order of change of catalytic activity of the silicates of different metals in the *cracking of cumene* was studied in ref. [603]. This reaction is often used both for modeling and for study of the principles of the cracking and shows the following sequence:

$MgO \cdot SiO_2 > Al_2O_3 \cdot SiO_2 > ZrO_2 \cdot SiO_2 > ZnO \cdot SiO_2 > PbO \cdot SiO_2 > CaO \cdot SiO_2 > SrO \cdot SiO_2$. Activity differences can be explained by reference to acidic properties. Magnesium silicate, however, does not fit into the general dependency because its acidity is lower than that of alumosilicate.

The acid and the catalytic properties of magnesium silicate are explained by the analogous scheme (35) for alumosilicate, replacing Si with Mg in the fourth coordination, which is required for the neutralization of two protons [218, 220]. Zirconium is tetravalent, as is silicon, and the same reasoning applies [220] that in the zirconium silicates its coordination number can be changed and also will lead to protonic acidity.

Maki [603] relates the order of change of activity of silicates to the value of ionic radius of the cation. In Fig. 92 is presented the dependence of the catalytic activity of the silicates on the polarization action of the metal ion, as determined by the value e/r. The activity increases with growth of e/r (i.e., decreases with increase of the cation radius). Magnesium silicate quite evidently does not fit into the correlation.

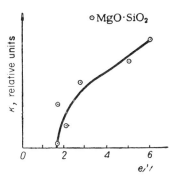

FIG. 92. The dependency of the silicate catalytic activity in the cumene cracking reaction [603] on the charge-to-radius ratio of the cation of the catalyst.

Substitution of the proton on the metal ion in alumosilicate as a rule decreases its catalytic properties in cracking and some other reactions of hydrocarbons (for example, isomerization). Bitepazh [604] studied the cracking of the butyl benzene on alumosilicate after ion exchange with different metal salts. It appears that alumosilicate is most active after exchange with Al^{3+}, Th^{4+}, or NH_4^+ ($\to H^+$). Zn^{2+} and Mg^{2+} give lower activity, and Ba^{2+} and Na^+ poisoned the catalyst. A similar study was made by Danforth [605]. The following order of the activity of the cation exchanged alumosilicate in the cracking of the cetane was found: $Ba^{2+} < Cs^+ < K^+ < Na^+ < Li^+$. In the study [606] on the cracking of cumene and decane it was found that Ca-alumosilicate

8.4. CRACKING AND ALKYLATION OF HYDROCARBONS

is more active than Na-alumosilicate. The least active, unexpectedly, appears to be H-alumosilicate. In this case treatment of the H form of alumosilicate with alkalies did not lead to deactivation, as is often the case, but instead to activation of the catalyst.

The catalytic activity of type X and Y zeolites in the cation form has been studied for *hexane cracking* [579] and cumene cracking [599, 607]. The following order of activity was found: BeY $\sim$ MgY $>$ CaY $>$ SrY $>$ MgX $>$ BaY $>$ CaX $>$ NaY $\sim$ NaX; and, according to the data [382], with alkali metals substituted for zeolites: LiNaX $>$ NaX $>$ KNaX $>$ PbNaX. Thus, here we also find a marked connection between catalytic activity and the polarization ability (e/r) of the cation. Divalent cations are more active than monovalent. The energy of activation for cumene cracking on CaX equals 38–40 kcal/mole, and on NaX—47 kcal/mole. The authors propose that in this case catalysis proceeds not on protonic but on the cationic centers, which form a strong electrostatic field. A molecule of hydrocarbon is polarized on them: $C^+-H^- \cdots Me^+$ or perhaps the hydride ion is split off with the formation of a carbonium ion.

The *cracking of heptane and octane* was studied [608] on the rare earth elements applied on Al_2O_3. Only $Al_2O_3 + Pr_2O_3$ appear to be more active than Al_2O_3. The rest of the rare earth elements: Nd_2O_3, Yb_2O_3, Sm_2O_3, and La_2O_3 led to somewhat lowered catalytic activity of Al_2O_3. Butane cracking to $CH_4 + C_3H_6$, was studied on lanthanide oxides [609] and showed very small differences in their activity compared on the basis of unit surface. The most active of these oxides is PrO_2.

The *reactions of alkylation* are to a considerable degree the reverse of the reaction of cracking. Therefore, the principles for catalyst selection for both types of reaction should be the same. Owing to the considerably lower temperatures of alkylation (from -30 to $+100°C$ as compared with that of cracking of 400–550°C), these reactions are conducted using homogeneous fluid phase protonic catalysts: H_2SO_4, HF, BF_3, and $AlCl_3$. The last two also act by the protonic mechanism (in the presence of H_2O, HCl, etc.).

The study of *benzene alkylation* by methanol [185] at 350°C and high pressure in the presence of HCl gave the following order for the catalytic activity of various salts applied on Al_2O_3: $ZnCl_2 > CdCl_2 > CuCl > HgCl_2$. According to the opinion of Gelbshtein *et al.* [185] proton catalysis also takes place, and the catalysts are complexes $H_n[MeCl_{m+n}]$ (where $n = 1, 2$).

The clearly evident connection between acidity and catalytic activity appears in the study of *alkylation of phenol by isobutylene* with the formation of di-*o,m*-isobutyl phenols [610]. The most active catalysts are alumosilicates and TiO_2. Simple oxides with less developed acidic properties: CuO, Al_2O_3, PbO, Fe_2O_3 were not active. It is interesting that in certain reactions of alkylation, for example in the alkylation of olefins with alkyl aromatic compounds in the side chain, alkali catalysts [595] exhibit catalytic activity.

9 • Polymerization Reactions

The reactions of polymerization have been studied very intensively in recent years in connection with the discovery, by Ziegler and Natta, of stereospecific polymerization of α-olefins. A large number of catalytic systems were studied and results of these investigations can be found mostly in the patent literature. There are only a few studies on the scientific basis for catalyst selection for polymerization. These studies have been carried out to an increasing extent in recent times. As a rule requirements for standardized conditions (measured surface, regime, etc.) were not established. A survey of the polymerization catalysts is given in the book of Gaylord and Mark [186]. Therefore, in the present chapter, only selected examples from the recent journals are discussed.

The selection of polymerization catalysts is a more involved task than the selection of catalysts for other classes of reactions. In addition to the usual requirements of activity and selectivity in the case of monomers, which are able to polymerize in several directions (for example diolefins and unsaturated aldehydes), there is also the requirement of obtaining polymers with the required molecular weight and arrangement of side groups, i.e., stereoregularity. The majority of the examples given in the following paragraphs are considered from the catalyst activity point of view. In special cases, problems of selectivity and stereoregularity are also considered.

In the presence of liquid acids (H_2SO_4 and H_3PO_4), polymerization of olefins proceeds readily with the formation of liquids and waxy products of low molecular weight (dimers, trimers, etc.). The polymerization of such substances was even studied in the nineteenth century and led to the development of industrial processes for polybenzenes, lubricant oils, etc. Processes on solid catalysts proceed in a similar way. The most widely used solid catalysts are alumosilicates, H_3PO_4 Kieselguhr, Al_2O_3, treated HF, etc.

In the case of *dimerization of isobutylene* [212], a relation between catalytic activity and acidity was observed for the following catalysts: Alumosilicates, zirconiumsilicate, magnesium silicate, TiO_2, Al_2O_3, $Al_2O_3 \cdot B_2O_3$, and ZrO_2. However, a linear dependency was not found. For a change of acidity of the order of five, the velocity of dimerization changes only five-fold.

The reaction of *propylene dimerization* was studied on a number of catalysts [611] usually employing alumosilicates as carrier. The activity was considerably increased when some oxides of transition metals (Cr_2O_3, Ce_2O_3, and Fe_2O_3) were applied on them. MgO, having basic properties, decreases the activity. According to the data of the study [612] in the polymerization of propylene to products of low molecular weight, a linear connection was observed between catalytic activity and protonic acidity of the silicate systems investigated. In the study [216], it was proven that the polymerization of ethylene on such catalysts as Al_2O_3 and alumosilicate proceeds on acid centers of the Bronsted type through the formation of carbonium ions:

$$C_2H_4 + H^+S \to (C_2H_5)^+S$$
$$(C_2H_5)^+S + C_2H_4 \to (C_4H_9)^+S, \quad \text{etc.}$$

where S is the catalyst site.

In a study of *dimerization of cyclopentadiene* on sulfides, Ingold and Wassermann [613] point out the connection between catalytic activity and the color of the catalysts investigated. The most active were CuS, followed by Cu_2S, Ag_2S, HgS, Tl_2S, PbS, SnS, BiS, FeS, and NiS. The sulfides of the metals of Group II (except HgS) were inactive or only slightly active. If these data are reliable, a question arises about the possibility of homolytic steps and single-electron transfer in the simpler reactions of polymerization. The activation of the alumosilicate catalysts by oxides of transition metals for the dimerization of propylene (as is also observed in a number of other cases of polymerization) indicates that there is some probability that such mechanisms exist. This fact complicates the rules for catalyst selection.

Ethylene polymerization for the production of polymers with molecular weights ranging from several thousand to several million has been carried out in several industrial processes during recent years. The U.S. company Phillips uses chromium oxide catalyst applied on alumosilicate for this purpose; Standard Oil employs oxides of Mo and other oxides of the transition metals. According to the data of Feller and

Field [614], the most active catalyst for ethylene polymerization is MoO_3. Other active catalysts appear to be V_2O_5, WO_3, Nb_2O_5, and Ta_2O_5. These, like CrO_3 and MoO_3, promote the formation of solid high molecular weight polymers. Promoters such as the reducing agents Na, CaH_2, and $LiBH_4$ were employed, and their effect is explained by the fact that the most active valence state for this process is not that of CrO_3 and MoO_3, but that of oxides with a lower degree of oxidation. As shown in Section 1.6, the active center for polymerization probably appears to be a cation with the electron configuration d^1. For example, it was shown by EPR [187] that the polymerization catalyst Cr has a +5 charge. Among other transition oxides of Groups V and VI the electron configuration d^1 (Mo^{3+}, W^{5+}, Nb^{4+}, and Ta^{4+}) is also possible. The other studied oxides of the transition metals (NiO, CoO, ZrO_2, ThO_2, CeO_2, and Fe_2O_3) were less active.

The study of polymerization of ethylene on the rare earth elements [615] has shown that, in general, the yield of solid polymer increases with an increase in the atomic weight of the metal (Fig. 93). Among

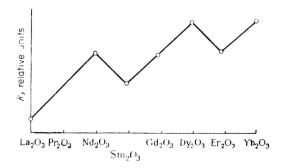

FIG. 93. The change of catalytic activity for ethylene polymerization in a series of rare earth elements.

a number of sulfides the most active catalysts appear to be the same as in the case of oxides [616]—the sulfides Group VI metals: W, Mo, Cr, and also the sulfides of V and Ti. For cations forming these sulfides, the d^1 configuration is also possible. On sulfides of the nontransition metals (Al_2S_3 and CdS) fluid products and low yields were obtained.

The second method for obtaining crystalline solid polyethylene was brought to industrial use as a result of the discovery by Ziegler and Natta of the complex catalysts $TiCl_n + MeAlk_m$ (where $n = 2$–4; MeAlk—metallo-organic compounds of alkali metals, most often metals of Groups I–III). The most widely used catalysts for polymerization are halides of Ti like soluble $TiCl_4$, or nonsoluble $TiCl_3$ and $TiCl_2$ with the cocatalysts—Al alkyls.

Natta [617] indicates that the best catalysts for polymerization are chlorides of the metals with a valence 3+ or less: Ti, V, Cr, and Zn with small work functions of $\varphi < 4$ eV. These metals also have low ionization potentials (< 7 eV). As catalysts for ethylene polymerization, the chlorides of Mn, Fe, Ni, W, Ta, Th, Hf, Co, Sc, and Pd [186] were also studied, though the maximal catalytic activity was obtained on chlorides of Ti, V, Zr, and Th. If we assume that the charge configuration of Ti and Zr in the catalyst complex is $+3$, their electron configuration should be d^1, i.e., the same as for oxides of Cr and Mo for polymerization. This indicates a possible generalization of the mechanisms for olefin polymerization on catalysts of these two different groups: oxides and chlorides. In the study by Smith and Zelmer [618], a plot is given for the velocities of polymerization of ethylene on oxides and chlorides in the presence of the cocatalyst $Al(C_2H_5)_3$. It was shown that the order of activity was as follows: $TiCl_4 > CrO_3 > V_2O_5$.

Natta reports that metallo-organic compounds and the halides of the transition metals form bridge complexes. For example, catalytically active complexes of the following type were detected:

$$\begin{array}{c} C_5H_5 \\ \diagdown \\ \diagup \\ C_5H_5 \end{array} Ti \begin{array}{c} R_1 \\ \diagdown \diagup \\ \diagup \diagdown \\ R_1 \end{array} Al \begin{array}{c} R_2 \\ \diagup \\ \diagdown \\ R_2 \end{array} \tag{78}$$

where R_1 is a halogen or alkyl group; R_2 is an alkyl group

In such complexes it is necessary to use a metal alkyl as cocatalyst [619] which is able to form a considerable localized electric field. For this purpose, the metals with small ionic radii (less than 1 Å) are especially convenient and especially those with strong electropositive character. The strong electropositive alkali and alkaline earth metals K, Rb, Cs, Ca, Ba, and Sr are not suitable because of their much too large radii, though in the form of metal alkyls they can polymerize ethylene and the other olefins by the anionic method [620]:

$$AlkMe + CH_2 = CHX \rightarrow Alk\text{—}CH_2\text{—}CHX\text{—}Me^+, \text{ etc.}$$

The most suitable cocatalysts appear to be the metallo-organic compounds Al, Be, and Li, because these metals are characterized by small ionic radii and electronegativity lower than 1.5. In the periodic system of the elements the "cocatalytic" activity of $MeAlk_m$ in general, decreases in going from left to right and from the bottom up. In Fig. 94

9. POLYMERIZATION REACTIONS

the change of cocatalytic activity of the alkyls of the metals of the second and third period is shown schematically for ethylene polymerization [621].

According Dawans and Teyssie [622], the cocatalytic activity decreases in the order $Alk_3B > Alk_3Al > Alk_2Be > Alk_2Mg > AlkLi$.

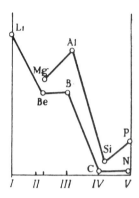

FIG. 94. A schematic representation of the change in activity of organic compounds of the elements of Groups I–V as cocatalysts in the reaction of polymerization of ethylene by the use of the catalyst $TiCl_2$.

Various authors studying olefin polymerization on complex catalysts reported that the polymer chain is bound during the process of growth into a type of complex (78) with the atoms of Al and Ti. Most reliable are the data by Cossee [188] and Arlman [331, 340] which were considered in Sections 1.6 and 3.1. According to these studies, the olefin polymerization proceeds according to the coordination scheme (29) on ions of transition metals with the structure d^1. This point of view explains the similar catalytic activity of oxides and chlorides with the same electron configuration of their transition metal. The alkyl of the nontransition metal serves as the agent which alkylates the transition metal and in this manner forms an initiating chain. The alkylation of the $ScCl_3$ surface by organic compounds was proven recently by a straightforward experiment using labeled atoms [623]. According to these observations, it follows that the cocatalyst should be selected on the basis of its ability to alkylate the halides of the transition metals, as the atom of nontransition metal itself does not participate directly in the polymerization.

In Fig. 95 is presented the periodic system of elements, showing the metals whose oxides or halides are active in the polymerization of olefins and metals whose alkyls are cocatalysts for polymerization and also elements for metallo-organic compounds which are themselves able to catalyze polymerization by the ionic or radical mechanism.

Everything said so far regarding the polymerization of ethylene is related to the polymerization of α-olefins with a number of C atoms greater than 2, but here there enters also the factor of stereospecificity.

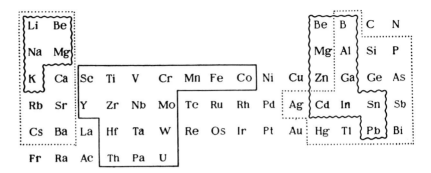

FIG. 95. The elements whose compounds are active for olefin polymerization: —— forming catalysts—oxides or halides; ～ forming cocatalysts—the metallo-organic compounds; ····· elements, the metallo-organic compounds of which, themselves catalyze the polymerization of olefins.

Among the oxides of the transition metals for the *reaction of propylene polymerization*, the most active are [9] the oxides of V, Cr, Mo, W, U, Mn, etc. The oxides of nontransition metals are not very active.

In order to obtain high molecular solid polymers of propylene, one uses, as in the case of ethylene, systems based on $TiCl + MeAlk_m$. The catalytic properties of pure chlorides have also been studied. It was, for example, shown [624] that the addition of the chlorides of Al, W, Mo, Mg, and other metals considerably increases the catalytic activity of $TiCl_3$ and $TiCl_2$ although the specific surface is decreased thereby. The molecular weight of the polymer obtained is, on the contrary, highest with pure $TiCl_3$ and $TiCl_2$, additives having the effect of lowering it. So far as the degree of stereoregularity is concerned, the maximum proportion of stereoregular "isotactic" polypropylene (85.3% of the overall yield) was observed with pure $TiCl_3$ and is somewhat lower (70–80%) on $TiCl_2$ and $TiCl_3$ with the addition of trivalent chlorides: $AlCl_3$, WCl_3, and $MoCl_3$. The addition of divalent chlorides ($MgCl_2$, $ZnCl_2$, and $SnCl_2$) result in lowered output of stereoregular product to 40–60%.

The catalytic activity of the halides entering into the complex of $MeX_n + Zn(C_2H_5)_2$ in the polymerization of propylene [625] decreases in the order: $\alpha\text{-}TiCl_3 > TiCl_2 > VCl_3 > TiBr_4 > NbCl_5 > CrCl_3$.

In the case of the complex catalysts $TiCl_n + MeAlk$, the highest stereospecificity leading to the formation of isotactic polypropylene is shown with crystal systems based on $TiCl_3$. The homogeneous catalysts seem to be inactive for stereospecific polymerization of α-olefins [619]. Polymerization on them leads to the formation of nonstereoregular atactic polyolefins. Other systems based on Ti have also been used.

In the series of halides, the output of the regular polymer is lowered in the following order [186]: $TiCl_3 \rightarrow TiBr_3 \rightarrow TiI_3$; in the series of other compounds of Ti: $TiCl_3 > TiCl_2 > TiCl_4 > TiCl_2(OC_4H_9)_2 > TiCl(OC_4H_9)_3 > Ti(OC_4H_9)_3 > Ti(OH)_4$ (in the presence of $Al(C_2H_5)_3$ as the cocatalyst). With halides of other metals a considerable output of isotactic polypropylene was observed on VCl_3, $CrCl_3$, VCl_4, and $VOCl_3$. According to the data [622], the percentage of polymer crystallinity which was obtained in the polymerization of propylene in the presence of these cocatalysts $Al(C_2H_5)_3$, i.e., output of isotactic product, decreased in the following order: $TiCl_3 > VCl_3 > ZrCl_4 > TiCl_4 > VCl_4 > TiB_4 > CrCl_3 > VOCl_3$. The higher halides ($TiCl_4$ and VCl_4) are partially reduced in the polymerization process to the lower ones, also the maximum stereoregularity can be obtained if one introduces into the reaction system relatively lower halides. In the presence of WCl_3, $FeCl_3$, and other halides the yield of the stereoregular product was not improved.

The overall catalytic activity, as is the case for isotactic polypropylene, depends on the nature of the cocatalyst $MeAlk_m$. For example, an increase in the ratio of Alk in the series: $Al(C_2H_5)_3$, $Al(C_3H_7)_3$, $Al(C_4H_9)_3$ and $Al(C_6H_{13})_3$ (in the presence of $TiCl_3$) decreases the catalytic activity and yield of isotactic polypropylene. The partial change of the alkyl by halogen decreases the catalytic activity of the complex but somewhat increases the proportion of isotactic product, and also the molecular weight of the polymer [186, 622, 626]. By polymerization of propylene in the case of the catalytic system $\alpha\text{-}TiCl_3 + Al(C_2H_5)_2X$ (where X is C_2H_5, F, Cl, Br, or I), the catalytic activity decreases in the order: $C_2H_5 > F > Cl > Br > I$, and the percentage of the isotacticity changes in the reverse direction [627]. In the presence of $Al(C_2H_5)_2I$, the isotactic product was formed almost completely. In the catalytic system where X is OC_6H_5, SC_6H_5, SeC_6H_5, or NC_5H_{10}, the catalysts were not very active. The smaller the radius of the ion of the metal in $MeAlk_m$ and the smaller its electronegativity, the higher will be the stereospecificity of the catalytic system based on $TiCl_3 + MeAlk_m$. In this case, it decreases in the following order: $Be(C_2H_5)_2 > Al(C_2H_5)_3 > Mg(C_2H_5)_2 > Zn(C_2H_5)_2$. It is interesting that, according to the data [628], the catalytic activity of these complexes changes in the same direction. A low activation energy was observed on the least active catalysts —$Zn(C_2H_5)_2 + \alpha\text{-}TiCl_3$ (8.2 kcal/mole) and a high one on the most active catalyst—$Be(C_2H_5)_2 + \alpha\text{-}TiCl_3$ (16.2 kcal/mole).

Almost all catalysts studied are chlorides having a lamellar lattice. Transfer from one modification to an other (for example from $\beta\text{-}TiCl_3$ to $\alpha\text{-}TiCl_3$) changes the activity and stereospecificity of the catalyst and the molecular weight of the catalyst. The violet crystal modification of α-TiCl$_3$ is especially active and stereospecific. This shows the special role of the solid surface in the formation of stereoregular poly-α-olefins. Natta [617] and with Paskuon [619] explains the stereospecificity of the surface of the epitaxial adsorption of the metal alkyl on the TiCl$_3$ surface and its relation to the formation of the corresponding stable metallo-organic complex surface compounds (Fig. 96). The order of the activity of the cocatalysts (see Fig. 94) is connected with the stability of their adsorption complexes. The mechanism of polymerization is connected with the breaking of bridge bonds of the complex type (78). The surface exhibits orienting action.

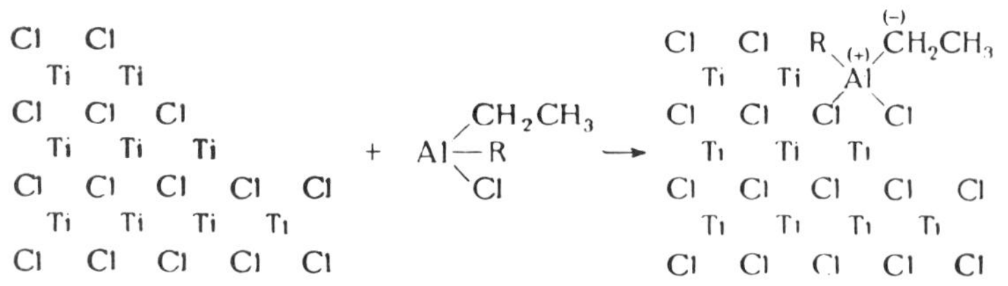

```
Cl   Cl                                          Cl   Cl   R_(+)  CH₂CH₃
  Ti   Ti                                          Ti   Ti    Al ⁽⁻⁾
Cl   Cl   Cl                 CH₂CH₃              Cl   Cl   Cl    Cl
  Ti   Ti   Ti      +   Al─R            →          Ti   Ti   Ti
Cl   Cl   Cl   Cl   Cl        Cl                 Cl   Cl   Cl   Cl   Cl
  Ti   Ti   Ti   Ti                                 Ti   Ti   Ti   Ti
Cl   Cl   Cl   Cl   Cl                           Cl   Cl   Cl   Cl   Cl
```

FIG. 96. The scheme of epitaxial adsorption of the alkyl Al on the surface of TiCl$_3$ with the formation of an active center for polymerization [617, 619].

A generally accepted theory of the mechanism of stereospecific action of the surface on the formation of the regular polymers is not yet established. In Section 3.1 another mechanism was presented [331] which is more probable from our point of view. According to this mechanism, the polymerization proceeds in the anion hole at the corners or ribs of the layer crystal TiCl$_3$; and the stereoregular effect is obtained by the asymmetric position of the ligands around the Ti^{3+} ion around this hole.

The study of the *copolymerization of ethylene with propylene* [629, 630] shows that the factors that can cause stereoregularity will provide a higher percentage of the propylene in the copolymer. The relative percentage of the propylene increases in the order: HfCl$_4$ < ZrCl$_4$ < TiCl$_4$ < VOCl$_3$ < VCl$_4$ (plus metallo-organic compounds). The overall catalytic activity varies perfectly in accordance with the ionic mechanism and is minimal on VCl$_4$.

The polymerization of isobutylene proceeds on catalysts of the same type as described above. For the processes on pure (soluble) halides of the elements or Groups III and IV the following order of activity was found [631]: BF$_3$ > AlBr$_3$ > TiCl$_4$ > TiBr$_4$ > SnCl$_4$ > BCl$_3$ > BBr$_3$. The molecular weight of the products changes in the same direction as the catalytic activity—from 150,000 in the presence of BF_3 to 12,000 for $SnCl_4$.

The polymerization of butadiene can proceed in different directions due to the presence of two double bonds in the molecule $H_2C=CH-CH=CH_2$. In ref. [632], the polymerization of butadiene in a solution of tetrahydrofuran was studied in the presence of the homogeneous catalysts: organic compounds of the nontransition metals of fourth period and cocatalysts $MgCH_3Br$. The most active were Ti-organic compounds: $Ti(C_2H_5)_4$, $Ti(CH=CH_2)_4$, and $Ti(C_6H_5)_4$, on which there were also obtained polymers having maximum molecular weights and containing 1,2-product. Close to them in activity, the compounds of chromium gave lower molecular weights and yields of 1,2-product. Co-organic compounds had medium catalytic activity, but produced a considerable proportion of 1,2-product. Finally, Fe- and Ni-organic compounds had low catalytic activity and led to the formation of the *trans*-1,4-polymer of low molecular weight.

As was shown [633] in the study of the polymerization in hydrocarbon media by the use of organic compounds of basic metals, the ratio of 1,2-polymer increases with increase in the polarity of the bonds in the order: $Li^+ < Na^+ < K^+$. In a polar medium the order is reversed because of the counteracting polarity of the complex formed.

In the presence of soluble halides it was found [634] that the following activity prevailed: $AlBr_3 > TiCl_4 > WCl_6 > VCl$. The stereospecificity of these halides, in contrast to their activity, does not depend on the cation. In all cases, *cis*-1,4-polymer with a yield of 90–94% was obtained.

The polymerization of styrene leads to production of crystalline polystyrene in the presence of the catalysts $TiCl_3 + Al(C_2H_5)_3$; and some related compounds. In studies of the polymerization on pure chlorides, it was found that here also $TiCl_3$ is the most active; then come $CrCl_3$ and VCl_3. The study of the same process on other chlorides [636] gave as the order of their catalytic activity: $SbCl_5 \gg SnCl_4 > BCl_3$. The polymerization of styrene can also be carried out in the presence of only metallo-organic catalysts [637], in agreement with the principle of catalyst selection, the following are ordered: $Si(C_6H_5)_4 > Pb(C_6H_5)_4 > Hg(C_6H_5)_2 > Al(C_6H_5)_3$. Note that the metallo-organic catalysts (excluding halides) catalyze the polymerization of other olefins, but only to polymers of low molecular weight.

Nakata *et al.* [638] studied the polymerization of styrene at low temperatures in benzene in the presence of inorganic peroxides. The order of the activity of the peroxides is as follows: $Ni > Zn > Co > Mg > Al > Cu > Cd > Mn$. In this case, the peroxides of the metals were not so much catalysts as initiators of the process of polymerization.

According to the data of Slinkin *et al.* [639] who studied the polymerization of styrene in the presence of organic semiconductors, chelate compounds of the type: $H[R>Me<]_nRH$ (where R is a quinizanine radical) the catalytic activity of changes in the following manner, depending on the type of chelate forming metals: $Cu^{2+} > Mn^{2+} > Ni^{2+} > Zn^{2+} > Co^{2+}$.

The polymerization of cyclobutene was studied on the following catalysts: acetyl acetonates and chlorides of the transition metals. $Al(C_2H_5)_2Cl$ or $Al(C_2H_5)_3$ served as the cocatalysts. The most active acetyl acetonates are those of V and Cr and the chlorides of Cr, V, Ti, and Mo. Low activity is exhibited by the chlorides of U, Mn, Fe, Co, Cu, Ru, and Rh. The principles of catalyst selection in this case are also similar to those for the polymerization of the other olefins. The heterogeneous systems are less active than homogeneous but are more stereospecific.

The polymerization of oxygen containing vinyls in a number of cases proceeds on the same catalysts as the polymerization of hydrocarbons.

The polymerization of isobutyl vinyl ether and similar vinyl ethers [641] proceeds with explosive speed at low temperatures in the presence of the catalyst—etherate of BF_3. Then follow, in order of decreasing activity, $TiCl_4$, $AlCl_3 \cdot O(C_2H_5)_2$, and halides of magnesium, both pure and in complex with ethyl ester. In the series of halides of Mg, the catalytic activity decreases with an increase in the ionic character of the bond from MgI_2 to MgF_2. The activation energy did not change with regularity, and was at a maximum in the case of MgI_2 and $MgBr_2$. Iwasaki *et al.* [641] indicate a relation between catalytic activity and Lewis acidity and report that the active catalysts must be compounds whose cations possess high values of e/r and high ionization potentials. For example, among metal halides, high activity should be displayed by the following: $BeCl_2$, BCl_3, $AlCl_3$, $GaCl_3$, and HCl; and low activity is displayed by chlorides of alkali and alkaline earth metals, respectively. In a number of cases this is in agreement with experiment. In a study of the polymerization of methyl vinyl ether [642] in the presence of the metal halides and $Al(C_2H_5)_3$, it was found that the most active are the halides of Bi, Mo, V, and Sn. The halides of Ta, Ti, and Cr were less active. Little activity was shown by those of Co, Zn, Ni, and Mn.

A number of different catalysts for the polymerization of vinyl esters, namely, the oxides of the transition metals and other oxides, both pure and in combination with C_2H_5MgBr, were also studied by Iwasaki *et al.*

[641]. The activity of BF_3 is highest; and their activity decreases in the following order: $NiO > V_2O_5 > SiO_2 > MgO > \gamma\text{-}Al_2O_3 > B_2O_3 > BeO > \alpha\text{-}Al_2O_3 > ZnO > CaO$.

Obviously, the oxides of the transition metals and acid oxides are more active than the basic oxides. The principles for change of catalyst stereospecificity, in this case, also does not agree with those for change in activity. For example, the etherate of BF_3 does not form a stereoregular polymer at room temperature. Catalysts like $AlCl_3$, $TiCl_4$, VCl_4, V_2O_5, WO_3, MoO_3, and NiO are also nonstereospecific. In the study [643] it is indicated that stereospecific catalysts for the polymerization of the vinyl ethers are only the ones whose coordination centers (cation) are in surrounding tetrahedra and in which the distance between two ligands does not exceed 3.5 Å. These requirements are met by $AlFCl_2$, CrO_2Cl_2, CrO_3, $AlR_3 + VCl_4$, sulfates of Fe, Al, Mg, $Cr + H_2SO_4$, and some others. The given conditions for stereospecificity are explored by a proposed mechanism in ref. [643]. According to this study, in order to obtain stereoregular polymers, it is necessary that the vinyl and other groups be connected with the active center of the catalyst.

The polymerization of acrylonitrile is easily carried out in the presence of acids or ionic coordinates [186, 620], metallo-organic compounds [644], alcoholates [645], and similar catalysts. Copolymerization of acrylonitrile with butadiene, styrene, and the other olefins is used for obtaining synthetic rubbers. A process of another type [646], which is basically different from the ones considered so far—the catalytic copolymerization of acrylonitrile into molecular condensed systems is also known.

The polymerization of acetylene and its product proceed most readily on copper catalysts which also catalyze other reactions of acetylene. The chlorides of copper in solution [647] are often used, as well as the pure heterogenous catalysts. For example, the polymerization of phenylacetylene from the gas phase at 250–350°C over Cu oxide was studied in ref. [648]. A solid polymer with a molecular weight of about 7000 was observed.

For the polymerization of 1-pentene in solution in the presence of the reducing agent ($NaBH_4$), the following order of activity of the halides and some other compounds of the transition metals was observed: $Co^{2+} > Ni^{2+} > Pt^{4+} > Os^{3+} > Pd^{2+} > Ru^{3+}$. The other salts studied: Zn^{2+}, Mn^{2+}, Hg^+, Sn^{2+}, Ti^{4+}, Zr^{4+}, and Na^+ did not catalyze the reaction. Thus the catalysts for this reaction appear to be only the compounds of the metals of Group VIII. It is possible that this is caused by the formation of donor π-complexes of the C=C bond with the metal.

In the polymerization of acetylene [650] and phenylacetylene, [651] a catalyst of the Ziegler–Natta type was also employed. The corresponding activity of chlorides in the presence of $Al(C_2H_5)_3$ was changed to the order: $TiCl_4 > VCl_4 > VOCl_3 > VCl_3 > CrCl_3 \sim MnCl_2 \sim FeCl_2 \sim CuCl_2$; in the case of acetylacetonate + $Al(C_2H_5)_3$ the order is: $V > Ti > Co$, Zr, and Cu. Thus, the catalytically active substances here appear to be the same systems which are active in the polymerization of olefins.

The polymerization of acetylene was also studied by Kuhn [41] on semiconductors with sphalerite structure. A relation between the catalytic activity and the conductivity type or the width of the forbidden zone cannot be established.

Consider now several examples of the selection of catalysts for polymerization involving rupture of the $-C=O$ bond.

The polymerization of acetaldehyde was studied [652] on numerous oxide catalysts. Its mechanism, it seems, is also of the coordination-ionic or Lewis acidity type; also the regularity of the change of catalytic activity differs from the corresponding regularity for olefin polymerization. For a number of oxides studied, the most active catalyst appears to be Al_2O_3. Next follow the oxides of the elements of Group VI: CrO_3 and MoO_3. Somewhat less active are B_2O_3, P_2O_5, MgO, ThO_2, and alumosilicate. The least active are oxides of Si, Cu, Ti, Zr, V, Zn, Mn, Fe, Ba, Pt, Pd, Ni, and Ca.

In a study of the polymerization of acetaldehyde in the presence of metallo-organic catalysts [653, 654] it was shown that the most active are compounds of the metals with a large value of e/r. The compounds of the earth alkali metal are not very active.

The order of the catalytic activity of the analogous metallo-organic compounds is the following: $Al > Fe > Li > Mg > Zn > Ti > Cd > Ba > Ca > Na$. It was proven that, in aldehyde media, the bond $Me-C$ in the metallo-organic compound is converted into $Me-O-C$, i.e., the atom of metal is set into an alcoholate group. In the presence of alkali and the catalysts $Al(C_2H_5)_3$, the order of activity is: $LiOH > NaOH > KOH$ [655].

The principles for the change of catalyst activity for acetaldehyde polymerization are very similar to the corresponding principles for olefin polymerization by the oxides which will be considered further in detail.

Metallo-organic compounds of Li and Al are also the most active for the polymerization of butylaldehyde [654]. The compounds of Zn, Cd, and B are less active.

The polymerization of dimethylketene [656] proceeds with the highest velocity in the presence of organic compounds of metals with low electronegativity. The order of the activity of the analogous metallo-organic compound was: $K > Na > Li > Mg > Al$ and Zn.

9. POLYMERIZATION REACTIONS

Formaldehyde is polymerized under the influence of acid catalysts. *The polymerization of trioxane* [657], its trimer, also proceeds easiest in the presence of the acid catalyst BF_3. Next follow $SnCl_4$ and $TiCl_4$. The catalyst (more precisely the initiator) in this case is on the surface of the trioxane crystals. In [658] the following order of the relative activity of the initiators for trioxane polymerization was obtained: $H_2SO_4 > HClO_4 > FeCl_3 > SbF_4 > SnCl_4 > TiCl_3 > BF_3O(C_2H_5)_2 > AlCl_3 > SbCl_5$. The protonic initiators appear to be more active than the nonprotonic. The highest molecular weight of polymer was obtained in the presence of $FeCl_3$.

Polymerization with the rupture of a three-member ring was accomplished in one case with protonic and in another with an aprotonic-acid mechanism. Let us consider the principles of catalyst selection for these reactions in more detail. The mechanism of *the polymerization of ethylene oxide* was considered in Section 2.2. According to scheme (41) the reaction proceeds based on the coordination mechanism by which, during the process of polymerization, every molecule of the catalyst is connected with an initial active center-cation which coordinates simultaneously with a molecule of monomer.

The principle for catalyst selection for the polymerization of ethylene oxide was considered in the studies of Krylov and Sinyak [280, 659]. The active catalysts appear to be the oxides and hydroxides of the metals of Group II. In Table VII are presented calculations based on the data of [280–282, 659]. The number of active centers on 1 cm² surface of hydroxides and the reaction velocity are related to the unit active center.

From Table VII, one can draw the conclusion that in the series of

TABLE VII

THE POLYMERIZATION OF ETHYLENE OXIDES ON THE HYDROXIDES OF GROUP II METALS

Catalyst	The number of active centers on 1 cm²	Specific activity related to a unit center, molecules/sec	
		25°C	90°C
$Be(OH)_2$	6.5×10^{13}	0.040	1.0
$Mg(OH)_2$	3.3×10^{13}	0.030	0.9
$Ca(OH)_2$	2.4×10^{13}	0.015	1.7
$Sr(OH)_2$	1.1×10^{13}	0.003	1.5
$Ba(OH)_2$	1.2×10^{13}	0.005	4.7

alkaline earth hydroxides with growth of the atomic weight of the metal the number of active centers on 1 cm² decreases. The specific activity, as related to the unit active center at low temperature, also decreases in the series from $Be(OH)_2 \rightarrow Ba(OH)_2$. At the high temperature, the specific activity has a tendency to increase.

Other active catalysts appear to be carbonates and oxalates of the metals of Group II. The activity of the catalysts for polymerization of ethylene oxides is determined, in general, by the cationic properties; the anion properties are of lesser importance.

The presence of the protonic acid centers on the catalysts studied (where, basically, the active centers seem to be Lewis acid sites) leads to a lowering of molecular weight. For example, the active catalyst $BeO \cdot CO_2 \cdot 5H_2O$ leads to the formation of a polymer with a molecular weight of only 2500, while at the same time on the oxides and hydroxides of Be, Mg, Ca, Sr, and Ba the molecular weight is 10^6–10^7. Polymers with a low molecular weight are formed on alumosilicate; and the catalytic activity of them is considerably lower than that of Al_2O_3 [282]. On the contrary, the exchange of H^+ ions of alumosilicate with Ca^{2+} and Sr^{2+} increases the activity [660]. The H^+ ions participate in the ring breaking reaction.

The oxides of Be, Mg, and Ca are considerably more active than the corresponding hydroxides, as a result of the presence of a larger number of active centers on their surfaces, i.e., of the metals not shielded by OH groups. In Fig. 97 is presented the dependency of the polymerization velocity of ethylene oxide based on 1 m² of the surface as related to the degree of $Mg(OH)_2$ dehydration.

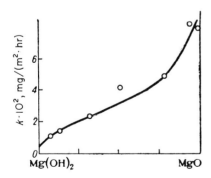

FIG. 97. The dependency of the polymerization velocity of ethylene oxide per 1 m² of surface on the degree of dehydration of $Mg(OH_2)$.

At higher temperatures (90–100°C), the compounds of Ca, Sr, and Ba are somewhat more active than the hydroxides of Be and Mg, and are close to the activity of BeO and MgO. This confirms a relationship corresponding to a straight line plot for activation energy, $\log k - 1/T$,

9. POLYMERIZATION REACTIONS

i.e., the presence of a compensating effect in the change of E and k_0. The activation energy on the alkaline earths as catalysts is larger than on the hydroxides Be and Mg, but k_0 also has a higher value (of order 2–3) It is possible that, just as in the case of the effect of changes on surface atoms and groups in the solid bases discussed in Section 2.1, in this case OH groups can participate in the reaction of initiation:

$$\overset{\delta+\delta-}{\text{CaOH}} + \overset{\text{CH}_2}{\underset{\text{CH}_2}{\text{O}\diagup\diagdown}} \rightarrow \cdots \text{Ca}-\text{O}-\text{CH}_2-\text{CH}_2-\text{OH} \qquad (79)$$

On other catalysts, the initiation of the chain is probably obtained according to the scheme:

$$\underset{\overset{\delta+\ \ \delta-}{\cdots \text{Me}\ \text{O}\ \text{Me}\cdots}}{\overset{\text{CH}_2}{\text{O}\diagup\diagdown\text{CH}_2}} + \rightarrow \underset{\cdots\text{Me}\ \ \text{O}\ \text{Me}\cdots}{\overset{\text{CH}_2}{\text{O}\diagup\diagdown\text{C}}} \rightarrow \underset{\cdots\text{Me}\ \text{O}\ \text{Me}\cdots}{\overset{\text{CH}=\text{CH}_2}{\underset{\text{H}}{\text{O}}}} \qquad (80)$$

Afterwards growth of the chain proceeds according to the mechanism (41) (Section 2.2), with the participation of the atom $\text{Me}^{\delta+}$.

Among the metals of the other groups, the most active were compounds of Al and Fe, i.e., again compounds containing cations with large values of the ratio e/r.

On treatment of Al_2O_3 with the HCl and BF_3 its catalytic activity for ethylene oxide polymerization [660] is tremendously decreased. However the energy of activation of the process of chain growth does not change and remains equal to 11.5 kcal/mole. In this case, it seems not only HCl, but also BF_3, exhibit protonic acid properties for the formation of the surface compound:

$$\text{Al}-\overset{\delta-}{\text{O}}\diagup\diagdown\underset{\text{H}^+}{\overset{\text{BF}_3}{}}$$

and increase of proton mobility connected with the surface OH-group. Obviously the protonic-acid centers are not the active centers for polymerization. The decrease in activity of the Al_2O_3 by treatment with BF_3 and HCl is due to the screening of active centers, which is greater in the case of the large molecules of BF_3.

The change of catalytic activity of the hydroxides of the transition metals was very interesting. The most active catalysts were FeOOH, Ca(OH)_2, i.e., compounds of the cations with the electron configuration

d^0 and d^5. These structures, as in the case of the oxidation–reduction reaction, fall into the same group of activity as the nontransition metals, but exhibit not a minimum but a maximum catalytic activity (Fig. 98). In this case, crystal field stabilization (Section 1.6) appears to be the factor not condusive to catalysis.

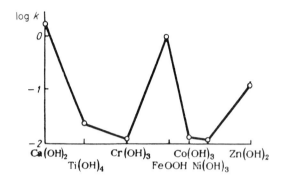

FIG. 98. The change in catalytic activity of the hydroxides of fourth period metals in the polymerization reaction of ethylene oxide.

From this point of view, high catalytic activity should be exhibited by compounds of Mn^{2+}, as Mn^{2+} has the electron structure d^5, as has also Fe^{3+}. However MnO appears to be only slightly active. The sample studied was obtained by the decomposition of $MnCO_3$ and it showed a low energy of activation for polymerization (9.7 kcal/mole), i.e., close to the most active Be-, Mg-, Fe-catalysts (8–12 kcal/mole). But it had a very low value of k_0 [10^{19} moles/(cm^2 · min.)], compared to that on Al_2O_3 with k_0 [10^{21} moles/(cm^2 · min)], and on BeO and FeOOH of 10^{22}–10^{23} moles/(cm^2 · min). It is possible that this is caused by the low number of active centers obtained by the stated method of MnO preparation.

Among the compounds of the transition metals of the other periods, catalytic activity is shown by hydroxides of Th^{4+} and Zr^{4+}, i.e., the compounds in which these metals have the electronic structure d^0 and are close in their property to the nontransition elements.

According to the data of the other investigations, analogous results were obtained. For example in ref. [661], polymerization of ethylene oxide was studied in the presence of soluble chlorides of metals. It proceeds with the maximum velocity in the presence of the metallic chlorides with high values of e/r: Sb^{5+}, Sn^{4+}, Al^{3+}, Be^{2+}, and B^{3+}. From the compounds of the transition elements, the only active compounds were those of Fe^{3+} and Ti^{4+}, with the cation structure d^5

9. POLYMERIZATION REACTIONS

and d^0. The other chlorides of the transition metals (FeCl$_2$, NiCl$_2$, NiBr$_2$, and CuCl$_2$) were nonactive.

Fukui et al. [662] and Kagiya et al. [663] studied the polymerization of ethylene oxide in the presence of different phosphates promoted by metal alkyls. Here also, catalytic activity is shown by the metal phosphates, cations of which have the electronic structure d^0 and d^5. The most active were FePO$_4$ promoted by the addition of Zn(C$_2$H$_5$)$_2$. The promoter in this case accepts the protons and thus prevents chain breakage. The other active catalysts are Zr$_3$(PO$_4$)$_4$ and Ti$_3$(PO$_4$)$_4$. There was no polymerization in the presence of V$_3$(PO$_4$)$_2$ (configuration d^3).

In the presence of partially hydrolized metal alkyls of Group II the order of activity was the following for the reaction of ethylene oxide polymerization [664]: Mg > Be > Zn > Cd > Hg, which is quite similar (with the exception of a reverse in position of Be and Mg) to the order of activity of the oxides and hydroxides of these metals [280–282, 659]. The polymerization proceeds with high velocity on the carbonates of alkaline earth metals [665]. Among them, the most active (in the presence of a small amount of H$_2$O) was Sr carbonate. A polymer of molecular weight of up to 10 million was formed on it.

The principles for catalyst selection for *the polymerization of propylene oxide* are similar to those for polymerization of ethylene oxide. According Krylov and Livshits [666], the active compounds are: oxides, hydroxides, and oxalates of Be, Mg, Al, and Fe. The absolute velocity of the reaction of propylene oxide polymerization is 1–2 orders lower than the velocity of polymerization of ethylene oxide on the same catalysts under the same conditions. This is explained by the loss of the free rotation of the methyl group of propylene oxide in the activated complex, and related to this is the additional decrease of activation entropy for propylene oxides as compared with ethylene oxide.

According to the data by Okazaki [667], the catalytic activity of dimer compounds for the reaction of polymerization of propylene in the presence of Zn(C$_2$H$_5$)$_2$ increases with the strength of ionic bond in the dimer compound. As a proof, there is presented the following order of increase of production of yM (where y is the percentage of polymer yield and M is the molecular weight of the polymer): Bi$_2$O$_3$ < MgO < CaO < BaO < CaF$_2$ < KF, which decreases with the order of increase of the difference in electronegativity Δx. Generally speaking, such regularity of principles for increase of the catalytic activity are possible if the reaction velocity is limited, e.g., by chain initiation, similar to the scheme of (80). In general, the application of the function yM for the characterization of the catalytic activity is not correct. If one extracts separately from this study the change of y or change of M, a less regular relationship is obtained.

The catalytic activity (yield of polymer y) for oxides in the presence of $Zn(C_2H_5)_2$, according to the data of Okazaki [667], decreases in the order: $ZnO > TiO_2 > SnO > WO_3 > Sb_2O_3 > MoO_3 > Cr_2O_3 > MnO_2$ and V_2O_5. Thus, the oxides of transition metals exhibit little activity.

In the case of dry powders of caustic alkali oxides, propylene is polymerized with the highest velocity on KOH [668]. Less activity was observed with NaOH, $NaNH_2$, and $NaOCH_3$. LiOH, $Ba(OH)_2$, and K_2CO_3 were only slightly active.

The transfer from the polymerization of ethylene oxide to propylene oxide is analogous to the transfer from the polyethylene to polypropylene. It is necessary to consider not only the catalytic activity but also the stereospecific effect of the catalyst. The best yield of crystalline isotactic polyoxypropylene has thus far been obtained [669] in the presence of the complex catalyst [$FeCl_3$ with the propylene oxide) which is possibly homogenous.

The polymerization of optically active propylene oxide in the presence of powdered KOH [670] or Mg oxalate [671] led to the formation of a crystalline optically active isotactic polymer.

On the surface of dehydrated MgC_2O_4 (hydrated Mg oxalate is nonactive) one can display regular adsorption in such a manner that the molecule of propylene oxide is oriented with the methyl group on the side opposite the surface (Fig. 99). Thus the atom of Mg binds the first monomer molecule by means of its sixth free coordination site (five coordination sites are used for the binding with the oxalate ions). It is easy to convince oneself from Fig. 99 that, in this case, the polymer

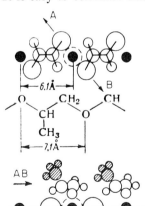

FIG. 99. The scheme for adsorption of molecules of propylene oxide on Mg oxalate. The lower part of the diagram gives a profile corresponding to section AB on the upper part.

9. POLYMERIZATION REACTIONS

molecule can be formed only by the adsorption of the next molecule of propylene oxide on the neighboring Mg atom. This mechanism accomplishes fixation of the configuration of the resulting asymmetric monomer atom. The dimensions of the C_3H_6O molecule and periodicity of the Mg ions involved are found to be in a sufficiently close geometric relation (5.9 and 6.1 Å) to constitute a satisfactory condition for chemisorption.

A second possible explanation, without assumption of regular adsorption, is the formation of stereoregular polymer in an anion hole on the surface. If one eliminates an oxalate-ion from the surface (Fig. 99), two neighboring Mg atoms gain two additional coordination valences. Therefore, every one of them can maintain a polymer chain and also attach to the underside monomer molecule in agreement with the coordination scheme (41) in Section 2.2. Such anion holes can place in a polymer chain either a d- or l-molecule of monomer. The first molecule which initiates a chain builds up asymmetrically inside of the hole and afterwards only molecules of the same sense are attached to the Mg atom. Therefore, from the racemic monomer, an optically nonactive polymer is formed, and from the optically active monomer an optically active polymer. This is confirmed by the experimentally found relations of the polymerization velocity constants for optically active and racemic propylene oxide $k_l/k_{dl} = k_d/k_{dl} = 2$ for a first order reaction.

Optical activity also occurs in the polymers if the propylene oxide is polymerized in the presence of the catalyst $Zn(C_2H_5)_2$ (it seems it is partially hydrolyzed) with the addition of l-borneol or menthol [672] or in the presence of an optically active catalyst—dehydrated Mg d-tartrate [673]. In the last case with racemic propylene oxide, in the initial step, the polymerization selectively proceeds only with l-propylene oxide, while d-propylene oxide remains in solution. As the reaction proceeds, the stereoselectivity of the catalyst decreases.

Figure 100 presents the geometric structure of the tartrate ion. The Mg ion is placed on the cross section of planes I and II. Its screening leads to the fact that the polymerization can proceed only in the anion holes. As these holes by themselves have inevitable asymmetry, therefore, with this catalyst, an optically active polymer is even formed from the racemic propylene oxide.

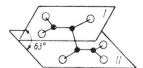

FIG. 100. The geometric structure of the tartrate ion.

The polymerization of the oxides of higher olefins proceeds in the presence of the carbonates, alcoholates, and metallo-organic compounds of Al, Zn, and Mg. Thus, these catalysts are identical to the catalysts for the polymerization of oxides of ethylene and propylene. The velocity of polymerization of the investigated [674, 675] oxides of butylene butadiene and styrene are smaller than the corresponding velocities for the polymerization of the oxides of lower olefins.

The polymerization of epichlorhydrin

$$CH_2\text{---}CH\text{---}CH_2\text{---}Cl$$
$$\diagdown O \diagup$$

was studied in the presence both of dehydrated chlorides and of their complexes with propylene oxide $MeCl_n \cdot 4(C_3H_6O)$. The order of the activity was the following: $SnCl_4 > TiCl_4 > AlCl_3 > FeCl_3$. The authors of the study [676] report that, on $SnCl_4$ and $TiCl_4$, polymerization proceeds by the cation mechanism with a split of charges; and on $AlCl_3$, $FeCl_3$ and their complexes with propylene oxide, by a coordination mechanism similar to the scheme of (41) for ethylene oxide (Section 2.2). According to their opinion, the activation energy of the reaction by the coordination mechanism is higher than that for the cation mechanism (the E values are not reported). In the presence of the complex $FeCl_3 \cdot 4(C_3H_6O)$, as in the case of propylene oxide, there is obtained a solid crystalline polyepichlorhydrin with a regular structure.

For the polymerization of fluorin substituted epoxides [677] $CF_3CC(CH_3)CH_2O$ and CF_3CHCH_2O the most effective catalysts are $AlCl_3$ and $FeCl_3$. In the presence of the best one, the velocity of polymerization is lower than in the presence of $AlCl_3$, but, with CF_3CHCH_2O a solid polymer is obtained. In order of decreasing activity, are the catalysts SO_2 and BCl_3. The polymerization of $CF_3C(CH_3)CHClO$ was studied on the same catalysts; the velocity of the polymerization appeared to be much slower.

One should also note studies on the polymerization of the cyclic analogs of olefinic oxides in which the atom of O is substituted by an S or NH group.

From a number of catalysts studied [678] for the *polymerization of styrene sulfide*, the only active catalysts were the Lewis acids: $ClCdSC_4H_9$, $Al(OC_3H_7)_3 + ZnCl_2$, $CuSC_4H_9$, and $(C_6H_5)_2Hg$. The bases, e.g., KOH, are only slightly active.

The polymerization of ethyleneimine

$$CH_2\text{---}CH_2$$
$$\diagdown NH \diagup$$

9. POLYMERIZATION REACTIONS   239

was studied [679] on numerous homogenous catalysts. The maximum yield of polymer was obtained in the presence of the protonic acids: HCl, H_2SO_4, and CH_3COOH. Ethyleneimine is also polymerized [660, 680] by heterogenous catalysts of protonic acid character: alumosilicates, incompletely dehydrated Al_2O_3, and also MoO_3. A large number of other oxides studied show that the transition metals are inactive though the adsorption of ethyleneimine proceeds on them with the formation of a firm bond. The treatment of Al_2O_3 by the protonic promoters BF_3 and HCl leads, in this case, not to decrease as in the case of polymerization of the ethylene oxide but to the increase of the catalytic activity.

In the polymerization of β-propiolactam [681] the connection between the activity of the homogenous to heterogenous catalyst and its basicity is quite clear. The salts of K were, as a rule, more active than the corresponding Na salts; and $Ba(OH)_2$ more active than $Ca(OH)_2$. Less activity was shown by phosphates of the transition metals. The most active among them were the phosphates of V, Ti, and Zr. The mechanism of the reaction on them, it seems, differs from that for polymerizations in the presence of alkalies. It follows that high polymer compounds are obtained only in the presence of the phosphates. On alkalies, a product with relatively low molecular weight is formed. The polymer with the highest molecular weight was obtained in the presence of $Ti_3(PO_4)_4$. One can assume that the mechanism of the polymerization on the phosphates of nontransition metals is connected with the participation of some form (or perhaps all) of it in the initiating step. If H_2O is completely removed from the phosphates reaction ceases.

This survey shows that, in spite of the great number of studies on polymerization catalysts, systematic data on the principles for catalyst selection are but few and they are contradictory. Only for the polymerization of the simpler olefins and olefin oxides are a sufficiently great number of studies of catalysts belonging to the different classes of chemical compounds under comparable conditions generally available.

Conclusions

The present monograph does not pretend to be a handbook on the principles of nonmetallic catalyst selection. Its purpose is concluded with an explanation of those properties of solids which can be employed for selection of a catalyst for a particular reaction and how experimental data on catalyst selection correlate with one or another theoretical assumption.

At the present time, there is no completely scientific theory for catalyst selection. This problem may be solved in the near future after a sufficient number of completely reliable data have been obtained. The survey given in this book of theoretical assumptions and experimental work can, to a certain degree, serve as the basis for building a future theory of catalyst selection.

As the data on catalyst selection for a number of particular reactions considered in Part II show, even today it is possible to formulate some selection principles with considerable reliability (though not to 100%). The reliability, as was shown, increases with the number of the investigations available for a given process, even though some of the required data (for example, measured surface, purity, kinetic regime) are not available.

A number of data indicate that the chemical constitution of the basic catalyst lattice appears to be an important factor, while impurities are of secondary importance. Thus we see the connection between catalytic activity and the position in the periodic table of elements which are the main constituents of basic catalyst structures. This does not exclude the possibility, in some cases, of the tremendous promoter or poisoning action of some impurities, especially for those catalysts which are semiconductors at low temperatures.

CONCLUSIONS

Consideration of catalyst selection principles emphasizes the importance of classifying catalytic processes into two types: electronic (oxidation–reduction, homolytic) and ionic (acid–basic, heterolytic). Characteristic catalysts of the first type, as was shown, are semiconductors; those of the second are acids and bases. Some reactions (for example, decomposition of alcohols and acids, polymerization) can proceed by both electronic and acid–basic mechanisms. In such cases, the principles of catalyst selection become more complicated.

Consider, on the basis of our survey, what properties one should employ to select catalysts for each of these two basic types of processes. First consider processes of the electronic type.

In spite of the opinion of a number of investigators who developed the electronic theory of catalysis, the type of catalyst (semiconductor) conductivity at elevated temperatures cannot serve as a sufficient criterion for selection. This is especially the case if we are dealing with different compounds (and not the same oxide having oxide impurities of different valences). As was shown, the correlation of the change of catalytic activity with the change of conductivity type is fortuitous or random in the majority of cases. In the case where one and the same catalyst is modified by donor or acceptor additives, the change in catalytic activity, as observed by experiments, appears to be insignificant and does not correspond to the change in semiconductor electroconductivity. In certain cases, this is due to the change of the catalysts to natural or migratory (from impurity to natural) conductivity. In other cases, it is due to the presence of a large number of surface electron levels, i.e., change of surface electron properties per unit volume. For some low-temperature catalytic reactions, the predominance of semiconductors with a specific type of conductivity (for example n-semiconductors in the decomposition of H_2O_2, p-semiconductors in the decomposition of N_2O and some other reactions of oxidation) was confirmed by statistical analysis of a large number of different catalysts. It is still unclear whether this regularity principle can serve for practical purposes in catalyst selection.

In those cases, involving semiconductors of a specific conductivity type, correlation of catalytic activity with the work function of the catalyst was observed. Especially satisfactory correlation can be obtained in the reactions where high catalytic activity of solid bases correlate with the catalytic activity of n-semiconductors as, e.g., in H_2O_2 decomposition.

In contrast to conductivity type, the width of the forbidden zone U of semiconductors can serve, in a number of cases, as a practical principle for selection. The decrease of catalytic activity with an increase in the width of the forbidden zone seems to be the principle which holds, in general, for the majority of oxidation–reduction reactions studied. The scattering of data, caused by the other factors (lattice type, electron structure of cation, and impurities), can sometimes distort this principle. However, the influence of U, as a rule, is stronger than that of the other factors. If one excludes the other factors by selecting catalysts from among semiconductors of the same crystal structure, and confining the change of metal or metalloid to those which form semiconductors in the range of one period or one group of the periodic system (as in the case of semiconductors of the isoelectronic series of germanium) the relation between catalytic activity and U can be represented uniquely.

The reasons for correlation of U with catalytic activity are, at the present time, still obscure. Probably, there is a connection between catalysis and conductivity. It is also possible that U influences catalytic activity by a relationship of U with the effective charge of cation. Further measurements of effective charge of cations on catalyst surfaces may clarify this.

In the case of impurity conductivity, a dependence of catalytic activity on U may possibly be observed if the concentrations of electrons and holes are used in the equation giving the velocity of a catalytic reaction. The width of the forbidden zone increases with increase in electronegativity difference. Therefore the latter, as was shown in a number of cases, appears to be a convenient criterion for catalyst selection, aside from the inaccuracy in the determination of electronegativity. This results from the fact that in one reaction type for semiconductors, the active catalysts are also solid acids but for the other type are solid bases. Both solid acids and solid bases can have large values of U, but the former have small Δx and the latter—large Δx values. Therefore, in the first case, the criterion of selection should be Δx; and in second, U. In the example considered for dehydrogenation of isopropyl alcohol, where both semiconductors and solid bases were active catalysts, the correlation of catalytic activity with Δx and with U were both satisfactory.

Concrete ideas regarding the relationship of catalytic activity with the number of d-electrons in the atom or metallic ions contained in the catalyst are given in crystal field theory and ligand field theory, discussed in Part I. For many oxidation–reduction reactions, these theories, as was shown in Part II, are confirmed. In particular, the minima for catalytic activity in the systems d^0 (for example, V_2O_5), d^5 (F_2O_3), d^{10} (ZnO), and the maxima with d^3 (MnO_2 and Cr_2O_3), d^7 (CoO) appear to be convincingly in agreement with principles in the case of the majority of reactions of oxidation, hydrogenation, dehydrogenation, and the decomposition of unstable oxygen-containing compounds. Note that the oxides of the fourth period were studied as catalysts to a much greater extent than any other solids.

At the present time the influence of two factors is still obscure and is not theoretically explainable: whether the width of the forbidden zone and the number of d-electrons, which were correlated simultaneously in a number of systems studied, are two different parameters or related to one and the same property. In Part I it was shown that in the series of oxides of the fourth period, the change of U is inversely related to changes in catalytic activity. Both these factors, it seems, are related to the connection between catalyst activity and the position of an element in the periodic system, according to studies by a number of authors.

The correlation of catalytic activity with color is related to the width of the forbidden zone and with electronic configuration of cations. Recent studies of catalytic activity of the rare earth elements in oxidation–reduction reactions indicates the absence of sharp distinctions between these properties. Earlier investigations in which larger differences were noted are probably not reliable.

So far as lattice parameter and type are concerned, the role of this factor in homolytic or oxidation–reduction reactions is, at present, far from clear. In the majority of examples studied, it does not correlate at all. This could be caused by the fact that, in the reactions studied, only very simple molecules were present. With increased complexity of the molecule, the probability of steric difficulties also increases and consequently there is more probability of the influence of lattice geometry on catalytic activity.

In cases where a correlation of activity with the lattice parameter was observed (dehydrogenation of alcohols) it was explained by a dual attachment of the reacting molecule with the catalyst to form the activated complex.

A compensation effect (linear relationship) between E and $\log k_0$ was not observed in the majority of cases with reactions on catalysts of different composition. On the contrary, it often occurs if some promoter or poisoning compound is added to one and the same compound. This is explained, in the general case, by the presence of two or several types of active centers.

The catalytic activity of solids in acid–base type reactions is related to the acid–base properties of solids. The reliability of these data is much lower than those for reactions of oxidation–reduction type.

There are not many reactions at the present time for which a reliable relationship between catalytic activity and proton acidity or Bronsted acidity has been established. Among them are some reactions of dehydration, alkylation, skeletal isomerization of olefins, and production of olefin polymers of low molecular weight. Some principles for selection of these catalysts can be explained by the fact that not only are the protons active, but also, so are the surface cations of the metals, i.e., Lewis acids.

In some cases, the connection with Lewis acidity or nonprotonic acidity is much more pronounced. This is the case for reactions of acetylene hydration or chlorbenzene hydrolysis, some reactions of halogenation and hydrohalogenation, numerous reactions involving polymerization of olefins and olefin oxides with the formation of high molecular weight products, etc. In all of these reactions, the catalytic activity increases with an increase in the polarizing action of the cation, determined by the ratio e/r. In some cases, the direction of the change in catalytic activity of compounds of the transition metals in acid reactions was opposite to the regularities found for oxidation–reduction reactions; and the systems d^0, d^5, d^{10} are more active than other compounds of the transition metals (bromination of benzene, polymerization of ethylene oxide, etc.) In other cases, a process of the heterolytic type proceeds, preferably in systems with a definite quantity of cation electrons. For example, a great number of reactions involving acetylene are catalyzed by cations in the electronic configuration d^{10} (Hg^{+2}, Zn^{+2}, Cd^{+2}, Cu^+, and others), olefinic polymerizations are catalyzed by the systems d^1 ($TiCl_3$, Mo_2O_5, chromia–alumina catalysts[1]).

As a first approximation for the evaluation of acid–base properties of oxide surfaces, one can use the difference in electronegativity Δx: the higher the Δx of metal and oxygen the stronger the basicity and the weaker the acidity.

The basicity of the catalyst, according to our survey, considerably increases the velocity of isomerization of olefins involving transfer of the double bond, condensation of many organic compounds with water elimination and some polymerization reactions. The active center for base catalysts in the majority of cases appears to be a surface ion of O^{2-} or $Me^{2+}O^{2-}$. Establishing of excess charge on these ions leads to the ability of solid bases, in some cases, to catalyze homolytic reactions, e.g., dehydrogenation of alcohols and hydrocarbons.

In catalysis by ionic pairs, indeed one can observe a relationship between catalytic activity and lattice parameter. However, in the majority of acid–base reactions the rate of the geometric factor is not completely understood. Very probably, though not completely confirmed,

---

[1] In chromia–alumina catalysts the Cr on the surface has a charge of $+5$.

the geometry of atoms and ions of a catalyst surface have an influence on the formation of stereoregular polymers.

A relationship between dielectric permeability of a solid and catalytic activity such as is observed in the majority of cases of the oxidation–reduction type, could not be established in the case of acid–base catalysis, due to the lack of data.

In a number of cases, rules for the selection of homogenous catalysts proved to be very similar to those for heterogenous catalysts.

The principle of energetic relations can help in catalyst selection because for the activation of reactants the presence of optimal values of bond energy are required. With either too loose or too strong binding of molecules to the catalyst, reactions will not occur. For example, molecules bound to the surface with the weak hydrogen bond or the strong covalent bond do not participate in the reactions of alcohol dehydration, and polymerization of ethylene oxide. The optimal bond energy, in this case, exists for molecules which have a coordination bond with the surface metal atoms. In our opinion however, at the present time, the theory available for treating these problems does not allow catalyst prediction for a particular reaction based on the values of the bond energy determined independently from catalytic reactions.

By a study of the principles involved, one can solve only partially the problem of catalyst selection and the closely related problem of the determination of mechanism and catalyst action. A number of conclusions regarding the mechanism for particular reactions was presented in Part II.

In the case of those reactions (not a great many) for which there are enough data on the catalyst, correlations for catalyst selection were developed by a statistical method of analysis. Correlation of catalytic activity was sought for those properties of solids, which, according to existing theories of catalysis considered in Part I, should influence catalytic activity. Methods of correlation analysis allow us to approach catalyst selection and also to explain the mechanism of catalytic action objectively, though it is a tedious task. The calculation procedure is still more involved if we use multiple regression techniques in order to determine the relationship between catalytic activity and several properties of a solid. In order to develop the rules for selection by these methods, it is expedient to use contemporary computer techniques.

Another way of applying computer techniques to catalyst selection is to determine regions of maximum activity by the use of sequential analysis. In the past, this method has been used only for the selection of mixed catalysts, the compositions of which were known, as, for example, the reaction of naphthalene oxidation [682]. The procedure [683] is terminated by applying a two-dimensional least squares approximation to a limited area and finding the maximum by the method of steepest ascent. Thus, referring to Fig. 61 (Section 7.1), (involving oxidation of CO) one obtains for the oxides Mg, Zn, Si, and Ge, points which mark out a central quadrilateral; for these oxides one obtains an equation of activity: $A = 0.50 - 0.75x + 0.16y$ (where x and y are the positions of metals in the period and group, respectively). The steepest ascent on the basis of this equation leads to the region of oxides between Ag and Cu which, indeed proved to have high activity for the oxidation of CO.

Consider the case of dehydrogenation of the alcohol iso-C_3H_7OH (Fig. 38 in Section 4.1) with three independent variables: the position of the metal in the period and group x and y, respectively; and the position of the metalloid in group z. If the initial point is to lie at the center of symmetry of the initial data, the choice may be limited, in this case, to four initial points: HgO, SnO, CdS, and PbS. The derived equation is $A = 2.53 + 0.60x - 1.68y + 1.16z$. The maximum ascent approaches HgS which, according to Table III, appears to be less active, however, than the other three compounds taken as the basis for calculation. In this case, the statistical approach does not lead to the compound with maximum activity.

There are some other ways of using statistics and computer techniques that help to plan experiments and limit the region of search when one does not have knowledge of the detailed mechanism of the catalyst action. This is especially useful when dealing with mixed and promoted catalysts where the number of independent parameters is considerably greater.

It is clear that the statistical approach does not yet solve the basic problem: development of a scientifically based theory for catalyst selection, applying contemporary developments in quantum physics, solid state physics, as well as heterogenous micro- and macro-kinetics. Nevertheless, as shown in Part I, the prerequisites for the development of such a theory do exist at the present time.

Appendix

Physical Properties of Some Nonmetallic Compounds Used as Catalysts

In the table below, data are given for the properties of the simplest compounds mentioned in the second part of this monograph as catalysts for one or another reaction. The entries in this table were extracted from a number of surveys and books[1] and are corrected according to new data from original references appearing from 1955–1964. If a compound exists in the form of several crystal modifications data are given for the most widespread modification, especially the form most stable under conditions of catalysis.

It should be noted that literature data are very contradictory for such properties as width of the forbidden zone U, work function φ and dielectric constant of some semiconductors especially for oxides of the transition metals. In these cases, priority was given to the most recent data obtained by the best contemporary methods for investigation. For example for V_2O_5 the following values are given for the width of the forbidden zone: 0.45 eV [687], 0.58 eV [688], 1 eV [689], 2.1 eV [690], and 3.0 eV [80]. From among these values the optional value of 2.1 eV was chosen since this was obtained in 1964. This is in good agreement with the values of U in the series of oxides of transition elements.

Wherever possible, data were presented for the following properties: distance between the atoms of metal and nonmetal Me—X; width of the forbidden zone obtained from optical data U_{opt}; work function, thermoelectronic φ_T; and dielectric permeability, "static" i.e., determined at zero frequency ϵ_0. In the absence of complete data, available values are given for Me—X, U, φ, and ϵ.

The values presented in the table were used for the plots in the text that involved catalytic activity.

[1] See refs. [52, 64, 70, 72–76, 79, 80, 88–107, 112, 113, 118, 119, 129, 131, 164, 169, 287, 288, 332–337, 684–686].

APPENDIX TABLE

Some Properties of Substances Used as Catalysts[a]

I	II[b]	III[c]	IV	V	VI	VII	VIII	IX
AgBr	cb.	NaCl	2.88	1.0	n, p	4.3	4.3	13.1
AgCl	cb.	NaCl	2.79	1.2	n, p	5.1	4.7	12.3
β-AgI	hex.	Wurtzite	2.32	0.8	p, n	3.0	4.0	
γ-AgI	cb.	Sphalerite		0.8	p	2.8	3.9	
AgO	mn.		2.03	1.7	p	0.7		
Ag$_2$O	cb.	Cuprite	2.05	1.7	p	1.5		
Ag$_2$S	rhomb.		2.87	0.7	n	1.2		
AlBr$_3$	mn.	AlCl$_3$	2.23	1.3				
AlCl$_3$	mn.	AlCl$_3$	2.15	1.5				
AlN	hex.	Wurtzite	1.87	1.5				
α-Al$_2$O$_3$	rhmbohed.	Corundum	1.85	2.0	n	4.0		8.5
β-Al$_2$O$_3$	hex.	Corundum		2.0				11.4–13.1
γ-Al$_2$O$_3$	cb.	Spinel	1.78	2.0	n	7.3	4.7	12.2
α-AlOOH	rhomb.	Diaspore	1.82					
γ-AlOOH	rhomb.	Boehmite	1.87					
Al(OH)$_3$	mn.	Hydrargillite	1.85					
Al$_2$S$_3$	hex.	Wurtzite	2.31	1.0		4.1		
AlSb	cb.	Sphalerite	2.64	0.6	p	1.57		
α-As	rhmbched.	α-As	2.80	0		1.2	5.1	
AsBr$_3$	rhomb.		2.35	0.8				
As$_2$O$_3$	cb.	Senarmontite	2.01	1.5		4.0		3.3
As$_2$S$_3$	mn.		2.20	0.5		2.5		
Au$_2$O$_3$				1.2	n		5.8	
B	tetrag.		1.94	0	p, n	1.5	4.7	12
BAs	cb.		2.04	0	p	2.9		
α-B$_2$O$_3$	hex.	Sphalerite	1.45	1.5	Ionic	9.0	4.7	32

248

I	II	III	IV	V	VI	VII	VIII	IX
BP	cb.	Sphalerite	3.15	0.1			4.6	
BaCl$_2$	rhomb.	PbCl$_2$	2.87	2.15				
BaH$_2$	rhomb.	SrH$_2$	2.76	1.25				
BaO	cb.	NaCl	2.68	2.65	Ionic	4.4	1.2	13
BaO$_2$	tetrag.	CaC$_2$	2.02	2.15	Ionic, p			
BaCl$_2$	rhomb.	SiS$_2$			n			
BeO	hex.	Wurtzite	1.65	2.0	Ionic	10.4	3.8	7.3
Be(OH)$_2$			1.63					
Bi	rhmbohed.	α-As	3.10	0		0.015	4.3	
α-Bi$_2$O$_3$	mn.	Valentinite	2.38	1.7		2.9		
α-Bi$_2$O$_3$	cb.	Mg$_3$P$_2$	2.45	1.7	p	1.4		
Bi$_2$S$_3$	rhomb.	Antimonite	2.60	0.7	n			
C	cb.	Diamond	1.54	0	p	7.0	4.4	5.7
C	hex.	Graphite	1.42	0	n	0.1	4.8	
CaBe$_2$	rhomb.	Deform. rutile	2.88	1.8				
CaCl$_2$	rhomb.	Deform. rutile	2.73	2.0	Ionic			10.5
CaF$_2$	cb.	Fluorite	2.38	3.0	Ionic	10.0		8.4
CaH$_2$	rhomb.	SrH$_2$	2.32	1.1				
CaO	cb.	NaCl	2.40	2.5	p, n	7.5	2.0	12
Ca(OH)$_2$	rhombohed.	CdI$_2$	2.36					
CdBr$_2$	rhombohed.	CdCl$_2$	2.84	1.3				
CdCl$_2$	rhombohed.	CdCl$_2$	2.74	1.5				

^{a} Column identifications:

  I   Chemical formula
  II   Type of syngony
  III Type of structure of crystal lattice
  IV Distance between atoms of metal and nonmetal, Å
  V   Difference in electronegativity
  VI Prominent type of conductivity at low temperature
  VII Width of forbidden zone, eV
  VIII Work function for pure compound, eV
  IX Dielectric constant

^{b} The following abbreviations are used: cb—cubic; hex—hexagonal; mn—monoclinic; rhomb—rhombic; rhombohed—rhombohedral; tetrag—tetragonal.

^{c} deform—deformed.

APPENDIX TABLE (*continued*)

I	II	III	IV	V	VI	VII	VIII	IX
CdO	cb.		2.35	2.0	n	2.5		
α-CdS	hex.	Wurtzite	2.52	1.0	n	2.4		
β-CdS	cb.	Sphalerite	2.53	1.0	n	2.2	4.4	11.6
CdSe	hex.	Wurtzite	2.62	0.8	n	1.7		
CdTe	cb.	Sphalerite	2.83	0.6	p	1.5		10.9
CeO$_2$	cb.	Fluorite	2.34	2.4	n			
Ce$_2$O$_3$	rhombohed.	La$_2$O$_3$	2.39	2.45				
CoCl$_2$	rhombohed.	CdCl$_2$	2.62	1.3				
CoO	cb.	NaCl	2.12	1.8	p	0.8		10.0
Co$_2$O$_3$	cb.	Spinel	1.75	1.8	n			9.6
Co$_3$O$_4$	cb.	Spinel	1.92	1.8	p	0.9	4.9	
Co(OH)$_2$	rhombohed.	CdI$_2$	2.10					
Co(OH)$_3$	hex.		2.13					
CoP	rhomb.		2.30	0.4		0		
CoS	hex.		2.56	0.8				
CrB$_2$	hex		2.18	0.1	n	0	3.4	
CrCl$_3$	rhombohed.		2.43	1.4				
CrN	cb.		2.07	1.4	n			
CrO	cb.	NaCl	2.24	2.0	p			
CrO$_2$	tetrag.	Rutile	1.90	0.3	n	0.3		
CrO$_3$	rhomb.		1.76	1.4	n	3.2		
Cr$_2$O$_3$	rhombohed.	Corundum	2.01	1.9	p, n	1.9	5.8	11.9–13.3
CrP	rhomb.	MnP	2.40	0.5				
CrS	hex.	NiAs	2.62	1.0	n	0.9		
Cr$_2$S$_3$	rhombohed.	Corundum	2.42	0.9				
CrSe	hex.	NiAs	2.74	0.8				
CrSi$_2$	tetrag.	FeSi$_2$	2.48	0.2	p	1.1		7.2
α-CsCl	cb.	CsCl	3.57	2.3				

β-CsCl	cb.	NaCl	3.52	2.3			
CsF	cb.	NaCl	3.01	3.3			
CsI	cb.	CsCl	3.96	1.9		6.1	
Cs$_2$O	rhombohed.	Anti-CdCl$_2$	2.86	2.8			
CuBr	cb.	Sphalerite	2.46	1.0	n, p	2.9	1.1
CuBr$_2$	mn.		2.76	0.8			8.0
CuCl	cb.	Sphalerite	2.34	1.2	n, p		8
CuCl$_2$	mn.		2.64	1.0		3.0	
CuI	mn.	Sphalerite	2.62	0.8			10.0
CuO	mn.	Tenorite	1.95	1.5	n, p	1.4	5.3	10
Cu$_2$O	cb.	Cuprite	1.84	1.5	p	1.9	5.2	18.1
Cu(OH)$_2$	rhombohed.	CdI$_2$	2.00					10.5
CuS	hex.	Covellite	2.54	0.5			
Cu$_2$S	cb.	Fluorite	2.42	0.7	p	1.7	
CuSe	hex.	Covellite	2.71	0.3			
CuTe	rhomb.		2.91	0.1			
Dy$_2$O$_3$	cb.	Mn$_2$O$_3$	2.28	2.4		3.1		11.1
Er$_2$O$_3$	cb.	Mn$_2$O$_3$	2.25	2.4		3.3		12.3
Eu$_2$O$_3$	cb.	Mn$_2$O$_3$	2.32	2.4		1.8		10.2
FeCl$_3$	rhombohed.	CrCl$_3$	2.46	1.2			
Fe$_2$N	rhomb.	ε-AgZn$_3$		1.2			
FeO	cb.	NaCl	2.16	1.85	p	0.4	3.85	14.2
α-Fe$_2$O$_3$	rhombohed.	Corundum	1.91	1.7	n	1.9		12
β-Fe$_2$O$_3$	cb.	Mn$_2$O$_3$	2.01	1.7			
γ-Fe$_2$O$_3$	cb.	Spinel	1.88	1.7	n	2.2	3.9	12
Fe$_3$O$_4$	cb.	Spinel	1.93	1.75	n	0.4	
α-FeOOH	rhomb.	Lepidocrocite	1.96				
γ-FeOOH	rhomb	Goethite	1.92				
Fe(OH)$_2$	rhombohed.	CdI$_2$	2.26				
Fe(OH)$_3$	cb.		2.14				
FeP	rhomb.	MnP	2.31	0.3			
FeS	hex.	NiAs	2.57	0.85	p	0.1	

251

APPENDIX TABLE (*continued*)

I	II	III	IV	V	VI	VII	VIII	IX
FeS$_2$	cb.	Pyrite	2.26	0.7	p, n	1.2		
FeS$_2$	rhomb.	Marcasite		0.7				
FeSi$_2$	tetrag.	FeSi$_2$	2.31	0.1	n	0.7		
GaAs	cb.	Sphalerite	2.45	0.4	n, p	1.4	4.6	12.5
GaAs·Ga$_2$Se$_3$	cb.	Sphalerite	2.38	0.6	p	1.7		
3 GaAs·Ga$_2$Se$_3$	cb.	Sphalerite	2.39	0.55	n	1.6		
GaN	hex.	Wurtzite	1.92	1.4		3.3		
Ga$_2$O$_3$	rhombohed.	Corundum	1.95	1.9		4.6		
GaP	cb.	Sphalerite	2.36	0.5		2.3		10.2
Ga$_2$S$_3$	cb.	Sphalerite	2.22	0.9	n	2.5		
GaSb	cb.	Sphalerite	2.63	0.2	p, n	0.7		14.8
GaSe	hex.			0.7	p	1.9		
Ga$_2$Se$_3$	cb.	Sphalerite	2.34	0.7	p	1.8		
GaTe	hex.			0.5	p	1.6		
Ga$_2$Te$_3$	cb.	Sphalerite	2.53	0.5	p	1.1		
GdB$_6$	cb.	CaB$_6$		0.9			2.1	
Gd$_2$O$_3$	cb.	Mn$_2$O$_3$	2.31	2.4		2.9		
Ge	cb.	Diamond	2.45	0	p, n	0.74	4.8	16.0
GeO$_2$	tetrag.	Rutile	1.87	1.65		6.0	5.5	
HfO$_2$	cb.	Fluorite	2.21	2.2	n, p	4.4	4.0	
HgCl$_2$	rhomb.	HgCl$_2$	2.25	1.1				
HgO	rhomb.		2.01	1.6	n	1.0		6.6
HgO	hex.		2.03	1.6	n	1.2		11.5
HgS	cb.	Sphalerite	2.53	0.6	n	1.0		18
Ho$_2$O$_3$	cb.	Mn$_2$O$_3$	2.26	2.4		2.8		25.6
InAs	cb.	Sphalerite	2.62	0.4	n, p	0.4		14.6
In$_2$O$_3$	cb.	Mn$_2$O$_3$	2.15	1.9	n	3.1		
InP	cb.	Sphalerite	2.55	0.5		1.3		15

Compound	System	Structure				
InSb	cb.	Sphalerite	2.79	0.2	0.2	
In$_2$Se$_3$	hex.	Wurtzite	2.51	0.7	1.2	
In$_2$Te$_3$	cb.	Sphalerite	2.64	0.6	1.1	
KBr	cb.	NaCl	3.29	2.0	1.1	16
KCl	cb.	NaCl	3.15	2.2	Ionic	6.8	4.9
KF	cb.	NaCl	2.67	3.2		7.6	5.0
KI	cb.	NaCl	3.53	1.8		10.4	6.1
K$_2$O	cb.	Fluorite	2.80	2.7		5.8	4.9
KOH	cb.	NaCl	2.89				
LaB$_6$	cb.	ThB$_6$	2.42	0.9		0.08	2.7
La$_2$O$_3$	rhombohed.	La$_2$O$_3$	2.46	2.4	n	2.6	20.8
La(OH)$_3$	hex.	UCL$_3$	2.75	1.8			
LiBr	cb.	NaCl	2.57	2.0			
LiCl	cb.	NaCl	2.00	3.0		10	11.0
LiF	cb.	NaCl	2.00	2.5		11.5	9.3
Li$_2$O	cb.	Fluorite	1.97				
LiOH	tetrag.	PbO	2.22	2.4			
Lu$_2$O$_3$	cb.	Mn$_2$O$_3$	2.69	1.6		3.9	12.9
MgBr$_2$	rhombohed.	CdI$_2$	2.54	1.8	Ionic		
MgCl$_2$	rhombohed.	CdCl$_2$	2.05	2.8	Ionic		
MgF$_2$	tetrag.	Rutile	2.10	2.3	n, p, ionic	8.7	3.5
MgO	cb.	NaCl	2.09				
Mg(OH)$_2$	rhombohed.	CdI$_2$	2.22	2.1	p	1.25	6.5
MnO	cb.	NaCl	1.87	1.8	n	0.3	10
β-MnO$_2$	tetrag.	Rutile	1.84	1.8	n	0.6	14.4
γ-MnO$_2$	rhomb.	Goethite	2.02	2.0			12.8
Mn$_2$O$_3$	cb.	Mn$_2$O$_3$	2.05	2.05			15
Mn$_3$O$_4$	cb.	Spinel	1.87				
Mn(OH)$_2$	rhombohed.	CdI$_2$	2.36	0.6	p	0	8
MnP	rhomb.	MnP	2.61	1.1			
α-MnS	cb.	NaCl	2.42	1.1			
β-MnS	cb.	Sphalerite					

APPENDIX TABLE (*continued*)

I	II	III	IV	V	VI	VII	VIII	IX
γ-MnS	hex.	Wurtzite	2.41	1.1				
Mo_2C	hex.	Mo_2C	2.01	1.1		0		
MoO_3	mn.	Deform. rutile	2.00	1.9				
MoO_3	rhomb.	MoO_3	1.75	1.4	p	2.9	4.25	7.6
MoS_2	hex.	Molybdenite	2.35	0.9	n	1.2		
Mo_2S_3	rhomb.		2.36	1.0	p			
$MoSi_2$	tetrag.	$MoSi_2$	2.62	0.4		0.1		
Mo_3Si	cb.	β-W		0.4		0		
NaBr	cb.	NaCl	2.98	1.9		6.5		6.1
NaCl	cb.	NaCl	2.81	2.1	Ionic	7.8		5.9
NaF	cb.	NaCl	2.31	3.1				
NaI	cb.	NaCl	3.24	1.7		5.4		6.6
$NaNH_2$			2.46					
Na_2O	cb.	Fluorite	2.41	2.6				
α-NaOH	rhomb.	NaCl	2.51					
NbC	cb.	NaCl	2.22	0.9	p	0		
NbN	cb.	NaCl	2.19	1.4	p	0		
NbO	cb.	NaCl	2.10	2.0		0		
NbO_2	tetrag.	Rutile	2.02	1.9				
Nb_2O_5	rhomb.	U_3O_8	2.03	1.9	n	3.3		19.7
Nd_2O_3	hex.	La_2O_3	2.48	2.4		2.2		
$Nd(OH)_3$	rhombohed.	UCl_3	2.49					
$NiBr_2$	rhombohed.	$CdCl_2$	2.74	1.1				
$NiCl_2$	rhombohed.	$CdCl_2$	2.43	1.3				
NiO	cb.	NaCl	2.09	1.8	p	1.95	5.55	12
Ni_2O_3	rhombohed.	Corundum	1.80	1.8	n			12.3
$Ni(OH)_2$	rhombohed.	CdI_2	2.09		p			
NiS	hex.	NiAs	2.56	0.8				

NiSe	hex.	NiAs	2.72	0.6		
PbBr$_2$	rhomb.	PbCl$_2$	3.01	1.2		
PLCl$_2$	rhomb.	PbCl$_2$	2.86	1.4		
PbI$_2$	rhombohed.	CdI$_2$	3.08	1.0		
PbO, red	tetrag.	PbO	2.30	1.9	n, p	3.1
PbO, yellow	rhomb.	PbO	2.24	1.9	p	2.3
β-PbO$_2$	tetrag.	Rutile	2.16	1.7	n	3.2
Pb$_3$O$_4$			2.18	1.8		1.8
PbS	cb.	NaCl	2.96	0.9	n, p	0.4	5–6	30
PdO	tetrag.		2.2	1.5				33.5
PrO$_2$	cb.	Fluorite	2.33	2.4		0.9		
Pr$_2$O$_3$	rhombohed.	La$_2$O$_3$	2.38	2.4	p	1.2		
Pr(OH)$_3$	hex.	UCl$_3$	2.35					
PtO	tetrag.	PbO						
PtO$_2$			1.9	1.4				
Pt$_3$O$_4$			2.19		n			
RbCl	cb.	CsCl	3.27	2.2	Ionic	9.0		22.5
RbF	cb.	NaCl	2.82	3.2				
Rb$_2$O	cb.	Fluorite	2.92	2.7				5.9
RbOH	rhomb.	TeI	3.00					6.0
ReO$_2$	mn.	Deform. rutile	1.99	1.4		1.2		
ReS$_2$	hex.	Molybdenite	2.34	0.4		1.1		
Rh$_2$O$_3$	rhombohed.	Corundum	2.01	1.5				7.8
RuO$_2$	tetrag.	Rutile	2.00	1.4		0.1		
Sb	rhombohed.	α-As	2.87	0		0.1	4.0	
SbBr$_3$	rhomb.		2.51	1.0				5.1
SbCl$_3$	rhomb.		2.36	1.2				5.4
Sb$_2$O$_3$	cb.	Senarmonite	2.22	1.7	p	4.2		12.8
Sb$_2$O$_4$	rhomb.			1.6				
Sb$_2$O$_5$	cb.			1.4	n			
Sb$_2$S$_3$	rhomb.	Antimonite	2.39	0.7	p, n	1.7		
ScN	cb.	NaCl	2.23	1.7		2.6		

APPENDIX TABLE (*continued*)

I	II	III	IV	V	VI	VII	VIII	IX
Sc$_2$O$_3$	cb.	Mn$_2$O$_3$	2.10	2.2	p, n	5.4	4.9	8.5
α-Se	rhombohed.	a-Se	2.32	0	p	1.8	4.8	12.5
Si	cb.	Diamond	2.35	0	n, p	1.1		5.5
α-SiO$_2$	cb.	α-quartz	1.61	1.7		11		4.4–4.6
β-SiO$_2$	hex.	β-quartz	1.59	1.7		8.0	5.0	
SiO$_2$	hex.	β-tridymite	1.54	1.7				
α-SiO$_2$	tetrag.	α-cristobalite	1.59	1.7				8.2
β-SiO$_2$	cb.	β-cristobalite	1.58	1.7				(silica gel)
Sm$_3$O$_3$	cb.	Mn$_2$O$_3$	2.22			2.3		18.3
γ-Sn	cb.	Diamond	2.80	0	n, p	0.09	4.4	42
SnCl$_2$	rhomb.	PbCl$_2$	2.78	1.35		3.6		
SnO	tetrag.	PbO	2.21	1.85	p, n			
SnO$_2$	tetrag.	Rutile	2.06	1.7	n	3.8		23.7
SnS	rhomb.	SnS	2.62	0.85	p	1.25		
SnS$_2$	rhombohed.	CdI$_2$	2.55	0.7				
SrCl$_2$	cb.	CaC$_2$	3.02	2.0	Ionic	5.7	1.4	12.4
SrO	cb.	NaCl	3.57	2.5	p, ionic	4.8		13.3
SrS	cb.	NaCl	3.00	1.5		0	3.1	
TaC	cb.	NaCl	2.23	1.1				
TaO$_2$	tetrag.	Rutile	2.02	2.1		3.6	4.6	
Ta$_2$O$_8$	rhomb.	U$_3$O$_8$	2.00		n	0.8		
Tb$_2$O$_3$	cb.	Mn$_2$O$_3$	2.29	2.4		0.3	4.8	5.2
Te	rhombohed.	α-Se	2.86	0	p	1.5		
TeO$_2$	tetrag.	Rutile	2.22	1.4	n	3.5	4.7	
ThO$_2$	cb.	Fluorite	2.42	2.4	n	0	3.9	
TiB$_2$	hex.	AlB$_2$	2.45	0.4	n	0	2.4	
TiC	cb.	NaCl	2.16	0.9	n			

256

Compound		Structure					
TiCl₂	rhombohed.	CdI₂	2.52	1.6			
α-TiCl₃	rhombohed.	AsI₃	2.50	1.5			
γ-TiCl₃	rhombohed.	BiI₃	2.50	1.5			
TiN	cb.	NaCl	2.12	1.5			
TiO	cb.	NaCl	2.12	2.1			
TiO₂	tetrag.	Anatase	1.95	1.9			
TiO₂	rhomb.	Brucite	1.87	1.9			
TiO₂	tetrag.	Rutile	1.94	1.9			2.9
Ti₂O₃	rhombohed.	Corundum	2.03	2.0	p	0	
TiP	hex.	NiAs	2.48	0.6	p	3.2	
TiS	hex.	NiAs		1.1		3.3	
TiS₂	rhombohed.	CdI₂	2.42	0.9		3.1	3.0 89–173
Tl₂O₃	cb.	Mn₂O₃	2.26	1.6	n	2.2	
Tl₂S	rhombohed.	CdI₂	3.23	1.0		0	
Tu₂O₃	cb.	Mn₂O₃	2.25	2.4		0	
UO₂	cb.	Fluorite	2.37	2.2	p	3.2	12.6
UO₃	cb.	ReO₃	2.08	1.9	n	0.3	
U₃O₈	rhomb.	U₃O₈	2.06; 2.31	2.0	n	2.6	
VC	cb.	NaCl	2.08	0.85		1.1	
VCl₃	rhombohed.	BiI₃	2.14	1.35	p	0	3.9
VN	cb.	NaCl	2.14	1.65		0.2	18–36
VO	cb.	NaCl	2.05	2.3		0.3	
VO₂	tetrag.	Rutile	1.76–2.05	1.85		0.3	
V₂O₃	rhombohed.	Corundum	2.05	2.15		2.1	
V₂O₈	rhomb.	V₂O₈	1.54–2.81		n		4.1
VP	hex.	NiAs	2.70	0.75		0	
VS	hex.	NiAs	2.66	1.3		0	
VSi₂	tetrag.	FeSi₂		0.2		0	
V₃Si	cb.	β-W		0.6		0	
WC	hex.		2.18	0.9		0	
W₂C	rhombohed.		2.06	1.1			4.6
WO₂	tetrag.	Rutile	2.00	1.9	n		5.0

APPENDIX TABLE (continued)

I	II	III	IV	V	VI	VII	VIII	IX
WO_3	rhomb.	WO_3	1.83	1.4	n	2.8	4.5	
WS_2	hex.	Molybdenite	2.48	0.9				
WSi_2	tetrag.	$MoSi_2$	2.63	0.4				
Y_2O_3	cb.	Mn_2O_3	2.27	2.3	n	0	5.5	14
Yb_2O_3	cb.	Mn_2O_3	2.25	2.4	n	2.9		6.6
$ZnCl_2$	rhombohed.	$CdCl_2$	2.31	1.5		3.1		
ZnO	hex.	Wurtzite	1.99	2.0	n	3.3	4.6	8–10.4
$Zn(OH)_2$	rhombohed.	CdI_2	1.95					
ZnS	hex.	Wurtzite	2.34	1.0	n	3.6		9–10.5
ZnS	cb.	Sphalerite	2.36	1.0	n	3.7		12–16
$ZnSe$	cb.	Sphalerite	2.43	0.6	p	2.7		8.1
$ZnTe$	cb.	Sphalerite	2.62	0.6	p	2.2		18.6
ZrC	cb.	NaCl	2.34	1.1		0	3.6	
ZrN	cb.	NaCl	2.32	1.6	p		2.9	
ZrO_2		Fluorite	2.20	2.1	n	2.3	4.5	
ZrO_2	mn.	Baddeleyite	2.04	2.1			5.0	
ZrS_2	rhombohed.	CdI_2	2.58	1.1				

References

1. S. Z. Roginskii, *Probl. Kinetiki i Kataliza, Acad. Nauk SSSR* **8**, 110 (1956).
2. G. K. Boreskov, *Probl. Kinetiki i Kataliza, Acad. Nauk SSSR* **10**, 123 (1969).
3. V. A. Roiter, "Introduction to Theory of Kinetics and Catalysis *Izv. Akad. Nauk. SSSR*, Kiev, 1962.
4. S. Berkman, G. Morell, and G. Egloff, "Catalysis in Organic and Inorganic Chemistry." Gostoptekhizdat, 1949. (in Russian)
5. A. A. Balandin, *Usp. Khim.* **13**, 465 (1944).
6. S. Z. Roginskii, *Zh. Fiz. Khim* **6**, 334, 1935.
7. S. Z. Roginskii, *Dokl. Akad. Nauk SSSR* **67**, 97 (1949).
8. F. F. Volkenshtein, Coll. "Scientific Basis of Catalyst Selection for Heterogenous Catalytic Reactions," p. 6. Izd. Nauka, 1966. (in Russian)
9. A. Clark, *Ind. Eng. Chem.* **45**, 1476 (1953).
10. D. A. Dowden and P. A. Reynolds, *Discussions Faraday Soc.* **8**, 184 (1959).
11. M. McD. Baker and G. I. Jenkins, *Advan. Catalysis* **7**, 1 (1955).
12. K. K. Ingold, "Mechanism of Reactions and Composition of Organic Compounds." IL, 1959. (in Russian)
13. A. F. Ioffe, "Reports on Scientific-Technical Investigations in the Commonwealth-Catalysis." p. 53. NKHTI, 1930. (in Russian)
14. S. Z. Roginskii and E. N. Shul, *Ukr. Khim. Zh.* **3**, 177 (1928).
15. C. Wagner and K. Hauffe, *Z. Elektrochem.* **44**, 172 (1938).
16. F. F. Volkenshtein, "Electronic Theory of Catalysis on Semi-Conductors." Fizmatgiz, 1960. (in Russian)
17. K. Hauffe, "Reactions on Solids and Their Surfaces." IL. 1962. (in Russian)
18. J. E. Germain, *J. Chim. Phys.* **51**, 691 (1954).
19. P. B. Weiss, *J. Chem. Phys.* **20**, 1483 (1952).
20. V. B. Kazanskii and G. B. Pariiskii, *Proc. 3rd Intern. Congr. Catalysis* **1**, 367 (1964). Amsterdam, 1965.
21. W. E. Garner, *Advan. Catalysis* **9**, 169 (1952).
22. W. E. Garner, F. S. Stone, and P. F. Tiley, *Proc. Roy. Soc., (London), Ser. A* **211**, 472 (1952).
23. K. Hauffe, *Angew. Chem.* **68**, 776 (1956).

24. G. M. Schwab, *Advan. Catalysis* **2**, 229 (1952).
25. K. Hauffe, R. Glang, and H. Engell, *Z. Phys. Chem.* **201**, 223 (1952).
26. R. M. Dell, F. S. Stone, and P. F. Tiley, *Trans. Faraday Soc.* **49**, 201 (1953).
27. H. J. Engell and K. Hauffe, *Z. Elektrochem.* **57**, 776 (1953).
28. V. L. Kuchaev and G. K. Boreskov, *Kinetika i Kataliz* **1**, 356 (1960).
29. J. Kubokava and O. Toyama, *J. Phys. Chem.* **60**, 833 (1956).
30. H. C. Powlinson and R. J. Cvetanovic, *Advan. Catalysis* **9**, 243 (1957).
31. G. M. Schwab, Sbornik Referatov 17-go Kongressa IUPAKA, Munchen, 1959.
32. G. M. Schwab and J. Block, *J. Chim. Phys.* **51**, 664 (1954).
33. G. Parravano, *J. Am. Chem. Soc.* **75**, 1148, 1452 (1953).
34. N. P. Keier, S. Z. Roginskii, and I. S. Sazonova, *Izv. Akad. Nauk SSSR, Ser. Fiz.* **22**, 183 (1957).
35. J. Coue, P. C. Gravell, R. E. Ranc, P. Rue, and S. J. Teichner, *Proc. 3rd Intern. Congr.* Catalysis **1**, 748 (1964). Amsterdam, 1965.
36. G. I. Chizhikova and N. P. Keier, *Probl. Kinetiki i Kataliza* **10**, 77 (1960).
37. L. N. Kuseva and N. P. Keier, *Probl. Kinetiki i Kataliza* **10**, 82 (1960).
38. W. E. Garner, D. A. Dowden, and J. F. Garcia de la Banda, *Anales Real. Soc. Espan. Fis. Quim.* **50B**, 35 (1954).
39. V. M. Frolov, O. V. Krylov, and S. Z. Roginskii, *Dokl. Akad. Nauk. SSSR* **III**, 623 (1956).
40. G. Parravano, and M. Boudart, *Advan. Catalysis* **7**, 509 (1955).
41. R. Kuhn, Inaug. Dissertation, Munchen, 1957.
42. Zh. Zhermen, "Heterogenous Catalysis." IL, 1961. (in Russian).
43. D. A. Dowden, *Bull. Soc. Chim. Belg.* **67**, 439 (1958).
44. K. Hauffe, *Advan. Catalysis* **7**, 213 (1955).
45. S. W. Weller and S. E. Voltz, *Advan. Catalysis* **9**, 215 (1957).
46. B. Arghiropoulos, F. Juillet, M. Prettre, and S. Teichner, *Compt. Rend.* **249**, 1895 (1959).
47. Y. Saito, Y. Yoneda, and S. Makishima, *Actes 2e Congr. Intern. Catalyse* **2**, 1937 (1962). Technip, Paris.
48. F. S. Stone, *Advan. Catalysis*, **13**, 1 (1962).
49. J. Rudolph, *Z. Naturforsch*, **14a**, 727 (1959).
50. R. Glemza and R. J. Kokes, *J. Phys. Chem.* **66**, 566 (1962).
51. A. Cimino, E. Molinari, and F. Cramarosa, *J. Catalysis* **2**, 315 (1963).
52. A. F. Ioffe, "Physics of Semiconductors." Izv. Akad. Nauk SSSR, 1957. (in Russian)
53. S. Z. Roginskii, *Kinetika i Kataliz* **1**, 15 (1960).
54. F. F. Volkenshtein and V. B. Sandomirskii, *Probl. Kinetiki i Kataliza* **8**, 189 (1955).
55. V. F. Synorov, Coll. "Problems of Metallurgy and Physics of Semiconductors," p. 120. Izv. Akad. Nauk. SSSR, 1959. (in Russian)
56. F. F. Volkenshtein and Sh. M. Kogan, *Zh. Fiz. Khim.* **34**, 166 (1960).
57. G. P. Lekhtinen, M. A. Rzaev, and L. S. Stilbans, *Zh. Fiz. Khim.* **27**, 1227 (1957).
58. A. N. Arseneva-Geil and Van Bao-kun, *Fiz. Tverd. Tela* **3**, 3621 (1961).
59. S. I. Chernyshev and V. E. Zgaevskii, *Fiz. Tverd. Tela* **2**, 2460 (1960).
60. E. Kh. Enikeev, L. Ya. Margolis, and S. Z. Roginskii, *Dokl. Akad. Nauk SSSR* **124**, 606 (1959).
61. S. R. Morrison, *Advan. Catalysis*, **7**, 259 (1955).
62. G. M. Zhabrova, V. I. Vladimirova, and O. M. Vinogradova, *Dokl. Akad. Nauk SSSR* **133**, 1375 (1960).
63. L. Ya. Margolis, E. Kh. Enikeev, O. V. Isaev, A. V. Krylova, and M. Ya. Kushnerev, *Kinetika i Kataliz* **3**, 181 (1962).

64. B. M. Tsarev, "Contact Difference of Potentials. G.T.T.I., 1955. (in Russian)
65. G. K. Boreskov and V. V. Popovskii, *Probl. Kinetiki i Kataliza* **10**, 67 (1960).
66. I. D. Morozova and V. V. Popovskii, *Kinetika i Kataliz* **3**, 445 (1962).
67. O. V. Krylov and S. Z. Roginskii, *Izv. Akad. Nauk. SSSR, Ser. khim. Nauk* **1**, 17 (1959).
68. O. V. Krylov and S. Z. Roginskii, *Dokl. Akad. Nauk. SSSR* **118**, 526 (1958).
69. M. A. Dalin and A. Z. Shikhmamedbekova, *Tr. Inst. Khim. Akad. Nauk Azer. SSR* **15**, 84 (1956).
70. G. Bush, *Usp. Fiz. Nauk* **47**, 258 (1960).
71. W. W. Piper and F. E. Williams, *Phys. Rev.* **84**, 659 (1952).
72. W. Ruppel, H. J. Gerritsen, and A. Rose, *Helv. Phys. Acta* **30**, 495 (1957).
73. G. Velker, G. Veise, Coll. "New Semiconducting Materials. p. 9. IL, 1958. (in Russian)
74. N. A. Goryunova, "Chemistry of Diamondlike Semiconductors." Publ. Leningrad State University, 1963. (in Russian)
75. V. P. Zhuze, V. M. Sergeeva, and A. Shelykh, *Fiz. Tverd. Tela* **2**, 2858 (1960).
76. N. A. Goryunova, V. S. Gregoreva, B. M. Konovalenko, and S. M. Ryvkin, *Zh. Tekhn. Fiz.* **25**, 1675 (1955).
77. O. V. Krylov and E. A. Fokina, *Zh. Fiz. Khim* **35**, 851 (1961).
78. O. V. Krylov and E. A. Fokina, *Kinetika i Kataliz* **5**, 284 (1964).
79. V. Denlen, "Introduction to Semiconductor Physics," IL, 1959. (in Russian)
80. V. Byub, "Photoconductivity of Solids." IL, 1962. (in Russian)
81. E. Cremer, *Advan. Catalysis* **7**, 303 (1955).
82. R. M. Dell, *J. Phys. Chem.* **64**, 1584 (1957).
83. G. M. Schwab, G. Greger, S. Krawczynski, and J. Penzkofer, *Z. Phys. Chem.* **15**, 363 (1958).
84. J. Penzkofer, Inaug. Dissertation, Munchen, 1957.
85. G. Greger, Inaug. Dissertation, Munchen, 1957.
86. G. M. Schwab, *in* "Semiconductor Surface Physics," p. 283. Univ. of Pennsylvania Press, Philadelphia, 1957.
87. A. F. Ioffe, *Izv. Akad. Nauk. SSSR, Ser. Fiz.* **15**, 477 (1951).
88. V. B. Pirson, *Zh. Bho* **5**, 493 (1961).
89. N. A. Goryunova, *Zh. VKHO* **5**, 522 (1961).
90. Zh. Syushe, "Physical Chemistry of Semiconductors." Pub. Metallurgiya, 1964. (in Russian)
91. O. G. Folberth and H. Welker, *J. Phys. Chem. Solids* **8**, 14 (1959).
92. O. H. Folberts, *Z. Naturforsch.* **13a**, 856 (1958).
93. B. Seraphin, *Z. Naturforsch.* **9a**, 450 (1954).
94. A. I. Gubanov, *Zh. Tekhn. Fiz.* **26**, 2170 (1956).
95. J. Adavi, *Phys. Rev.* **105**, 789 (1957).
96. V. P. Zhuze, *Zh. Tekhn. Fiz.*, **25**, 2079 (1955).
97. D. P. Belotskii, Coll. "Problems of Metallurgy and Semiconductor physics," p. 18. Izv. Akad. Nauk. SSSR, 1961. (in Russian)
98. G. H. L. Goodman, *J. Electron.* **1**, 115 (1955).
99. L. Pauling, "Nature of the Chem. Bond." IL, 1946. (in Russian)
100. C. F. Cole, *Proc. I.R.E.* **50**, 1856 (1962).
101. H. Mott and R. Gerni, "Electronic Processes in Ionic Crystals." IL, 1959. (in Russian)
102. B. F. Ormont, Coll. "Problems of Metallurgy and Semiconductor Physics," p. 5–103, 1961.
103. W. Ruppel, A. Rose and H. J. Gerritsen, *Helv. Phys. Acta* **30**, 238 (1957).

104. P. Manca, *J. Phys. Chem. Solids* **20**, 268 (1961).
105. M. V. Fok, *Fiz. Tverd. Tela* **5**, 1491 (1963).
106. H. Pfister, *Z. Naturforsch* **10a**, 79 (1955).
107. G. B. Bokii, "Introduction to Crystal Chemistry." Ivz. Moscow State Univ., 1960.
108. H. P. Philiph, *Phys. Rev.* **111**, 440 (1958).
109. M. Balkanski, *Proc. Intern. Conf. Semicond. Phys.*, Prague p. 932, (1960), 1961.
110. B. B. Sandomirskii, *Zh. Eksp. Teor. Fiz.* **43**, 2309 (1962).
111. V. L. Boich-Bruevich, Coll. "Solid State Physics," Vol. 2. Ivz. Akad. Nauk. SSSR, **2**, 1959. p. 177.
112. A. A. Vorobev and E. K. Zavadovskaya, "Electric Stability of Solid Dielectrics." Gostekhteoret, 1956. (in Russian)
113. E. C. McIrvine, *J. Phys. Chem. Solids* **15**, 356 (1960).
114. V. L. Boich-Bruevich, *Zh. Fiz. Khim.* **27**, 622, 960 (1953).
115. E. K. Putseiko and A. N. Terenin, *Dokl. Akad. Nauk. SSSR* **70**, 401 (1950).
116. Coll. "Surface Phenomena on Semiconductors." IL, 1958. (in Russian)
117. V. Heine, *Phys. Rev.* **138**, 1689 (1965).
118. M. Haissinsky, *J. Phys. Radium* **7**, 7 (1946).
119. S. S. Batsanov, "Electronegativity of Elements and Chemical Bond." Izv. Akad Nauk SSSR, Novosibirsk, 1962. (in Russian)
120. Ya. K. Syrkin, *Usp. Chim.* **31**, 397 (1962).
121. R. S. Mulliken, *J. Chem. Phys.* **2**, 782 (1934).
122. C. H. L. Goodman, *Proc. Phys. Soc.* **B67**, 258 (1954).
123. R. T. Sanderson, *J. Inorg. Nucl. Chem.* **7**, 288 (1958).
124. C. H. L. Goodman, *Nature* **187**, 590 (1960).
125. J. Hinze, H. H. Jaffe, *J. Phys. Chem.* **67**, 1501 (1963).
126. Z. V. Zvonkova, *Prob. Fiz. Khim.* **2**, 97 (1959).
127. Z. V. Zvonkova, *Kristallografiya* **4**, 668 (1959).
128. G. B. Slater, *Phys. Rev.* **69**, 404 (1946).
129. S. S. Shalyt, Coll. "Semiconductors in Science and Technology," Vol. 1, p. 7. Izv. Akad. Nauk SSSR, 1957. (in Russian)
130. V. Pirson, Coll. "Semiconducting Substances. Problems of the Chemical Bond," p. 249. IL, 1960. (in Russian)
131. P. Aigrain and M. Balkanski, "Constantes Selectionnées Relatives aux Semiconducteurs. Table de Constantes UICPA." Oxford — London — Paris — New York — Los Angeles, 1961.
132. B. Szigetti, *Trans. Faraday Soc.* **45**, 155 (1949).
133. G. Picus, E. Burstein, B. W. Henvis, and M. Hass, *J. Phys. Chem. Solids* **8**, 282 (1959).
134. M. Hass and B. W. Henvis, *J. Phys. Chem. Solids* **23**, 1099 (1962).
135. G. L. Barinskit and E. G. Nadzhakov, *Dokl. Akad. Nauk. SSSR* **129**, 1279 (1959).
136. E. V. Vainshtein, Yu. F. Kopelev, and B. I. Kotlyar, *Dokl. Akad. Nauk. SSSR* **137**, 1117 (1961).
137. Ya. A. Ugar, E. L. Domashevskaya, and T. A. Marshakova, Coll. "Chemical Bond in Semiconductors and solids." p. 347. Pub. Nauka i Tekhnika, Minsk, 1965. (in Russian)
138. H. Kimmel, *Z. Naturforsch* **18a**, 650 (1963).
139. V. I. Orvil-Tomas, *Usp. Khim.* **28**, 731 (1958).
140. J. P. Suchet, *J. Phys. Chem. Solids* **21**, 156 (1961).
141. A. A. Medvinskii, *Zh. Strukt. Khim.* **3**, 584 (1962).
142. V. V. Maikevich, *Fiz. Tverd. Tela* **5**, 3500 (1963).

143. D. E. O'Reilly, *Advan. Catalysis* **12**, 31 (1960).
144. D. E. O'Reilly and C. P. Poole, *J. Phys. Chem.* **67**, 1762 (1963).
145. J. J. Lander and J. Morrison, *J. Chem. Phys.* **37**, 729 (1962).
146. J. J. Lander and J. Morrison, *J. Appl. Phys.* **34**, 1403 (1963).
147. N. D. Sokolov, *Probl. Kinetiki i Kataliza* **8**, 141 (1955).
148. Z. L. Nagaev, *Zh. Fiz. Khim* **35**, 327 (1961).
149. Z. L. Nagaev, *Kinetica i Kataliz* **3**, 907 (1962).
150. S. Z. Roginskii, *Probl. Kinetiki i Kataliza* **10**, 5 (1960).
151. F. J. Morin, *Bell System Tech. J.* **37**, 1047 (1958).
152. R. R. Heikes, W. D. Johnston, *J. Chem. Phys.* **26**, 582 (1937).
153. R. R. Heikes, *Phys. Rev.* **99**, 1232 (1955).
154. V. A. Ioffe and G. A. Smolenskii, "Symposium on Dielectrical Spectroscopy," p. 5. IL 1960. (in Russian)
155. J. H. DeBoer and E. J. W. Verwey, *Proc. Phys. Soc.* **49**, 59 (1957).
156. I. T. Sheftel, Ya. V. Pavlotskii and G. P. Tekster-Proskuryakova, "Collection of Thesis Reports of the All Union Conference on Semiconductor Compounds" (18–27.XII.1961), Izv. Akad. Nauk. SSSR, 1961, p. 65. (in Russian)
157. J. B. Goodenough, *J. Appl. Phys.* **31**, 359 (1960).
158. N. V. Geld and V. A. Tskhai, *Zh. Strukt. Khim.* **4**, 235 (1963).
159. F. J. Morin, *J. Appl. Phys.* **32**, 2195 (1961).
160. I. I. Ioffe, Z. I. Ezhakova, and A. G. Lyubarskii, *Dokl. Akad. Nauk. SSSR* **154**, 903 (1964).
161. I. I. Ioffe and L. Ya. Margolis, *Preprints 3rd Intern. Congr. Catalysis (1964)*. Amsterdam, 1965.
162. V. P. Zhuze and A. I. Shelykh, *Fiz. Tverd. Tela* **5**, 1756 (1963).
163. V. N. Bogomolov and V. P. Zhuze, *Fiz. Tverd. Tela* **5**, 3285 (1963).
164. L. Orgel, "Introduction to Chemistry of Transition Metals (Ligand Field Theory)." Pub. Mir, 1964. (in Russian)
165. K. Balkhauzen, "Introduction to Ligand Field Theory." Pub. Mir, 1964.
166. I. B. Bersuker and A. V. Ablov, "Chemical Bond in Complex Compounds." Pub. Akad. Nauk. Mold. SSR, Kishinev, 1962. (in Russian)
167. C. K. Jorgensen, "Absorption Spectra and Chemical Bonding in Complexes." Addison-Wesley, Reading Massachusetts, 1962; C. K. Jorgensen, *Mol. Phys.* **6**, 43 (1957).
168. J. F. Dunitz and L. E. Orgel, *J. Phys. Chem. Solids* **3**, 318 (1952).
169. J. D. Dunitz and L. E. Orgel, *Advan. Inorg. Chem. Radiochem.* **2**, 1 (1960).
170. Coll. "Contemporary Chemistry of Coordination Compounds." IL, 1963. (in Russian)
171. H. A. Bethe, *Ann. Phys.* (5), **3**, 133 (1929).
172. J. A. Van Vleck, *J. Chem. Phys.* **3**, 803, 807 (1935).
173. D. A. Dowden, D. Wells, *Actes 2e Congr. Intern. Catalyse*, **2**, 1489. Technip, Paris, 1961.
174. D. A. Dowden, N. MacKenzie, and B. M. Trapnell, *Proc. Roy. Soc.* **A 237**, 245 (1956).
175. G. M. Dixon, D. Nicholls, and H. Steiner, *Proc. 3rd Intern. Congr. Catalysis* **2**, 815 (1964). Amsterdam, 1965.
176. F. Basolo and R. G. Pearson, "Mechanisms of Inorganic Reactions." Wiley, New York, 1958.
177. M. Dikson and E. Vebb, "Fermentation." IL, 1961. (in Russian)
178. G. W. Watt, *J. Electrochem. Soc.* **108**, 423 (1961).

179. M. I. Temkin, N. M. Morozov, V. M. Pyshev, L. O. Apelbaum, L. I. Lukyanova, and V. A. Demidkin, *Probl. Fiz. Khim, Trudy NIFKHI im. Kaprova,* **2**, 14 (1959).
180. D. S. McIver and H. H. Tobin, *J. Phys. Chem.* **65**, 1665 (1961).
181. K. S. De, M. J. Tossiter, and F. S. Stone, *Proc. 3rd Intern. Congr. Catalysis* **1**, 520 (1964). Amsterdam, 1965.
182. K. S. De and F. S. Stone, *Nature,* **194**, 570 (1962).
183. K. Klier, *Kinetika i Kataliz* **3**, 65 (1962).
184. R. M. Flid, *Kinetika i Kataliz* **2**, 66 (1961).
185. A. I. Gelbshtein, M. I. Siling, G. G. Sheglova, and I. B. Vasileva, *Kinetika i Kataliz* **5**, 460 (1964).
186. N. Gaylord, G. Mark, "Linear and Stereoregular Polymers." IL, (1962). (in Russian)
187. F. M. Bukaneva, Yu. I. Pecherskaya, V. B. Kazanskii, and V. A. Dzisko, *Kinetika i Kataliz* **3**, 358 (1962).
188. P. Cossee, *J. Catalysis* **3**, 80 (1964).
189. S. Yu. Elovich, G. M. Zhabrova, L. Ya, Margolis, and S. Z. Roginskii, *Dokl. Akad. Nauk SSSR* **52**, 421 (1946).
190. G. M. Zhabrova, S. Z. Roginskii, and E. A. Fokina, *Zh. Obshch. Khim.* **21**, 10 (1956).
191. S. Z. Roginskii, *Probl. Kinetiki i Kataliza* **6**, 9 (1949).
192. K. S. Pitzer and J. H. Hildebrand, *J. Am. Chem. Soc.* **63**, 2472 (1941).
193. O. W. Florke, *Math. Naturwiss. Unterricht.* **15**, 193 (1962).
194. D. Morris and L. Arcus, Symposium "Semiconducting Substances, Problems of Chemical Bond," p. 57. IL, 1960. (in Russian)
195. Ch. N. R. Fao, "Electron Spectra in Chemistry." Publ. Mir, 1964. (in Russian)
196. A. I. Shatenshtein, "Theory of Acids and Bases." Goskhimizdat, 1949. (in Russian)
197. V. I. Vernadskii, "Outline of Geochemistry," pp. 94–98, 107–119. Gostoptehizdat, 1934. (in Russian)
198. I. I. Ioffe and S. Z. Roginskii, *Zh. Fiz. Khim.* **31**, 612 (1957).
199. K. V. Topchieva, *Uch. Zap. Mosk. Gos. Univ.* **174**, 75 (1955).
200. K. V. Topchieva, *Probl. Kinetiki i Kataliza* **10**, 247 (1960).
201. A. G. Oblad, T. H. Milliken, and G. A. Mills, *Advan. Catalysis* **3**, 199 (1951).
202. O. V. Krylov and E. A. Fokina, *Probl. Kinetiki i Kataliza* **8**, 248 (1955).
203. O. V. Krylov and E. A. Fokina, *Izv. Akad. Nauk. SSSR, O Kh N.* **3**, 266 (1958).
204. O. V. Krylov and E. A. Fokina, *Probl. Kinetiki i Kataliza* **9**, 304 (1957).
205. Yu. A. Eltekov and S. Aktanova, "Thesis Reports of the 5th Union conference on Colloidal Chemistry." (Odessa, May, 1962) Pub. Akad. Nauk. SSSR, 1963, p. 158. (in Russian)
206. L. P. Hammet, "Physical Organic Chemistry." McGraw-Hill, New York, 1940.
207. A. I. Shatenshtein, "Isotope Exchange and Hydrogen Exchange in Organic Substances." Izv. Akad. Nauk. SSSR, 1960.
208. C. Walling, *J. Am. Chem. Soc.* **72**, 1164 (1950).
209. M. W. Tamele, *Discussions Faraday Soc.* **8**, 270 (1950).
210. H. A. Benesi, *J. Am. Chem. Soc.* **21**, 5490 (1956).
211. O. Johnson, *J. Phys. Chem.* **59**, 827 (1955).
212. V. A. Dzisko and M. S. Borisova, *Kinetika i Kataliz* **1**, 144 (1960).
213. V. A. Dzisko, M. S. Borisova, I. S. Kotsarenko and E. V. Kuznetsova, *Kinetika i Kataliz* **3**, 728 (1962).
214. A. L. Makarov, G. K. Boreskov, and V. A. Dzisko, *Kinetika i Kataliz* **2**, 84 (1961).
215. V. A. Dzisko, M. S. Borisova, N. V. Akinova, and A. D. Makarov, *Kinetika i Kataliz* **5**, 681, 689 (1964).

REFERENCES

216. V. A. Dzisko, *Proc. 3rd Intern. Congr. Catalysis* **1**, 422 (1964). Amsterdam, 1965.
217. N. M. Chirkov, *Probl. Kinetiki i Kataliza* **10**, 255 (1960).
218. C. L. Thomas, *Ind. Eng. Chem.* **41**, 2564 (1949).
219. A. P. Ballad and K. V. Topchieva, *Usp. Khim.* **20**, 161 (1951).
220. R. C. Hansford, *Advan. Catalysis* **4**, 1 (1952).
221. Y. Trambouze, M. Perrin, and L. de Mourgues, *Advan. Catalysis* **9**, 544 (1957).
222. Z. A. Markova, *Kinetika i Kataliz* **2**, 435 (1961).
223. L. G. Karakchiev, V. A. Barachevskii, and V. F. Kholmogorov, *Kinetika i Kataliz* **5**, 630 (1969).
224. I. D. Chapman and M. L. Hair, *Proc. 3rd Intern. Congr. Catalysis* **1**, 396 (1964). Amsterdam, 1965.
225. M. Sato, T. Aonuma, and T. Shiba, *Proc. 3rd Intern. Congr. Catalysis* **1**, 396 (1964). Amsterdam, 1965.
226. J. Amenoniya, J. H. B. Chenier, and R. J. Cvetanovic, *ibid.*, **2**, 1135 (1965).
227. S. E. Tung and E. McInninch, *ibid.*, **1**, 698 (1965).
228. A. Clark, V. C. F. Holm, and D. M. Blackburn, *J. Catalysis* **1**, 244 (1962).
229. N. Ohta, *J. Chem. Soc Japan, Ind. Chem. Sect.* **51**, 16, 138 (1948); *Chem. Abstr.* **44**, 9224 (1950).
230. A. N. Terenin, *Probl. Kinetiki i Kataliza* **10**, 214 (1960).
231. A. N. Terenin, *Vestn. Leningr. Univ.* **8**, 11, 143 (1953).
232. M. Robin and K. N. Trueblood, *J. Am. Chem. Soc.* **79**, 5138 (1957).
233. A. Terenin and V. Filimonov, "Hydrogen Bonding," p. 545. London — New York — Paris — Los Angeles, 1957.
234. M. B. Basila, *J. Phys. Chem.* **66**, 2223 (1962).
235. W. K. Hall, H. P. Leftin, F. J. Chesselske, and D. E. O'Reilly, *J. Catalysis* **2**, 506 (1963).
236. E. P. Parry, *J. Catalysis* **2**, 371 (1963).
237. O. V. Krylov and E. A. Fokina, *Kinetika i Kataliz* **1**, 421, 542 (1960).
238. P. Y. Hsieh, *J. Catalysis* **2**, 211 (1962).
239. Y. Kubokawa, *J. Phys. Chem.* **67**, 769 (1963).
240. M. Misono, G. Saito, Y. Yoneda, *Proc. 3rd Intern. Congr. Catalysis* **1**, 408 (1964). Amsterdam, 1965.
241. S. Levy and M. Folman, *J. Phys. Chem.* **67**, 1278 (1963).
242. N. G. Yaroslavskii and A. N. Terenin, *Dokl. Akad. Nauk. SSSR* **66**, 885 (1949).
243. O. V. Krylov, Z. A. Markova, I. I. Tretyakov, and E. A. Fokina, *Kinetika i Katal.* **6**, 128 (1965).
244. O. V. Krylov, S. Z. Roginskii, and E. A. Fokina, *Izv. Akad. Nauk. SSSR, OkhN*, **6**, 668 (1956).
245. O. Glemser and E. Harterct, *Z. Anorg. Chem.* **283**, 111 (1956).
246. C. Cabannes-Ott, *Ann. Chim.* **5**, 905 (1960).
247. C. C. Pimentel, A. C. McClellan, "The Hydrogen Bond," pp. 279, 334, Freeman, San Francisco, 1960.
248. H. Lux, *Z. Elektrochem.* **45**, 303 (1939).
249. L. A. Shvartsman and I. A. Tomilin, *Usp. Khim.* **26**, 554 (1957).
250. O. V. Krylov and E. A. Fokina, *Dokl. Akad. Nauk. SSSR* **120**, 333 (1958).
251. G. K. Miesserov, *Usp. Khim.*, **22**, 279 (1953).
252. R. C. Hansford, *Advan. Catalysis* **4**, 1 (1952).
253. J. D. Danforth, *Advan. Catalysis* **9**, 556 (1957).
254. R. J. Haldeman and P. H. Emmett, *J. Am. Chem. Soc.* **78**, 2917 (1956).

255. J. Turkevich, F. Nozaki, and D. Stamires, *Proc. 3rd Intern. Congr. Catalysis* 1, 586 (1964). Amsterdam, 1965.
256. H. P. Leftin and W. K. Hall, *Actes 2e Congr. Intern. Catalyse, Technip*, 1353. Paris, 1961.
257. H. P. Leftin and E. Hermana, *Proc. 3rd Intern. Congr. Catalysis* 2, 1064 (1964). Amsterdam, 1965.
258. R. S. Mulliken, *J. Chem. Phys.* **19**, 514 (1951).
259. R. S. Mulliken, *J. Am. Chem. Soc.* **72**, 600 (1950).
260. R. S. Mulliken, *J. Am. Chem. Soc.* **74**, 811 (1952).
261. A. E. Martell and M. Calvin, "The Chemistry of Metal–Chelate Compounds." Prentice-Hall, Englewood Cliffs, New Jersey, 1959.
262. A. A. Frost and R. G. Pearson, "Kinetics and Mechanism. A Study of Homogenous Chemical Reactions." Wiley, New York, 1961.
263. R. Steinberger and F. H. Westheimer, *J. Am. Chem. Soc.* **73**, 429 (1951).
264. K. J. Pedersen, *Acta. Chem. Scand.* **2**, 385 (1948).
265. A. I. Sidorova and A. N. Terenin, *Izv. Akad. Nauk. SSSR, OkhN* **2**, 152 (1950).
266. J. E. Mapes and R. P. Eischens, *J. Phys. Chem.* **58**, 1059 (1954).
267. E. I. Kotov, *Opt. i Spektroskopiya* **3**, 115 (1957).
268. G. Blyholder and E. A. Richardson, *J. Phys. Chem.* **65**, 2597 (1962).
269. Y. Amenomiya, J. H. B. Chenier, and R. J. Cvetanovic, *J. Phys. Chem.* **68**, 52 (1966).
270. V. N. Filimonov and D. S. Bystrov, *Opt. i Spektroskopiya* **3**, 480 (1957).
271. N. Ya. Turova, K. N. Semenenko, and A. V. Novoselova, *Zh. Neorgan. Khim.* **8**, 882 (1963).
272. A. N. Sidorov, *Zh. Fiz. Khim.* **30**, 995 (1956).
273. V. I. Kvlividze, N. M. Ievskaya, T. S. Egorova, V. F. Kiselev, and N. D. Sokolov, *Kinetika i Kataliz* **3**, 91 (1962).
274. A. Weyl, "A New Approach to the Chemistry of the Solid State," Pennsylvania, 1958.
275. A. J. de Rossett, C. G. Finstrom, and C. J. Adams, *J. Catalysis* **1**, 235 (1962).
276. V. A. Barachevskii, V. E. Kholmogorov, and A. N. Terenin, Coll. "Methods of Investigations of Catalysts and Catalytic Reactions," p. 3. Izd. Sib. Otd. Akad. Nauk. SSSR, Novosibirsk, 1965. (in Russian)
277. Y. Trambouze, *J. Chim. Phys.* **51**, 723 (1954).
278. K. Tanabe, M. Katayama, *J. Res. Inst. Catalysis, Sapporo*, **7**, 106 (1959).
279. J. B. Peri, *Proc. 3rd Intern. Congr. Catalysis* **2**, 1100 (1964). Amsterdam, 1965.
280. O. V. Krylov and Yu. E. Sinyak, *Neftekhimiya* **2**, 688 (1962).
281. O. V. Krylov, M. Ya. Kushnerev, and E. A. Fokina, *Neftekhimiya* **2**, 697 (1962).
282. O. V. Krylov, M. Ya. Kushnerev, Z. A. Markova, and E. A. Fokina, *Vysokomolekul. Soedin.* **7**, 984 (1965).
283. S. G. Hindin, A. G. Oblad, and G. A. Mills, *J. Am. Chem. Soc.* **77**, 535 (1955).
284. H. P. Leftin, M. C. Hobson, *Advan. Catalysis* **14**, 115 (1963).
285. G. Kortum and J. Vogel, *Ber.* **93**, 706 (1960).
286. I. S. Sazonova, G. M. Sushentseva, T. P. Khokhlova, and N. P. Keier, *Kinetika i Kataliz.* **31**, 751 (1962).
287. G. M. Schwab, *Proc. 3rd Intern. Congr. Catalysis* **2**, 985 (1964). Amsterdam, 1965.
288. A. A. Vorobev, "Physical Properties of Ionic Crystal Dielectrics," Book 1. Pub. TGU, Tomsk, 1960. (in Russian)
289. Ya. M. Ksendov, *Izv. Akad. Nauk SSSR, Ser. Fiz.* **22**, 237 (1958).
290. M. S. Kosman and I. A. Gesse, *Izv. Akad. Nauk SSSR, Ser. Fiz.* **22**, 315 (1958).

291. O. V. Krylov, *Kinetika i Kataliz* **2**, 674 (1961).
292. C. Glasstone, K. Laidler, and G. Eyring, "The Theory of Rate Processes." IL, 1948. (in Russian)
293. G. Parravano, *J. Chem. Phys.* **20**, 342 (1952).
294. L. E. Orgel, *Discussions Faraday Soc.* **26**, 138 (1958).
295. N. S. Hush, *Discussions Faraday Soc.* **26**, 145 (1958).
296. A. A. Balandin, *Zh. RFKhO* **61**, 909 (1929).
297. A. A. Balandin, "Multiplet Theory of Catalysis," Chapter 1. Pub. M.G.U., 1963; Ch. 2, 1966. (in Russian)
298. A. M. Rubinshtein, *Usp. Khim.* **21**, 1287 (1952).
299. A. A. Balandin and N. P. Egorova, *Dokl. Akad. Nauk. SSSR* **57**, 255 (1947).
300. A. A. Balandin, N. P. Sokolova, and Yu. P. Simanov, *Izv. Akad. Nauk SSSR, OKhN* **3**, 415 (1961).
301. A. M. Rubinshtein, S. G. Kurikov, and B. A. Zkharov, *Izv. Akad. Nauk SSSR, OkhN*, 587 (1956).
302. O. V. Krylov, S. Z. Roginskii, and E. A. Fokina, *Probl. Kinetiki i Kataliza* **10**, 117 (1960).
303. A. A. Balandin and I. D. Rozhdestvenskaya, *Zh. Fiz. Khim.* **34**, 872 (1960).
304. A. A. Balandin and I. I. Brusov, *Zh. Opshch. Khim.* **7**, 18 (1936).
305. O. V. Krylov and E. A. Fokina, *Zh. Fiz. Khim.* **33**, 2555 (1959).
306. E. F. G. Herington and E. K. Rideal, *Proc. Roy. Soc. (London)*, Ser. A **190**, 289, 309 (1947).
307. A. A. Balandin and G. V. Isagulyants, *Dokl. Akad. Nauk SSSR* **64**, 207 (49).
308. A. A. Balandin and V. S. Fedorov, *Dokl. Akad. Nauk SSSR* **30**, 21 (1964).
309. R. Griffith, *Advan. Catalysis*, **1**, 91 (1948).
310. E. F. G. Herington and E. K. Rideal, *Proc. Roy. Soc. (London)*, Ser. A **184**, 435 (1945).
311. A. F. Plate, *Probl. Kinetiki i Kataliza* **6**, 239 (1949).
312. A. A. Balandin, *Kinetika i Kataliz* **1**, 5 (1960).
313. M. I. Temkin, *Zh. Fiz. Khim.* **31**, 3 (1957).
314. G. I. Golodets, V. A. Roiter, *Ukr. Khim. Zh.* **7**, 667 (1963).
315. W. M. H. Sachtler and J. Fahrenfort, *Actes 2e Congr. Intern. Catalyse* **1**, 831 (1960). Technip, Paris, 1961.
316. S. Makishima, Y. Yoneda, and Y. Saito, *Actes 2e Congr. Intern. Catalyse* **1**, 617 (1960). Technip, Paris, 1961.
317. S. L. Kiperman, "Introduction to Kinetics of Heterogenous Catalytic Reactions." Pub. "Nauka," 1964. (in Russian)
318. M. Green, *Ann. N. Y. Acad. Sci.* **101**, 1001 (1963).
319. E. A. Shilov, *Dokl. Akad. Nauk SSSR* **18**, 643 (1938).
320. A. Eucken and K. Heuer, *Z. Phys. Chem.* **A196**, 41 (1950).
321. A. Eucken, *Naturwiss.* **34**, 274 (1947).
322. W. Wicke, *Z. Elektrochem.* **53**, 279 (1949).
323. J. Turkevich and R. K. Smith, *Nature* **157**, 874 (1946).
324. J. Turkevich and R. K. Smith, *J. Chem. Phys.* **16**, 466 (1948).
325. M. A. Kaliko, *Zh. Fiz. Khim.* **33**, 2517 (1959).
326. H. Noller and K. Ostermeier, *Z. Electrochem.* **60**, 921 (1956).
327. H. Noller and K. Ostermeier, *Z. Elektrochem.* **63**, 191 (1959).
328. P. Andréu, R. Letterer, W. Low, H. Noller, and E. Schmitz, *Proc. 3rd Intern. Congr. Catalysis* **2**, 857 (1964). Amsterdam, 1965.
329. E. C. Pitzer and J. C. Frazer, *J. Phys. Chem.* **45**, 761 (1941).

330. E. I. Klabunovskii, *Khim. Nauka i Prom.* **2**, 197 (1957).
331. E. J. Arlman and P. Cossee, *J. Catalysis* **3**, 99 (1964).
332. E. Mooser and W. B. Pearson, *J. Electron.* **1**, 629 (1956).
333. E. Mooser and W. B. Pearson, *J. Chem. Phys.* **26**, 893 (1957).
334. E. Mooser and W. P. Pearson, *Phys. Rev.* **101**, 1608 (1956).
335. A. R. Hippel, "Dielectrics and Wavelengths." IL, 1961. (in Russian)
336. N. A. Goryunova, *Probl. Kinetiki i Kataliza* **10**, 96 (1960).
337. G. Krebs, Sb. "Semiconducting Substances." p. 132, IL, 1961. (in Russian)
338. H. G. Backman, F. R. Achmed, and W. H. Barnes, *Z. Krist.* **115**, 110 (1961).
339. D. Brennan, *Discussions Faraday Soc.* **28**, 219 (1959).
340. E. J. Arlman, *J. Catalysis* **3**, 89 (1964).
341. S. Brunauer, "Adsorption of Gases and Vapors." IL, 1948. (in Russian)
342. A. P. Karnaukhov, *Kinetika i Kataliz* **3**, 583 (1962).
343. M. I. Temkin, *Kinetika i Kataliz*, **3**, 509 (1962).
344. F. F. Volkenshtein, *Probl. Kinetiki i Kataliza* **8**, 202 (1955).
345. O. Neunhoeffer, *Chem. Techn.* **9**, 3 (1957).
346. I. V. Dunih-Barkovskii and N. V. Smirnov, "Theory of Probability and Mathematical Statistics in Technology," 1955. (in Russian)
347. A. A. Tolstopyatova, and A. A. Balandin, *Probl. Kinetiki i Kataliza* **10**, 351 (1960).
348. G. A. Gaziev, O. V. Krylov, S. Z. Roginskii, G. V. Samsonov, E. A. Fokina, and M. I. Yanovskii, *Dokl. Akad. Nauk SSSR* **140**, 863 (1961).
349. A. A. Tolstopyatova and A. A. Balandin, *Dokl. Akad. Nauk SSSR* **138**, 1376 (1961).
350. V. A. Komarov, *Zh. Fiz. Khim.* **27**, 1748 (1953).
351. V. A. Kamarov and N. P. Timofeeva, *Zh. Obshch. Khim.* **26**, 3306 (1956).
352. O. V. Krylov, S. Z. Roginskii, and E. A. Fokina, *Izv. Akad. Nauk SSSR, OKHN*, **4**, 421 (1957).
353. O. V. Krylov, Coll. "Heterogenous Catalysis in the Chemical Industry," p. 437. Goskhimizdst, 1955. (in Russian)
354. D. J. Wheeler, P. W. Darby, and C. Kemball, *J. Chem. Soc.*, 1960, 322.
355. A. A. Tolstopyatova and A. A. Balandin, Coll. "Rare Earth Elements," p. 307. Pub. Akad. Nauk SSSR, 1958. (in Russian)
356. O. V. Krylov, M. Ya. Kushnerev, and E. A. Fokina, *Izv. Akad. Nauk SSSR, OKHN* **12**, 1413 (1958).
357. V. A. Komarov, S. A. Absulaeva, and E. A. Chernikova, *Kinetika i Kataliz* **3**, 320 (1962).
358. G. V. Tsitsishvili, Sh. I. Sidamonidze, and Sh. A. Zedgenidze, *Dokl. Akad. Nauk SSSR* **153**, 1395 (1963).
359. A. M. Rubinshtein, N. A. Pribytkova, V. M. Akimov, L. D. Kretalova, and A. L. Klicho-Gurvich, *Izv. Akad. Nauk SSSR, OKHN* **9**, 1552 (1961).
360. A. Simon and C. Oehme, *Z. Anorg. Chem.* **317**, 230 (1962).
361. L. G. Maidemovskaya and I. A. Kirovskaya, *Tr. Tomsk. Gos. Univ.* **158**, (1964).
362. A. M. Rubinshtein, S. G. Kulikov, A. A. Dulov, and N. A. Pribytkova, *Izv. Akad. Nauk SSSR, OKHN*, 596 (1956).
363. A. A. Tolstopyatova and A. A. Balandin, *Dokl. Akad. Nauk SSSR* **138**, 1365 (1961).
364. V. A. Komarov, *Uch. Zap. Leningr. Gos. Univ. Khimiya*, **13**, 41 (1953).
365. S. Landa, O. Weisser, and J. Mostecky, *Coll. Czech. Chem. Commun.* **24**, 1036 (1958).
366. E. Cremer, *Z. Phys. Chem.* **A144**, 231 (1929).
367. S. K. Bhattacharrya and N. D. Ganguiy, *J. Appl. Chem.* **12**, 97 (1962).

368. K. V. Topchieva, I. A. Zenkovich, and F. M. Bukanava, *Vestn. Mosk. Univ., Ser. Khim.* **1**, 34 (1961).
369. I. R. Konenko, *Autoreferat Kand. Diss.*, IOKH Akad. Nauk SSSR, 1964.
370. I. Hatta, S. Borcsok, F. Solymosi, and Z. G. Szabo, *Proc. 3rd Intern. Congr. Catalysis* **2**, 1340 (1964). Amsterdam, 1965.
371. O. V. Krylov, *Kinetika i Kataliz* **3**, 502 (1962).
372. G. Rienäcker, *Z. Chem.* **3**, 121 (1963).
373. P. L. Gale, J. Haber, and F. S. Stone, *J. Catalysis* **1**, 32 (1960).
374. R. G. Greenler, *J. Chem. Phys.* **37**, 2094 (1962).
375. O. V. Krylov, *Zh. Fiz. Khim.* **39**, 2911 (1965).
376. P. Handler, "Semiconductor Surface Physics," p. 3. Univ. of Pennsylvania Press, 1957.
377. V. M. Frolov and E. K. Radshabli, *Azerb. Khim. Zh.* **6**, 47 (1962).
378. H. Pines and W. O. Haag, *J. Am. Chem. Soc.* **82**, 2488 (1960); H. Pines and C. N. Pillai, *J. Am. Chem. Soc.* **83**, 3230 (1961).
379. O. V. Krylov, *Zh. Fiz. Khim.* **39**, 2656 (1965).
380. A. J. Lundeen and R. Van Hoozer, *J. Am. Chem. Soc.* **85**, 2180 (1960).
381. A. M. Rubinshtein, V. A. Zakharov, N. A. Pribytkova, and V. A. Afanasev, *Dokl. Akad. Nauk SSSR* **103**, 83 (1955).
382. P. N. Galich, I. T. Golubchenko, V. S. Gutyrya, V. G. Ilin, and I. E. Neimark, *Dokl. Akad. Nauk SSSR* **161**, 627 (1965).
383. A. M. Rubinshtein and E. P. Gracheva, *Zh. Fiz. Khim.* **8**, 725 (1936).
384. L. Kh. Freidlin, V. Z. Sharf, and Z. T. Tukhtamuradov, *Neftekhimiya* **2**, 730 (1962).
385. M. I. Shemishchin, S. I. Burmistrov, and A. A. Ivznov, *Izv. vuzov. Khimiya i Khim. Tekhnol.* **4**, 837 (1961).
386. Z. G. Szabo, F. Solymosi, and J. Batta, *Z. Phys. Chem.* **17**, 125 (1958).
387. V. A. Komarov, E. A. Chernikova, G. V. Komarov, and Z. I. Leonchik, *Vestn. Leningr. Univ. No. 16, Ser. Fiz. Khim.* **3**, 120 (1960).
388. P. Mars, "The Mechanism of Heterogenous Catalysis," p. 52. Elsevier, Amsterdam, 1960.
389. V. A. Komarov, E. A. Chernikova, G. V. Komarov, and Z. I. Leonchik, *Zh. Fiz. Khim.* **36**, 2577 (1962).
390. J. J. F. Scholten, P. Mars, P. G. Menon, and R. Van Hardeveld, *Proc. 3rd Intern. Congr. Catalysis* **2**, 881 (1966). Amsterdam, 1965.
391. P. Mars, J. J. F. Scholten, and P. Zwietering, *Advan. Catalysis* **14**, 35 (1963).
392. A. M. Rubinshtein and V. I. Yakerson, *Kinetika i Kataliz* **2**, 118 (1961).
393. A. M. Rubinshtein and V. I. Yakerson, *Zh. Fiz. Khim.* **30**, 2789, 3153 (1960).
394. V. I. Yakerson, E. A. Fedorovskaya, and A. M. Rubinshtein, *Dokl. Akad. Nauk SSSR* **140**, 626 (1961).
395. V. I. Yakerson, E. A. Fedorovskaya, A. L. Klyachko-Gurvich, and A. M. Rubinshtein, *Kinetika i Kataliz* **2**, 907 (1961).
396. V. I. Yakerson, L. I. Lafer, and A. M. Rubinshtein, Coll. "Scientific Basis of Catalyst Selection for Heterogenous Catalytic Reactions," p. 142. Pub. "Nauka," 1966. (in Russian)
397. G. M. Schwab and V. Leute, *J. Catalysis* **1**, 191 (1962).
398. G. D. Lyubarskii, *Usp. Khim.* **27**, 316 (1958).
399. S. Voltz and S. Weller, *J. Am. Chem. Soc.* **75**, 5227 (1953).
400. J. M. Bridges and G. Houghton, *J. Am. Chem. Soc.* **81**, 1334.
401. A. A. Balandin, A. A. Tolstopyatova, and A. A. Ferapontov, *Izv. Akad. Nauk SSSR, OKHN* **10**, 1761 (1960).

402. H. S. Taylor and R. A. Briggs, *J. Am. Chem. Soc.* **63**, 2500 (1941).
403. L. Wright and S. Weller, *J. Am. Chem. Soc.* **76**, 5302, 5305 (1954).
404. Kh. M. Minechev, M. A. Markov, and O. K. Schukina, *Neftekhimiya* **1**, 489 (1961); Kh. M. Minechev, M. A. Markov, and V. I. Bogomolov, *Neftekhimiya*, **2**, 144 (1962).
405. C. B. McGeough and G. Houghton, *J. Phys. Chem.* **65**, 1887 (1961).
406. V. A. Shvets and V. B. Kazanskii, *Dokl. Akad. Nauk SSSR* **167**, 1331 (1966); *Kinetika i Kataliz* **7**, 712 (1966).
407. L. L. Van Reijen, W. M. H. Sachtler, P. Cossee, and D. M. Brouwer, *Proc. 3rd Intern. Congr. Catalysis* **2**, 829 (1966). Amsterdam, 1965.
408. L. Yu. Tazhunkhuei, S. Z. Roginskii, G. V. Samsonov, and M. I. Yanovskii, *Neftekhimiya* **3**, 843 (1963).
409. G. V. Samsonov, *Kinetika i Kataliz* **6**, 424 (1965); Coll. "Scientific Basis of Catalyst Selection for Heterogenous Catalytic Processes," p. 236. Pub. "Nauka," 1966.
410. E. E. Vainshtein, E. A. Zhurakovskii, and I. B. Staryi, *Zh. Neorg. Khim.* **4**, 245 (1959).
411. E. A. Zhurakovskii and V. P. Dzeganovskii, *Dokl. Akad. Nauk SSSR* **150**, 1260 (1963).
412. J. J. Rooney, *J. Catalysis* **2**, 53 (1963).
413. Kh. M. Minachev, M. A. Markov, and O. K. Shukina, *Neftekhimiya*, **1**, 610 (1961).
414. I. R. Konenko, A. A. Tolstopyatova, and V. A. Ferapontov, *Izv. Akad. Nauk SSSR, OKHN*, 1899.
415. H. Steiner, *Discussions Faraday Soc.* **8**, 266 (1950).
416. H. S. Taylor and H. Fehrer, *J. Am. Chem. Soc.* **63**, 1385 (1941).
417. J. Turkevich, H. Fehrer, and H. S. Taylor, *J. Am. Chem. Soc.* **63**, 1129 (1961).
418. Y. Takegami, T. Ueno, and K. Kawajiri, *J. Chem. Soc., Japan, Ind. Chem. Sect.* **66**, 1068 (1963), cited in *R Zh. Khim.* 1964, 13B612.
419. B. D. Polkovnikov, A. A. Balandin, and A. M. Taber, Coll. "Catalytic Reactions in the Fluid Phase," p. 25. Izv. Akad. Nauk Kaz. SSSR, Alma-Ata, 1963. (in Russian)
420. A. A. Rubinshtein, A. A. Dulov, S. G. Kulikov, and B. A. Zakharov, *Izv. Akad. Nauk SSSR, OKHN*, **5** (1956).
421. A. V. Cozovoi, S. A. Senyavin, and A. B. Vol-Epshtein, *Zh. Prikl. Khim.* **28**, 175 (1955).
422. I. V. Kelechits, *Probl. Kinetiki i Kataliza* **10**, 121 (1960).
423. A. V. Lozovoii and S. A. Senyavin, Coll. "Chemical Treatment of Fuels," p. 180. Pub. Izd. Akad Nauk SSSR, 1957. (in Russian)
424. S. Landa and J. Masak, *Coll. Czech. Chem. Commun.* **25**, 761, 766 (1960).
425. J. Varga, J. Szebenyi, and E. Kocsis, *Acta Chim. Hungar.* **15**, 133 (1958); J. Varga and J. Szebenyi, *Acta Chim. Hungar.* **16**, 193 (1958).
426. A. K. Roebuck and B. L. Everihg, *Ind. Eng. Chem.* **50**, 1135 (1958).
427. M. F. Sloar, A. S. Matlack, and D. S. Breslow, *J. Am. Chem. Soc.* **85**, 4014 (1963).
428. F. P. Ivanovskii, V. A. Kontsova, and G. S. Beskova, *Zh. Fiz. Khim.* **32**, 2569 (1959).
429. A. N. Bashkirov, S. M. Loktev, and G. V. Sabirova, *Tr. Inst. Nefti Akad. Nauk SSSR* **8**, 168 (1956).
430. H. Kurita and J. Tsutsumi, *J. Chem. Soc., Japan, Pure Chem. Sect.* **82**, 1461 (1961); cited in *Zh. Khim.* 1962, 16B385.
431. R. B. Anderson, *Advan. Catalysis*, **5**, 355 (1953).
432. J. Changfoot and F. Sebba, *Actes 2e Congr. Intern. Catalyse* **2**, 1905 (1961). Technip, Paris.
433. G. B. Lotz and F. Sebba, *Trans. Faraday Soc.* **53**, 1246 (1957).

434. E. G. Vrieland and P. W. Selwood, *J. Catalysis* **3**, 539 (1964).
435. O. V. Krylov, Yu. N. Rufov, E. A. Fokina, and V. M. Frolov; "Kataliza i Kinetyka Chemiana," p. 100. Warszawa, 1969. (in Polish)
436. Khuan Yu-Mei, N. P. Keier, and S. Z. Roginskii, *Dokl. Akad Nauk SSSR* **133**, 413 (1960).
437. N. P. Keier, G. K. Boreskov, L. F. Rubtsova, and E. G. Rukhadze, *Kinetika i Kataliz* **3**, 680 (1962).
438. V. M. Frolov, O. V. Krylov, and S. Z. Roginskii, *Probl. Kinetiki i Kataliza, Izv. Akad. Nauk SSSR*, **10**, 502 (1960).
439. M. Szwarc, *Proc. Roy. Soc. (London) Ser. A.* **198**, 667 (1949).
440. O. V. Krylov, *Dokl. Akad. Nauk SSSR* **130**, 1063 (1960).
441. I. E. Adadurov, *Zh. Fiz. Khim.* **2**, 735 (1931).
442. H. S. Taylor, *Can. J. Chem.* **33**, 838 (5955).
443. D. A. Dowden, N. Mackenzie, and B. M. W. Trapnell, *Advan. Catalysis* **9**, 65 (1957).
444. G. K. Boreskov and V. L. Kuchaev, *Dokl. Akad. Nauk SSSR* **119**, 302 (1956).
445. J. L. Sandler, *Actes 2e Congr. Intern. Catalyse* **1**, 227 (1961). Technip, Paris.
446. V. C. F. Holm and R. W. Blue, *Ind. Eng. Chem.* **44**, 107 (1952).
447. R. L. Wilson, C. Kemball, and A. K. Galway, *Trans. Faraday Soc.* **58**, 583 (1962).
448. H. W. Kohn and E. H. Taylor, *J. Catalysis* **2**, 32 (1963).
449. P. R. Ashmead, D. D. Eley, and R. Rudham, *J. Catalysis* **3**, 280 (1964).
450. J. Halpern, *Advan. Catalysis* **11**, 301 (1959).
451. G. M. Schwab, O. Jenkner, and W. Leitenberger, *Z. Elektrochem.* **63**, 461 (1959).
452. S. Malinowski and T. Kobylinski, *Roczniki Chem.* **35**, 917 (1961).
453. E. Wigner, *Z. Physik. Chem.* **B23**, 28 (1933).
454. P. Selwood, "Magnetochemistry," IL, 1958. (in Russian)
455. R. A. Buyanov, *Kinetika i Kataliz* **1**, 306, 418, 617 (1960).
456. Y. L. Sandler, *Can. J. Chem.* **32**, 249 (1954).
457. D. R. Ashmead, D. D. Eley, and R. Rudham, *Trans. Faraday Soc.* **59**, 207 (1963).
458. N. Wakao, J. M. Smith, and P. W. Selwood, *J. Catalysis* **1**, 62 (1962).
459. V. V. Voevodskii and G. K. Lavrovskaya, *Zh. Fiz. Khim.* **25**, 1050 (1951).
460. J. C. Greaves and J. W. Linnett, *Trans. Faraday Soc.* **55**, 1338, 1355 (1959).
461. J. W. Linnett and D. G. Marsden, *Proc. Roy. Soc. (London) Ser. A* **234**, 489, 504 (1956).
462. W. V. Smith, *J. Chem. Phys.* **11**, 110 (1943).
463. R. Gowlard, *Jet Propulsion* **28**, 737 (1958).
464. G. Mannella and P. Harteck, *J. Chem. Phys.* **34**, 2177 (1961).
465. R. A. Young, *J. Chem. Phys.* **34**, 1292, 1295 (1960).
466. S. Z. Roginskii and A. B. Shekhter, *Acta Physicochim. URSS* **1**, 318 (1934); **6**, 401 (1937).
467. N. Ya Bubcn and A. B. Shekhter, *Acta Physicochim. URSS* **10**, 371 (1939).
468. P. G. Dickens and M. B. Sutcliff, *Trans. Faraday Soc.* **60**, 1272 (1964).
469. G. K. Boreskov and V. V. Popovskii, *Kinetika i Kataliz* **2**, 657 (1961).
470. G. T. Bakumenko, *Kinetika i Kataliz*, **6**, 74 (1965).
471. Y. Yoneda and S. Makishima, *Actes 2e Congr. Intern. Catalyse* **2**, 2103 (1961). Technip, Paris.
472. G. Rienäcker, R. Buhrmann, and M. Birckenstaedt, *Z. Anorg. Chem.* **258**, 280 (1949); **262**, 82 (1950).
473. G. Rienäcker, *Acta Chim. Hungar.* **14**, 179 (1958).
474. B. Neumann, C. Kroger, and R. Iwanowski, *Z. Electrochem.* **37**, 121 (1931).
475. L. J. E. Hofer, P. Gussey, and R. B. Anderson, *J. Catalysis* **3**, 451 (1964).

476. G. Rienäcker and G. Schneeberg, *Z. Anorg. Chem.* **282**, 222 (1955).
477. C. I. Engelder and M. Blumer, *J. Phys. Chem.* **36**, 1353 (1932).
478. R. Rudham and F. S. Stone, *in* "Chemisorption," Butterworth, London, 1957.
479. G. Rienäcker and E. Buchholz, *Z. Anorg. Chem.* **290**, 320 (1957).
480. J. Horacek and V. Pechanek, *J. Korbl. Coll. Czech. Chem. Commun.* **27**, 1254 (1962).
481. L. A. Sazonov and M. G. Logvinenko, *Kinetika i Kataliz* **3**, 761 (1962).
482. J. Mooi and P. W. Selwood, *J. Am. Chem. Soc.* **74**, 2461 (1952).
483. R. P. Eischens and W. A. Pliskin, *Advan. Catalysis* **10**, 1 (1958).
484. G. K. Boreskov, "Catalysis in the Production of Sulphuric Acid," Goskhimizdat, 1954. (in Russian)
485. B. Neumann, *Z. Elektrochem.* **74**, 734 (1928).
486. J. K. Dixon and J. E. Longfield, Collection, "Catalysis in the Petrochemical and Oil Refinery Industries," p. 302. Gostoptekhizdat, 1963. (in Russian)
487. G. M. Schwab and E. Kaldis, *Z. Phys. Chem.* **42**, 72 (1964).
488. S. I. Papko, *Zh. Fiz. Khim.* **34**, 161, 518 (1960).
489. S. I. Papko, *Zh. Fiz. Khim.* **38**, 2491 (1964).
490. W. Krauss, *Z. Elektrochem.* **53**, 320 (1949).
491. N. P. Kurin and M. S. Zakharov, *Izv. Tomsk. Politekhn. Inst.* **92**, 34 (1960).
492. N. P. Kurin and M. S. Zakharov, *Izv. Vuzov. Khimiya i Khim. Tekhnol.* **3**, 141 (1960).
493. N. P. Kurin and M. S. Zakharov, collection, "College Catalysis" Vol. 2, p. 234. Pub. Moscow State Univ., 1962. (in Russian)
494. M. M. Karavaev, *Autoreferat. Rand. Diss.*, *1 KHTI*, Ivanovo, 1959.
495. I. Komuro and Yamamoto, T. Kwan, *Bull. Chem. Soc. Japan* **36**, 1532 (1963).
496. V. V. Popovskii and G. K. Boreskov, *Kinetika i Kataliz* **1**, 566 (1960).
497. E. R. S. Winter, *in* "Chemisorption," p. 188. Butterworth, London, 1957.
498. A. P. Dzisyak, G. K. Boreskov, and L. A. Kasatkina, *Kinetika i Kataliz* **3**, 81 (1962).
499. S. M. Karpacheva and A. M. Rozen, *Dokl. Akad. Nauk SSSR* **75**, 239 (1956).
500. G. K. Boreskov and V. V. Popovskii, *Kinetika i Kataliz* **2**, 657 (1961).
501. G. K. Boreskov, A. P. Dzisyak, and L. A. Kasatkina, *Kinetika i Kataliz* **4**, 388 (1964).
502. L. Ya. Margolis, "Heterogenous Catalytic Oxidation of Hydrocarbons (monomer synthesis)." Gostoptekhizdat, 1962. (in Russian)
503. Kh. M. Minachev and G. V. Autoshin, *Dokl. Akad. Nauk SSSR* **161**, 122 (1965).
504. E. Greenhalgh and B. M. W. Trapnell, *Advan. Catalysis* **9**, 238 (1957).
505. P. P. Clopp and G. M. Parravano, *J. Phys. Chem.* **62**, 1055 (1958).
506. G. M. Zhabarova, S. Z. Roginskii, and E. A. Fokina, *Zh. Org. Khim.* **24**, 10 (1954).
507. A. B. Hart, J. MacFayden, and R. A. Ross, *Trans. Faraday Soc.* **59**, 1458 (1963).
508. V. A. Zhuravlev and A. A. Kazhelyuk, *Zh. Fiz. Khim.* **3**, 271 (1964).
509. S. Z. Roginskii, *Khim. Nauka i Promy.* **2**, 138 (1957).
510. W. Wolski, *Roczniki Chem.* **30**, 733 (1956).
511. A. B. Hart and R. A. Ross, *J. Catalysis* **2**, 121 (1963).
512. G. M. Schwab and H. Schultes, *Z. Phys. Chem.* **B9**, 265 (1930).
513. G. M. Schwab and N. Keller, *Naturwiss.* **37**, 43 (1950).
514. G. M. Schwab, R. Staeger and H. H. Von Baumbach, *Z. Phys. Chem.* **B21**, 65 (1933).
515. F. Stone, Coll. "Solid State Chemistry," p. 525. IL, 1961. (in Russian)
516. J. M. Fraser and F. Daniels, *J. Phys. Chem.* **62**, 215 (1958).
517. T. M. Yureva, V. V. Popovskii, and G. K. Boreskov, *Kinetika i Kataliz* **6**, 1041 (1965).
518. G. M. Schwab and G. Hartman, *Z. Phys. Chem.* **6**, 56, 72 (1956).

REFERENCES

519. G. I. Emalyanova, Autoref. Kand. Diss., Moscow State Univ., 1962.
520. I. A. Kurshunov and A. V. Savitskii, "Trudypokhimii i Khimicheskoi Tekhnologii," Vol. 3, p. 507. Gorkii, 1959.
521. F. Solymosi and N. Krix, *Acta Khim. Hungar.* **34**, 241 (1962).
522. A. Yu. Prokopchik and N. V. Yanitskii, *Zh. Fiz. Khim.* **28**, 1999 (1954).
523. P. K. Norkus, A. Yu. Prokopchik, *Tr. Akad. Nauk Lit. SSR* **B3**, 43 (1960).
524. A. M. Lunetskas and A. Yu. Prokopchik, *Tr. Akad. Nauk Lit. SSR* **B3**, 53 (1960).
525. A. P. Kazragis, *Autoreferat. Kand. Diss., VGU*, Vilnyus, 1963.
526. S. Z. Roginskii, L. A. Sapozhnikov, and N. A. Kucherenko, *Ukr. Khim. Zh.* **4**, 99 (1929).
527. A. Hermoini (Makowski), and A. Salmon, *Chem. Ind.* **41**, 1264 (1960).
528. S. Patai, M. Albeck, and H. Cross, *J. Appl. Chem.* **12**, 217, 225, 230 (1962).
529. H. Eggert, *Inorg. Chem.* **2**, 304 (1963).
530. R. B. Anderson, K. C. Stein, J. J. Feenan, and L. J. E. Hofer, *Ind. Eng. Chem.* **53**, 809 (1961).
531. T. V. Andrushkevich, V. V. Popovskii, and G. K. Boreskov, *Kinetika i Kataliz* **6**, 860 (1965).
532. B. Dmuchovski, M. C. Freerks, and F. B. Zienty, *J. Catalysis* **4**, 577 (1965).
533. M. Paget, R. M. Utrilla, J. C. Balacennu, and M. D. Leachieur, *Bull. Soc. Chim. France* **5**, 1013 (1962).
534. M. Ya. Rubanik, K. M. Kholyavenko, A. V. Gershingorina, and V. I. Lazukin, *Kinetika i Kataliz* **5**, 116 (1966).
535. T. Takeuchi, M. Honma, and H. Takeda, *Bull. Chem. Soc., Japan*, **39**, 2101 (1966).
536. Y. Morooka and A. Ozaki, *J. Catalysis* **5**, 116 (1966).
537. O. V. Isaev, L. Ya. Margolis, and S. Z. Roginskii, *Zh. Org. Khim.* **29**, 1522 (1959).
538. L. N. Kutseva and L. Ya. Margolis, *Zh. Org. Khim.* **32**, 102 (1962).
539. K. C. Stein, J. J. Feenan, G. P. Thompson, J. F. Shutz, L. J. E. Hofer, and R. B. Anderson, *Ind. Eng. Chem.* **52**, 671 (1960).
540. L. Ya. Margolis and O. M. Todes, *Probl. Kinetiki i Kataliza* **6**, 281 (1949).
541. Z. I. Ezhkova, I. I. Ioffe, V. B. Kazanskii, A. V. Krylova, A. G. Lyubavskii, and L. Ya. Margolis, *Kinetika i Kataliz* **5**, 861 (1964).
542. S. K. Bhattacharyva and N. Venkataraman, *J. Appl. Chem.* **8**, 737 (1958).
543. G. Charlot, *Bull. Soc. Chim. France* **51**, 1007 (1932); **53**, 572 (1933); *Compt. Rend.* **194**, 374 (1932).
544. I. I. Ioffe and N. V. Klimova, *Kinetika i Kataliz* **4**, 779 (1963).
545. Dzh. Uetington, G. Dzheks, U. Kristi, and A. Brozi, Coll. "Production and Application of Isotopes." Trudy 2-i Konferentsii Romirnom Ispolzovaniyu Atomnoi Energii (Zheneva, 1958) t. 10, 1959.
546. V. A. Nekrasova and N. A. Rumyantseva, *Khim. Nauka i Promy.* **2**, 806 (1957); V. A. Nekrosova, Coll. "College Catalysis," p. 290. Vol. 2. Pub. Moscow State Univ., 1962.
547. N. N. Lebedev and I. I. Baltadzhi, *Kinetika i Kataliz* **4**, 886 (1963).
548. B. V. Tronov and L. A. Pershina, *Zh. Org. Khim.* **24**, 1608 (1954).
549. M. E. Winfield, from the collection "Catalysis in the Petrochemical and Oil Refinery Industry," Vol. 3, p. 106. Gospromtekhizdat, 1963. (in Russian)
550. G. A. Mills, S. Weller, S. G. Hindin, and T. H. Milliken, *Z. Elektrochem.* **60**, 823 (1956).
551. H. Noller, E. Hantke, and G. M. Schwab, *Z. Phys. Chem.* **3**, 103 (1963).
552. P. Andréu, E. Bussmann, H. Noller, and S. R. Sim, *Z. Elektrochem.* **66**, 739 (1962).

553. T. Enqvist, *Suomen Kemistisauran Tiedonantoja* **67**, 45 (1958). Cited in *R. Zh. Khim.*, 1960, 35867.
554. F. Patat and P. Weidlich, *Helv. Chim. Acta* **32**, 783 (1949).
555. Yu. A. Gorin and I. K. Gorn, *Zh. Org. Khim.* **28**, 2328 (1958).
556. Yu. A. Gorin and I. K. Gorn, Coll. "Monomer Synthesis for SK," p. 232, Goskhimizdat, 1960.
557. I. K. Gorn and Yu. A. Gorin, *Zh. Org. Khim.* **29**, 2125 (1959).
558. Yu. A. Gorin, Coll. "Monomer Synthesis for SK," p. 216, Goskhimizdat, 1960.
559. J. Marton, G. Zollner, G. Level, A. Tatraaljai, and G. Balint, *Acta Chim. Hungar.* **21**, 375 (1959).
560. E. N. Rostovskii, I. A. Arbuzova, *Izv. Akad. Nauk SSSR, UKhN* **2**, 317 (1960).
561. I. B. Vasileva, A. I. Gelbshtein, I. N. Tolstikova, and Dao Van Tyong, *Kinetika i Kataliz* **5**, 144 (1964).
562. L. Kh. Freidlin, *Usp. Khim.* **23**, 581 (1954).
563. L. Chalkley, *J. Am. Chem. Soc.* **51**, 2489 (1929).
564. J. D. F. Marsch, W. B. S. Newling, and J. Rich, *J. Appl. Chem.* **2**, 681 (1952).
565. E. Bamann and M. Melsenheimer, *Chem. Ber.* **71**, 1986, 2086, 2233 (1938).
566. S. Malinowski, H. Jedrzejewska, S. Basinski, Z. Lipski, and J. Moszczenska, *Roczniki Chem.* **30**, 1129 (1956).
567. S. Malinowski and S. Basinski, *Roczniki Chem.* **36**, 821 (1962).
568. S. Malinowski, S. Basinski, S. Szczepanska, and W. Kiewlicz, *Proc. 3rd. Intern. Congr. Catalysis* **1**, 441 (1964). Amsterdam, 1965.
569. P. M. Scheidt, *J. Catalysis* **3**, 372 (1964).
570. K. Fukui and M. Takei, *Bull. Inst. Chem. Res., Kyoto Univ.* **26**, 81 (1951); cited in *Chem. Abs.* 5426e (1955).
571. K. Nodzu, S. Kunichika, S. Oka, and K. Kusinose, *J. Chem. Soc. Japan, Ind. Chem. Sect.* **57**, 914 (1956); cited in *Chem. Abs.*, 793b (1956).
572. T. Isoshima, *Bull. Inst. Res., Kyoto Univ.* **32**, 168 (1954); cited in *Chem. Abs.* 793d (1956).
573. R. Inoue, A. Ichikawa, and K. Furukawa, *Chem. High Polymers* **15**, 136 (1958); cited in *R Zh. Khim*, 1959, 526.
574. S. K. Bhattacharyya and B. N. Avasti, *Ind. Eng. Chem.* **55**, 45 (1963).
575. R. Inoue, A. Ichikawa, and K. Furukawa, *Chem. High Polymers* **15**, 136 (1958); cited in *R. Zh. Khim.*, 1959, 23527.
576. L. Rand, J. V. Swisher, and C. J. Cronin, *J. Org. Chem.* **27**, 3505 (1962).
577. J. C. Minor, and R. V. Lawrens, *Ind. Eng. Chem.* **50**, 1127 (1958).
578. A. Striegler, *Chem. Tech.* **10**, 79 (1958).
579. M. A. Popov and N. I. Shuikin, *Izv. Akad. Nauk SSSR, OKEN* **4**, 645 (1961).
580. H. G. Viche, E. Frenchimont, and P. Valange, *Chem. Ber.* **96**, 426 (1963).
581. P. Mastagli, P. Lambert, and D. Baladie, *Compt. Rend.* **255**, 2978 (1963).
582. M. A. Popov and N. S. Lobanova, *Zh. Prikl. Khim.* **36**, 856 (1963).
583. B. N. Dolgov, Yu. I. Khudobin, and N. P. Kharitonov, *Izv. Akad. Nauk SSSR, OKHN* **1**, 113 (1958).
584. S. Malinowski, J. Kehl, and T. Gora, *Roczniki Chem.* **36**, 1039 (1962).
585. A. V. Volokhina and G. I. Kudryavtsev, *Vysokomolekul Soedin.* **1**, 1724 (1959).
586. A. A. Blagonravora, G. A. Levkovich, and N. A. Pronina, *Intern. Symp. Macromol. Chem. (June 1960, Moscow)* **1**, 255 (1960). (in Russian)
587. J. W. Britain and P. G. Gemeinhardt, *J. Appl. Polymer Sci.* **4**, 207 (1960).
588. O. A. Osipov, V. I. Minkin, and O. E. Kashirenikov, *Vysokomolekul. Soedin.* **3**, 1774 (1961).

589. O. E. Kashirenikov, *Autoreferat Kand. Diss. RGU, Rostovna Donu*, 1962.
590. A. A. Petrov, "Catalytic Isomerization of Hydrocarbons," Izv. Akad. Nauk SSSR, 1960. (in Russian)
591. R. M. Roberts, *J. Phys. Chem.* **63**, 1400 (1959).
592. N. W. Foster and R. Cvetanevic, *J. Am. Chem. Soc.* **82**, 4274 (1960).
593. R. J. Cvetanevic and N. F. Foster, *Discussions Faraday Soc.* **28**, 219 (1959).
594. A. Ozaki and K. Kimura, *J. Catalysis* **3**, 395 (1964).
595. H. Pines and L. Schaap, *J. Am. Chem. Soc.* **79**, 2956 (1957).
596. I. V. Gostunskaya and B. A. Kazanskii, *Zh. Org. Khim.* **76**, 407 (1951).
597. B. A. Kazanskii, A. L. Liberman, M. Yu. Lukina, and I. V. Gostunskaya, *Khim. Nauka i Promy.* **2**, 172 (1957).
598. Ya. T. Eidus and B. I. Nefedov, *Usp. Khim.* **32**, 1025 (1963).
599. P. E. Pickert, J. A. Rabo, E. Dempsey, and V. Schomaker, *Proc. 3rd. Intern. Congr. Catalysis* **1** (1964). Amsterdam, 1965.
600. V. Sh. Feldblyum, S. I. Kryukov, M. I. Farberov, A. V. Golovko, I. Ya. Tyuryaev, and A. G. Pankov, *Neftekhimiya* **3**, 20 (1963).
601. M. A. Mosfoslavskii, *Ukr. Khim. Zh.* **29**, 1276 (1963).
602. A. A. Spryskov and Yu. G. Erykalov, *Zh. Org. Khim.* **31**, 292 (1961).
603. M. Maki, *J. Fuel Soc. Japan* **32**, 308 (1953); cited in *Chem. Abs.*, 11167c (1954).
604. Yu. A. Bitepazh, *Zh. Org. Khim.* **17**, 199 (1947).
605. J. D. Danforth, *J. Phys. Chem.* **58**, 1030 (1954).
606. V. J. Frillet, P. B. Weisz, and P. L. Golden, *J. Catalysis* **7**, 114 (1959).
607. E. G. Boreskova, K. V. Topchieva, and L. I. Piquzova, *Kinetika i Kataliz* **5**, 903 (1964).
608. K. V. Topchieva, I. A. Zenkovich, F. M. Bukanaeva, *Vestn. Mosk. Univ., Ser. Khim.* **15**, 3 (1960).
609. Kh. M. Minachev and Yu. S. Khodakov, *Kinetika i Kataliz* **6**, 89 (1965).
610. Ya. M. Paushkin, A. V. Topchiev, and M. V. Kurashov, *Dokl. Akad. Nauk SSSR* **130**, 1033 (1960).
611. I. N. Nazarov, E. I. Klabunovskii, and N. A. Kravchenko, *Izv. Akad. Nauk SSSR, OKHN* **1**, 73 (1960).
612. O. Johnson, *J. Phys. Chem.* **59**, 827 (1958).
613. E. H. Ingold and A. Wassermann, *Trans. Faraday Soc.* **35**, 1922 (1939).
614. M. Feller and E. Field, *Ind. Eng. Chem.* **50**, 155 (1958); **51**, 113 (1959).
615. C. P. Brown and J. Saunders, *J. Polymer Sci.* **43**, 579 (1960).
616. C. P. Brown and J. Saunders, *J. Polymer Sci.* **43**, 580 (1960).
617. G. Natta, *Angew. Chem.* **68**, 393 (1956).
618. W. E. Smith and R. C. Zelmer, *J. Polymer Sci.* **A1**, 2587 (1963).
619. D. Natta and I. Paskuon, *Kinetika i Kataliz* **3**, 805 (1962).
620. T. Tsuruta, "Reactions Producing Synthetic Polymers," Goskhim Izdat, 1963. (in Russian)
621. J. Furukawa and T. Tsuruta, *J. Polymer Sci.* **36**, 275 (1959).
622. F. Dawans and P. Teyssie, *Bull. Soc. Chim. France* **10**, 2376 (1963).
623. E. J. Arlman, *Proc. 3rd Intern. Congr. Catalysis* **1** (1964). Amsterdam, 1965.
624. V. Bochek, *Intern. Symp. Macromol. Chem. (Moscow, June 1960)* **2**, 230. Pub. Akad. Nauk SSSR, 1960.
625. J. Boor, *J. Polymer Sci.* **C1**, 237 (1963).
626. H. Schnecko, M. Reinmoller, K. Weilrauch, and W. Kern, *J. Polymer Sci.* **C4**, 71 (1963).
627. I. Pasquon, A. Tambelli, and G. Gatti, *Makromol. Chem.* **61**, 116 (1963).

628. A. P. Firsov, B. N. Kashparov, and N. M. Chirkov, *Visokomolekul. Soedin.* **6**, 352 (1964); A. P. Firsov and N. M. Chirkov, *Izv. Akad. Nauk SSSR, Ser, Khim.* 11, 1964, (1964).
629. W. L. Carrick, F. J. Karol, G. J. Karapinka, and J. J. Smith, *J. Am. Chem. Soc.* **82**, 1502 (1960).
630. F. J. Karol and W. L. Carrick, *J. Am. Chem. Soc.* **83**, 2654 (1961); W. L. Garrick, *J. Am. Chem. Soc.* **80**, 6455 (1958).
631. P. H. Plesch, M. Polanyi, and H. A. Skinner, *J. Chem. Soc.*, 257 (1947).
632. M. Tsutsui and J. Ariyoshi, *J. Polymer Sci.* **A3**, 1729 (1965).
633. R. V. Basova, A. A. Arest-Yakubovich, D. A. Solovykh, N. V. Desyatova, A. R. Gantmakher, and S. S. Medvedev, *Dokl. Akad. Nauk SSSR* **143**, 1067 (1963).
634. B. D. Babitskii, B. A. Dolgoplosk, V. A. Kormer, M. I. Lobach, E. I. Tinyakova, N. N. Chesnokova, and V. A. Yakovlev, *Vysokomolekul Soedin.* **6**, 2202 (1964).
635. K. S. Minsker, G. T. Fedoseeva, and G. A. Razuvaev, *Vysokomolekul. Soedin.* **5**, 655 (1963).
636. D. C. Pepper, *Quart. Rev.* **8**, 88 (1954).
637. L. N. Nazarova and G. M. Aleksandrov, *Vysokomolekul. Soedin.* **3**, 1823 (1961).
638. T. Nakata, T. Otsu, and M. Imoto, collection "Chemistry and Technology of Polymers," No. 10, p. 3, 1964. (in Russian)
639. A. A. Slinkin, A. A. Dulov, and A. M. Rubinshtein, *Izv. Akad. Nauk SSSR, OKHN* **6**, 1140 (1963).
640. G. Natta, G. Dall'Asta, G. Mazzanti, and G. Motroni, *Makromol. Chem.* **69**, 163 (1963); G. Natta, G. Dall'Asta, and L. Porri, *Makromol. Chem.* **81**, 253 (1965).
641. K. Iwasaki, H. Fukutani, and S. Nakano, *J. Polymer Sci.* **A1**, 1937 (1963); K. Iwasaki, H. Fukutani, Y. Tsuchida, and S. Nakano, *J. Polymer Sci.* **A1**, 2371 (1963).
642. E. J. Vandenberg, *J. Polymer Sci.* **C1**, 207 (1963).
643. S. Nakano, K. Iwasaki, and H. Fukutani, *J. Polymer Sci.* **A1**, 3277 (1963).
644. S. Inoue, T. Tsuruta, and J. Furukawa, *Makromol. Chem.* **42**, 12 (1960).
645. Ben Ami Feit and A. Zilkha, *J. Polymer Sci.* **A7**, 287 (1963).
646. V. A. Kargin, V. A. Kabanov, V. P. Zubov, and I. M. Panisov, *Vysokomolekul. Soedin.* **3**, 426 (1961).
647. O. A. Chaltykyan, "Copper Catalysis," Pub. Aipetrat, Erevan, 1963. (in Russian)
648. M. I. Cherkashin and Yu. G. Aseev, *Izv. Akad. Nauk SSSR, Ser. Khim.* **2**, 388 (1964).
649. L. B. Luttinger and E. C. Colthup, *J. Org. Chem.* **27**, 3752 (1962).
650. S. Kambara, M. Hatano, and T. Hoshoe, *J. Chem. Soc. Japan, Ind. Chem. Sect.* **65**, 720 (1962); cited in *R. Zh. Khim.* 1963, 7C26.
651. H. Noguchi and S. Kambara, *J. Polymer Sci.* **B1**, 553 (1963).
652. J. Furukawa, T. Saegusa, T. Tsuruta, H. Fujii, A. Kawasaki, and T. Tatano, *Makromol. Chem.* **33**, 32 (1959).
653. Kenkiusho, *Plastics* **28**, No. 305, 11, 20, 21, 22 (1963).
654. H. Sobue and H. Kubota, Coll. "Chemistry and Technology of Polymers," No. 1, p. 13, 1964 (in Russian); *J. Polymer Sci.* **C4**, 147 (1964).
655. H. Tani, T. Aoyagi, and T. Araki, *J. Polymer Sci.* **2**, 921 (1964).
656. Y. Yamashita and S. Nunomoto, *Makromol. Chem.* **58**, 244 (1962).
657. S. Okamura, T. Higashimura, and K. Takeda, *Makromol. Chem.* **51**, 217 (1962).
658. W. Jaacks and W. Kern, *Makromol. Chem.* **62**, 1 (1963).
659. O. V. Krylov and Yu. E. Synyak, *Vysokomolekul. Soedin.* **3**, 898 (1961).
660. O. O. Bartan, O. V. Krylov, and E. A. Kokina, *Kinetika i Kataliz* **7**, 289 (1966).

661. R. O. Colclough, G. Gee, W. C. Higginson, J. B. Jackson, and M. Lith, Coll. "Chemistry and Technology of Polymers," No. 1, p. 74, 1953. (in Russian).
662. K. Fukui, T. Tagitani, T. Shimizu, and B. Sano, Coll. "Chemistry and Technology of Polymers," No. 1, p. 132, 1963. (in Russian)
663. T. Kagiya, T. Sano, T. Shimizu, and K. Fukui, *J. Chem. Soc. Japan, Ind. Chem. Sect.* **66**, 1148, 1893 (1963); *R. Zh. Khim.*, 1964, 14C22, 18C99.
664. J. Furukawa, T. Saeguss, T. Tsuruta, and G. Kakogawa, *Makromol. Chem.* **36**, 25 (1949).
665. F. N. Hill, F. E. Bailey, and J. T. Fitzpatrick, *Ind. Eng. Chem.* **50**, 5 (1958).
666. O. V. Krylov and V. S. Livshits, *Neftekhimiya* **5**, 40 (1965).
667. K. Okazaki, *Makromol. Chem.* **43**, 84 (1961).
668. L. E. St. Pierre and C. C. Price, *J. Am. Chem. Soc.* **78**, 3432 (1956).
669. L. A. Bakalo and B. A. Krentsel, *Usp. Khim.* **31**, 657 (1962).
670. C. C. Price and M. Osgan, *J. Am. Chem. Soc.* **78**, 4787 (1959).
671. V. S. Livshits, O. V. Krylov, and E. I. Klabunovskii, *Dokl. Akad. Nauk SSSR* **161**, 633 (1965).
672. S. Inoue, T. Tsuruta, and J. Furukawa, *Makromol. Chem.* **53**, 215 (1962); J. Furukawa, Coll. "Chemistry and Technology of Polymers," No. 3, p. 28, 1963. (in Russian).
673. V. S. Livshits, Coll. "Methods for Catalysis Investigation," Izd. Sib. Otd. Akad. Nauk SSSR, Novosibirsk, **3**, 237 (1965).
674. E. J. Vandenberg, *J. Polymer Sci.* **47**, 486, 489 (1960).
675. F. E. Bailey and H. G. France, *J. Polymer Sci.* **45**, 43 (1960).
676. L. A. Bakalo, B. A. Krentsel, and A. V. Topchiev, *Neftekhimiya* **3**, 207 (1963).
677. D. D. Smith, R. M. Murch, and O. R. Pierce, *Ind. Eng. Chem.* **49**, 1241 (1957).
678. A. Noshay and C. C. Prince, *J. Polymer Sci.* **54**, 533 (1961).
679. G. D. Jones, A. Langsjoen, M. M. Neumann, and J. Zomfeler, *J. Org. Chem.* **9**, 125 (1944).
780. O. O. Bartan and O. V. Krylov, *Izv. Akad. Nauk SSSR, Ser. Khim.* **11**, 2053 (1965).
681. K. I. Fukui, T. Kagitani, T. Shimizu, and B. Sano, Coll. "Chemistry and Technology of Polymers," p. 118, 1963. (in Russian)
682. N. L. Franklin, P. H. Pinchbeck, and F. Popper, *Trans. Inst. Chem. Eng.* **36**, 259 (1958).
683. V. V. Nelimov, *Usp. Khim.* **29**, 1362 (1960).
684. B. F. Ormont, "Crystal Structures of Inorganic Compounds," Gostekhteoret, 1950. (in Russian)
685. V. S. Fomenko, "Emission Properties of Elements and Chemical Compounds," Izd. Akad. Nauk USSR, Kiev, 1961. (in Russian)
686. W. Noddack and H. Walch, *Z. Electrochem.* **63**, 269 (1959).
687. J. Boros, B. Jeszenszky, "Festkörperphysik und Physik," p. 117. Leuchtstoffe, Berlin, 1958.
688. I. B. Patrina and V. A. Ioffe, *Fiz. Tverd. Tela* **6**, 3227 (1964).
689. A. I. Manikov, O. A. Esin, and B. M. Leninskikh, *Zh. Fiz. Khim.* **36**, 2734 (1962).
690. A. Companion, *J. Phys. Chem., Solids*, **25**, 357 (1964).

Subject Index

A

Acetaldehyde
  condensation with furfural, 208
  crotonic condensation of, 208
  polymerization, 230
Acetone synthesis from acetaldehyde, 208
Acetylene
  hydration of, 204
  hydrohalogenation of, 204
  participation, reactions with, 204
  polymerization, 229
Acidity, measurement of, 70
Acids, decomposition of, 134ff
Acrylic acid, nitriles of, 204
Acrylonitrile polymerization, 229
Activation, σ-, 51
Alcohol
  dehydration of, 94, 99, 115ff
  dehydrogenation of, 115ff
Aldol reactions, 207
Alkaline centers, 77
Alkyl chlorides, dehydrohalogenation of, 199
Alkylation of hydrocarbons, 215, 217
Allyl alcohol with ammonia, 210
Aluminum oxide, black, 68
Aminoenanthic acid polycondensation, 211
Ammonia
  decomposition of, 152
  oxidation to nitric oxide, 174
Anions, sp-orbitals, 49
Antiferromagnetism, 48
Aromatics, hydrogenation of, 150
Arsine, decomposition of, 156
Asymmetric synthesis, 101
Atomic polarization, 43
Atomization, energy of, 33

B

Balandin's multiplet theory, 93ff
Band spacing, d-, 127
Basicity, measurement of, 70
Benzaldehyde, by oxidation of toluene, 183
Benzene
  alkylation by methanol, 217
  bromination of, 185
  chlorination of, 185
  oxidation of
    complete, 181, 182
    to maleic anhydride, 183
Benzylchloride polycondensation, 212
2-Benzylidene-3-keto-2, 3-hydronaphthene isomerization, 215
Bonds, σ and π, 45
Brønsted acids and bases, 70
Butadiene
  from ethyl alcohol, 209
  hydrogenation, 150
  polymerization, 227

Butane dehydrogenation to butylenes, 140
1-Butene isomerization to 2-butene, 213
tert-Butyl hydroperoxides, decomposition of, 188

C

Calcium hypochlorite decomposition, 194
Carbon disulfide, hydrogenation of, 151
Carbon monoxide
  conversion to hydrocarbons, 154
  hydrogenation to methane, 149
  oxidation of, 169
Carbonium ions, 76
Catalyst selection principles, summary of, 241
Charge transfer spectra, 58
Charges, effective, 40, 44
Chlorates, reaction with polyvinylbenzene, 196
Chlorobenzene hydrolysis, 205
2-Chlorobutane dehydrochlorination, 203
Chloroform condensation with acetone, 210
α-Chloromethyl naphthalene polycondensation, 212
Cinnamic aldehyde hydrogenation, 150
Compensation effect, 19, 25
Complex stability, 81
Condensation reactions, 207
Conduction
  hybrid, 26
  metallic, 48
Conductivity, electrical, 6
Conductors, intrinsic, 22
Coordination, changes of, 60, 62
Coordination complexes, 81
Correlation coefficients, 122, 123
Cracking of hydrocarbons, 215
Crotonic aldehyde hydrogenation, 150
Crotonic condensation of acetaldehyde, 208
Crystal field theory, 51, 55, 60
Crystal lattice parameter, 93ff
Cumene cracking, 215
Cyclobutene polymerization, 228
Cyclohexane
  dehydrogenation, 141
  isomerization to methyl cyclopentane, 213
  oxidation of, complete, 181
Cyclohexanone condensation, 208

Cyclohexene hydrogenation, 150
Cyclopentadiene
  dimerization, 220
  hydrogenation, 150
Cyclopropane isomerization to propylene, 213

D

Decarbonylation, 83
Degeneracy, 55
Decomposition
  of inorganic hydrides, 152
  of oxygen-containing substances, 185
Dehydrocondensation of trialkyl and triaryl silanes, 211
Dehydrogenation of hydrocarbons, 140
Dehydrohalogenation of alkyl chlorides, 199
p-Dichlorobenzene isomerization, 215
Dichloroethane dehydrochlorination, 204
Dielectric constant, 89
Dielectric permeability, 32, 90
Diethylaniline from aniline and ethanol, 210
Differential-isotope methods, 77
2,3-Dimethylbutane, complete oxidation of, 181
Dimethylketene polymerization, 230
Donor–acceptor bond, 82
Donor–acceptor interaction, 45
Doped levels, 36
Duplet, 94, 96, 97

E

Electroaffinity constant, 105
Electroconductivity, 6
Electron(s)
  d-, 45, 47
  free, 9
  s-, p-, 45
  transfer, 9
Electronegativity difference, 31, 40, 127
Electronic polarizability, 90
Electronic theory, 12
Energy
  of ionization, 36
  levels, 7
  of single bond, 33
  specific surface, 33

SUBJECT INDEX 281

Epichlorhydrin polymerization, 238
EPR, 44, 88
Escape, work of, 16
Ethylaniline from aniline and ethanol, 210
Ethylchloride dehydrochlorination, 203
Ethylene
  copolymerization with propylene, 226
  hydration, 198
  oxidation of, 179
  polymerization, 220
Ethyleneimine polymerization, 238
Ethylene oxide polymerization, 231

F

Ferroelectrics, 92
Ferromagnetism, 48
Fermi level, 7
Fluorine substituted epoxides, polymerization of, 238
Forbidden zone, width of, 21, 28
  measurement of, 35
Formaldehyde condensation, 211
Formaldoxime decomposition, 139

G

Geometry, 93ff.
Germanium hydride, decomposition of, 156
Glyceryl phosphate splitting, 206

H

Hall effect, 37
Heptane cracking, 217
Heterolytic reactions, 80
Hexafluoroethylene, oxidation of, 184
Hexane
  chlorination of, 185
  cracking on zeolites, 217
  dehydrogenation, 141
  oxidation of, complete, 181
Holes, free, 9
Homomolecular exchange of oxygen, 176
Homolytic reactions, 80
Hybridization, 42
Hydrazine, decomposition of, 153
Hydrides, inorganic, decomposition of, 152
Hydrocarbons
  alkylation of, 215, 217
  cracking of, 215

  dehydrogenation of, 140
Hydrogen
  atoms, recombination of, 163
  bonding, 72
  ortho–para conversion, 157, 160
  oxidation of, 168
  transfer reactions between organic compounds, 160
Hydrogen bromide, decomposition of, 156
Hydrogen–deuterium exchange, 157
Hydrogen halides
  addition of, 198
  removed of, 198
Hydrogen peroxide, decomposition of, 186
Hydrogenation, 148
Hydroxide radius, 78

I

Indicator method, 73, 86
Inorganic hydrides, decomposition of, 152
Insulators, 22
Iodates, reaction with polyvinylbenzene, 196
Ionic crystals, 105, 106
Ionic polarizability, 82
Ionization
  constants, 82
  degree of, 41
  energy of, 36
IR spectra, 76, 87
Isobutane dehydrogenation, 141
Isobutylene
  dimerization, 220
  polymerization, 226
Isobutyl vinyl ether polymerization, 228
Isoelectronic series, 23
Isomerization reactions, 212
Isooctane, oxidation of, complete, 182

J

Jahn–Teller effect, 107

K

K-spectra, 43
Knoevenagel condensation, 210

L

Lattice defects, 95
Lattice parameter, 93
LEED, 98
Lewis acidity, determination of, 84
Lewis acids and bases, 75, 81
Ligand field theory, 53
Ligands, 52

M

Madelung, constant of, 33
Maleic anhydride, by oxidation of benzene, 183
Mercury oxides, decomposition of, 194
Methallyl alcohol, dehydrogenative dimerization of, 210
Methane, oxidation of, 179
2-Methylbutane, complete oxidation of, 181
Methylchloride dehydrochlorination, 203
2-Methyl-1-pentene isomerization, 214
Mobility of current carrier, 48
Molecular orbitals, 55
Monolayer formation, 72
Multiplet theory, 93ff.

N

Naphthalene, complete oxidation of, 183
Nitric acid synthesis, 175
Nitric oxide, decomposition of, 191
Nitrobenzene, reduction of, 150
Nitrogen atoms, recombination of, 163
Nitroparaffins, decomposition of, 195
Nitrous oxide, decomposition of, 188
Nuclear magnetic resonance, 45

O

Octane cracking, 217
Octet rule, 103
Olefins
  isomerization, 100
  polymerization of, 65, 100
Optical electronegativity, 58
Orbitals, 55
Organic compounds, oxidation of, 178
Ortho–para hydrogen conversion, 157
Oxidation
  of organic compounds, 178
  reactions, simple, 168
Oxides
  of higher olefins, polymerization of, 238
  of mercury, decomposition of, 194
Oxygen
  atoms recombination of, 163
  chemisorption, 178
  homomolecular exchange of, 176
Oxygen-containing substances, decomposition of, 185
Ozone decomposition, 192

P

Paramagnetic resonance, electronic, 44
Parameter, crystal lattice, 93
Partial coefficient of correlation, 124
Pentane
  dehydrogenation, 141, 148
  oxidation of, complete, 181, 182
1-Pentene
  isomerization, 214
  oxidation of, complete, 181
Periodic table dependence, 14, 34
Permanganates, decomposition of, 193
Permeability, dielectric, 32
Phenol alkylation by isobutylene, 217
pK, 73
Polarizability, 43, 89
Polarization, 61
  atomic, 43
Polarizing action of cation, 81
Polycondensation reactions, 207
Polymerization, 87
  of olefins, 65
of propylene oxide, 235
Polyurethanes by reaction of polyisocyanates with polyesters, 211
Potassium azide decomposition, 197
Potassium chlorate decomposition, 194
Potassium perchlorate decomposition, 194
Potassium permanganate decomposition, 193
Propane dehydrogenation, 141
β-Propiolactam polymerization, 239
Propionic aldehyde condensation, 208

SUBJECT INDEX 283

Propylene
  dimerization, 220
  hydration, 198
  hydrogenation, 149
  polymerization, 224
Proton acids and bases, 70
Prussic acid hydrolysis, 206

Q

Quasi-insulated surface, 80
Quaternary coordination, 76

R

Reactions
  acceptor, 8
  donor, 8
  oxidation–reduction, 6
Refraction, index of, 43
Regression, 123, 124
Rosin, esterification of, 210

S

Semiconductors, 9, 22
Silanes, dehydrocondensation of, 211
Sodium metaphosphate conversion to orthophasphate, 206
Solids, color of, 66
Spin, 53
Stabilization, crystal field, 54, 60
Statistical weight, 117
Stereospecificity, 102
Stibine, decomposition of, 156
Styrene polymerization, 227
Styrene sulfide polymerization, 238
Sugars, via formaldehyde condensation, 211
Sulfur dioxide, oxidation to sulfur trioxide, 174
Symmetry, 55

T

Tetralin dehydrogenation, 148
Thermoelectric force, 37
Toluene
  bromination of, 185
  oxidation
    to benzaldehyde, 183
    complete, 183
Transition
  conduction, 26
  d–d, 58
Trioxane polymerization, 231

V

Vinylacetate synthesis from acetylene, 204

W

Water
  addition of, 198
  removal of, 198
Work
  of escape, 16
  of polarization, 91
  of removal, 91
Work function, 15

X

X ray K-spectra, 43

Z

Zeolites, hexane cracking on, 217
Ziegler–Natta catalysts, 65